LONDON MATHEMATICAL SOCIETY LECTURE NOTE SERIES

Managing Editor: PROFESSOR I.M. James, Mathematical Institute, 24-29 St.Giles, Oxford

Already published in this series

1. General cohomology theory and K-theory, PETER HILTON.
4. Algebraic topology: A student's guide, J.F. ADAMS.
5. Commutative algebra, J.T. KNIGHT.
7. Introduction to combinatory logic, J.R. HINDLEY, B. LERCHER and J.P. SELDIN.
8. Integration and harmonic analysis on compact groups, R.E. EDWARDS.
9. Elliptic functions and elliptic curves, PATRICK DU VAL.
10. Numerical ranges II, F.F. BONSALL and J. DUNCAN.
11. New developments in topology, G. SEGAL (ed.).
12. Symposium on complex analysis Canterbury, 1973, J. CLUNIE and W.K. HAYMAN (eds.).
13. Combinatorics, Proceedings of the British combinatorial conference 1973, T.P. McDONOUGH and V.C. MAVRON (eds.).
14. Analytic theory of abelian varieties, H.P.F. SWINNERTON-DYER.
15. An introduction to topoligical groups, P.J. HIGGINS.
16. Topics in finite groups, TERENCE M. GAGEN.
17. Differentiable germs and catastrophes, THEODOR BRÖCKER and L. LANDER.
18. A geometric approach to homology theory, S. BUONCRISTIANO, C.P. ROURKE and B.J. SANDERSON.
19. Graph theory, coding theory and block designs, P.J. CAMERON and J.H. VAN LINT.
20. Sheaf theory, B.R. TENNISON.
21. Automatic continuity of linear operators, ALLAN M. SINCLAIR.
22. Presentations of groups, D.L. JOHNSON.
23. Parallelisms of complete designs, PETER J. CAMERON.
24. The topology of Stiefel manifolds, I.M. JAMES.
25. Lie groups and compact groups, J.F. PRICE.
26. Transformation groups: Proceedings of the conference in the University of Newcastle upon Tyne, August 1976, CZES KOSNIOWSKI.
27. Skew field constructions, P.M. COHN.
28. Brownian motion, Hardy spaces and bounded mean oscillation, K.E. PETERSEN.
29. Pontryagin duality and the structure of locally compact abelian groups, SIDNEY A. MORRIS.
30. Interaction models, N.L. BIGGS.
31. Continuous crossed products and type III von Neumann algebras, A. VAN DAELE.
32. Uniform algebras and Jensen measures, T.W. GAMELIN.
33. Permutation groups and combinatorial structures. N.L. BIGGS and A.T. WHITE.
34. Representation theory of Lie groups, M.F. ATIYAH.
35. Trace ideals and their applications, BARRY SIMON.
36. Homological group theory, edited by C.T.C. WALL.

Partially Ordered Rings and Semi-Algebraic Geometry

Gregory W. Brumfiel

CAMBRIDGE UNIVERSITY PRESS
CAMBRIDGE
LONDON NEW YORK NEW ROCHELLE
MELBOURNE SYDNEY

CAMBRIDGE UNIVERSITY PRESS
Cambridge, New York, Melbourne, Madrid, Cape Town, Singapore, São Paulo

Cambridge University Press
The Edinburgh Building, Cambridge CB2 8RU, UK

Published in the United States of America by Cambridge University Press, New York

www.cambridge.org
Information on this title: www.cambridge.org/9780521228459

First published 1979
Re-issued in this digitally printed version 2007

A catalogue record for this publication is available from the British Library

ISBN 978-0-521-22845-9 paperback

Contents

Preface

This text represents an attempt to formulate foundations and rudimentary results of a type of geometry and topology in purely algebraic terms. I feel that the approach taken here is very natural and that it is only coincidental that the point of view I advocate did not emerge fifty years ago.

The mathematics itself is most similar to elementary commutative algebra and algebraic geometry. The level of difficulty is about like that of the texts on commutative algebra by Zariski - Samuel or Atiyah-Macdonald. Although, strictly speaking, the text might be read without any previous knowledge of basic commutative algebra, essential motivation would probably be lacking. On the other hand, I see no reason why a student couldn't simultaneously read this text and some classical commutative algebra.

In the final two chapters, I assume the reader is familiar with, or can read elsewhere, basic results of Artin-Schreier theory, Krull valuation theory, and algebraic geometry. The basic algebra texts listed in the bibliography as references [63] - [68] contain more than adequate background material in the appropriate sections. The final two chapters of this text are, in fact, somewhat independent of the first six chapters. I recommend that after looking at the introduction, the reader look through Chapters VII and VIII in order to gain motivation for the foundational material of Chapters I through VI.

Introduction

It is my hope that the methods developed in this text will lead to an interesting embedding of algebraic topology in a purely algebraic category, namely, some category of partially ordered rings. At the same time, the theory provides a convenient abstract setting for the theory of real semi-algebraic sets, quite analogous to commutative algebra as a setting for modern algebraic geometry.

I might motivate the study of partially ordered rings (somewhat frivolously) as follows. One observes that the integers, together with their ordering, is an initial object for a lot of mathematics. On the one hand, consideration of order properties leads to the topology of the real line, then to Euclidean spaces, and eventually to abstract continuity and general point set topology. On the other hand, consideration of arithmetic properties leads to the abstract theory of rings, fields, ideals, and modules.

Following either route, one can go too far. Completely general topological spaces and continuous maps are uninteresting. Completely general rings and modules are uninteresting. Thus the mainstream in topology concentrates on nice spaces (for example, polyhedra and manifolds) and the mainstream in algebra concentrates on nice rings (for example, finitely generated rings over fields, subrings of the complex numbers and their homomorphic images.) The two theories seem to intersect eventually in category theory and semi-simplicial homotopy theory. The topologists put back in some algebra and the algebraists put back in some topology. On the other hand, algebraic geometry works best over an algebraically closed field and such concepts as manifolds with boundary, homotopy of maps and mapping cones, which are extremely useful to topologists, are not readily available in pure algebra. A simple observation is that such concepts are easily described by algebraic equalities and inequalities in real affine space, however. Certainly all of geometry is deeply rooted in the study of equalities and inequalities of functions on real affine space. Even in point set topology real functions and inequalities play a key role, for example, in the theory of paracompact spaces.

The real number field is a formally real, real closed field. In fact, the Artin-Schreier theory of formally real fields is precisely an abstract algebraic treatment of inequalities in field theory. The real closed fields are the analogues of algebraically closed fields-they admit no proper algebraic extensions in which inequalities still make sense.

It thus seems to me that a true understanding of the relations between algebraic geometry and topology must stem from a deeper understanding of real algebraic geometry, or, actually, semi-algebraic geometry. Moreover, real algebraic geometry should not be studied by attempting to extend classical algebraic geometry to non-algebraically closed ground fields, nor by regarding the real field as a field with an added structure of a topology. Instead, the abstract algebraic treatment of inequalities originated by Artin and Schreier should be extended from fields to (partially ordered) algebras, with real closed fields replacing the algebraically closed fields as ground fields. It is obvious that such a category of partially ordered algebras provides an abstract setting for semi-algebraic geometry (study of sets defined by finitely many real polynomial equalities and inequalities), and it seems plausible that such a category would allow a natural development of algebraic topology and homotopy theory.

It is essential that the reader understand that algebraic topology (at least the homotopy category of finite simplicial complexes and the study of reasonable functors on this category) is known to be completely independent of topology, that is, independent of limits, continuity, the infinite arithmetic of open and closed sets; even the completeness of the real numbers is irrelevant. The most highly developed reduction of homotopy theory to pure algebra is the semi-simplicial, or combinatorial, approach, developed by D. M. Kan, J. C. Moore, M. M. Postnikov, and others in the 1950's. The problem with this reduction (ignoring inefficiency) is that it seems unmotivated without first developing the point set topology of finite simplicial complexes, which, in turn, is founded on the topology of the real line. Also, differential topology seems unnatural in this setting.

My philosophy is that the derivation of homotopy theory from point set topology is an historical accident. Basically, I regard the real goal to be a mathematization of our experience, sensation, and perception of space, time, and matter. This experience is inherently finite, but involves counting, hence algebra, and order relations, hence inequalities. Our *immediate* perception of boundaries of objects, and spatial and temporal order relations justifies a more structured approach than the mathematical reduction of all experience to simply counting small or big finite sets. In fact, it seems to me that a reasonable first approximation of our perception is the set theory of sets in affine space defined by finite collections of algebraic equalities and inequalities, along with the Boolean set theoretic operations of finite unions and intersections, and differences. This sort of set theory has much in common with topology, but is fundamentally very different. Thus we will often use the language of open sets, closed sets and so on, but it is to be understood that a set in the plane like $y > x^2$ is not open because it is a union (necessarily infinite) of open balls but rather because it is the set of points where the single algebraic function $y - x^2$ is positive. In general, functions can be replaced by their graphs, hence admissible functions from one semi-algebraic set to another can be thought of as certain types of semi-algebraic subsets of the product. Thus morphisms also avoid the infinite definitions of point set topology.

According to the Hilbert Basis Theorem, any set in affine space over a field defined by algebraic equalities $f_\alpha(x_1 \dots x_n) = 0$ is already defined by a finite subset of the equations $f_\alpha = 0$. Over an ordered field, subsets of affine space defined by inequalities $g_\beta(x_1 \dots x_n) \geq 0$ as well as equalities $f_\alpha(x_1 \dots x_n) = 0$ are clearly not always representable by finitely many equalities and inequalities. It is perhaps this fact which has led to the divergence of the fields of algebra and topology.

Topology is a good example of a subject which produces answers of interest before the real problems are fully clarified. As several examples of such answers, which come up in more than one context in mathematics, I would list Lie theory, the theory of compact surfaces, the Bott periodicity theorem and K-theory, the classification of differentiable structures on spheres, the theory of cohomology operations (even cohomology theory itself), the computations of the classical group bordism rings, and the emerging classification of singularities of maps. These subjects deal with topological concepts, but also turn out to be related to problems in algebra and number theory. Thus one feels certain that it is good stuff.

On the other hand, much of geometric topology is concerned with analyzing just what pathology can and cannot occur, using infinite definitions and constructions. Thus one has space filling curves, but a very strong regularity theorem about simple closed curves in the plane. I regard this as evidence that arbitrary continuous curves tend to be uninteresting, but it is not evidence that simple closed curves are interesting. In fact, in three space one has wild embedded arcs and spheres and their classification is not regarded as a mainstream problem. Under the assumption of topological local flatness, it is known that every n-sphere in $n+1$ space bounds a topological ball. More important, however, is the question of whether a smooth n-sphere in $n+1$ space bounds a smooth ball. Provocatively, if $n \neq 3$ the answer is known to be yes, but the proof is harder than the corresponding topological theorem. If $n = 3$, it is possible that some smooth copy of S^3 in $\mathbb{R}^4$ bounds a topological ball which is not diffeomorphic or piecewise linearly equivalent to the standard D^4.

Pathology can involve morphisms, as do these examples, or absolute properties of spaces. Thus topological manifolds of dimension one and two are classified, manifolds of dimension three are known to possess unique piecewise linear and differentiable structures, but are not yet classified, while it is still unknown if manifolds of dimension four and higher can always be triangulated.

There is perhaps a widespread feeling that if attention is restricted to, say, differentiable manifolds, pathology disappears. It is known (Whitehead) that smooth manifolds admit a unique compatible piecewise linear structure and it is also known (Nash, Tognoli) that compact smooth manifolds are diffeomorphic to non-singular real algebraic varieties. Also (Milnor, Serre and many other contributors) any compact manifold admits only finitely many distinct differentiable structures. But the differentiable category allows too many morphisms. Any closed set in Euclidean space can be realized as the zeros of a smooth (C^∞) function. The studies of singularities, diffeomorphisms, flows, and foliations in recent years have produced many pathological phenomena, as well as regularity theorems under suitable hypotheses. Another area in which there are strong regularity theorems for the objects is the study of smooth compact Lie group actions on compact manifolds. The compact Lie groups are algebraic groups and Palais has extended the Nash-Tognoli theorem to an equivariant version which roughly says all compact Lie group actions on compact manifolds are algebraic, up to isomorphism.

My own interest is exactly the reverse of this tradition of seeking regularity theorems in topological situations. Instead, I advocate beginning with algebra and working toward geometry, in an attempt to discover just what geometric phenomena are realizable by finite algebraic constructions. It is not so much a question of one type of mathematics being superior to another, but simply a question of how best to understand the dividing line between algebra and topology, and I think this dividing line should be approached from both directions. The notion of inequalities is very close to this dividing line, but essentially on the algebraic side. From the algebraic point of view it is more or less clear that the real algebraic numbers are just as useful as, or even preferable to, the real numbers. I will return to this philosophy from time to time in the course of this introduction.

In any event, in this book, I first develop systematically an abstract theory of partially ordered rings. The models I have in mind are rings of real valued algebraic functions on certain semi-algebraic sets in affine space and their quotients by various allowable ideals. My axioms thus reflect properties of these models, and aside from a certain amount of curiosity, the abstract theory interests me only insofar as it contributes to the eventual goals of better understanding semi-algebraic geometry and algebraic topology.

The second half of the book is devoted to a systematic introduction to real semi-algebraic geometry via Artin-Schreier Theory and the language of partially ordered rings. For the most part, the actual results can be found in the existing literature. In particular, the papers by Dubois, Efroymson, Lang, and Stengle, referred to in the bibliography are very similar in spirit to (and, in fact, influenced greatly) my philosophy. Also, there are excellent introductory accounts of Artin-Schreier theory in the algebra texts of Lang, Jacobson, and van der Waerden.

What, then, is a partially ordered ring? Generally, the definition found in the literature is a ring, together with a subset of elements called positive such that sums and products of positive elements are positive and such that if x and $-x$ are positive, then $x = 0$. (The most efficient terminology is to call 0 positive and to refer to non-zero positives as strictly positive.) *I make the further assumption that all squares are positive.* Also, of course, all rings are commutative with unit. Given such a set of positives, a partial order relation is defined on the ring by $x \geq y$ if $x - y$ is positive. The definition is purely algebraic.

The assumption that squares are positive is justified, first, because it is true in all the examples I want to fall within the scope of the theory and, secondly, because it seems to be a very useful assumption for proving analogues of the basic results in commutative algebra.

It is easy to see that a ring admits such a partial order if and only if the following condition holds: whenever $\sum_{i=1}^{n} a_i^2 = 0$, then each $a_i^2 = 0$, $1 \leq i \leq n$. The set of all finite sums of squares is then an allowable set of positives, in fact, clearly the smallest such. Nilpotent elements are decidedly permitted by this condition.

The morphisms between partially ordered rings which are important are the order preserving ring homomorphisms. Kernels of such morphisms are called convex ideals and are characterized by the property that if a sum of positive elements belongs to the ideal, then so does each summand. This gives a category, (POR).

The category (POR) turns out to be not the best approximation to the ultimate goals. A useful subcategory is the category (PORNN), partially ordered rings with no nilpotent elements. However, nilpotent elements are actually useful, just as in modern algebraic geometry. A compromise is the intermediate category (PORCK), partially ordered rings with convex killers. The added axiom is the following condition: whenever $\left(\sum_{i=1}^{n} p_i\right)x = 0$, p_i positive, then each $p_i x = 0$, $1 \leq i \leq n$. Throughout this introduction, I will indicate advantages of (PORCK). To begin, in (POR) one can have $nx = 0$, but $x \neq 0$. Since $n = 1 + \cdots + 1$, such pathology is avoided in (PORCK). Secondly, in (PORCK) the associated primes of a convex ideal are automatically convex, as are the isolated primary components. Thirdly, a ring of polynomials in finitely many indeterminates over a ring A in (PORCK) can be regarded faithfully as a ring of A-valued functions on affine space over A. A more subtle advantage is that the defining condition for (PORCK) makes sense if x takes values in a module over A. Thus, in (PORCK), a ring is an admissible module over itself, in a certain useful sense.

In any category of partially ordered rings, only ideals which are kernels of morphisms in the category should be considered at all. Thus, in (POR) one sees the convex ideals, in (PORNN), one sees the radical convex ideals $I = \sqrt{I}$, and in (PORCK) one sees what I call absolutely convex ideals. Namely, I is absolutely convex if $\left(\sum_{i=1}^{n} p_i\right)x \in I$, p_i positive, implies $p_i x \in I$, $1 \leq i \leq n$.

Much of this first volume is concerned with the partially ordered analogues of basic results on ideals in commutative algebra. Each such result must be checked, but, as a rule, slight extensions of classical arguments work for convex ideals in partially ordered rings. As examples, we mention:

1. Every proper convex ideal is contained in a maximal convex ideal. (Note maximal convex $\neq$ convex maximal.)
2. Maximal convex ideals are prime.
3. The intersection of all prime convex ideals is the (convex) ideal of all nilpotent elements.
4. The intersection of all prime convex ideals containing a given convex ideal is the nil radical of the ideal.

These results hold uniformly in the categories (POR) and (PORCK) since a convex prime ideal Q is necessarily absolutely convex. (Proof: if $(\Sigma p_i)x \in Q$, then $(\Sigma p_i)x^2 = \Sigma p_i x^2 \in Q$, hence $p_i x^2 \in Q$, hence $p_i x \in Q$.)

5. In all the categories, residue partially ordered rings are defined as the cosets of a convex or absolutely convex ideal. The positive cosets are those containing a positive element. The usual fundamental isomorphism lemmas and correspondences between ideals in residue constructions are established.
6. Localizations are defined in all the categories and the desired basic properties established. Of importance here is the concept of a concave multiplicative set, which, when it contains a positive element also contains all larger elements. This is natural, since if one wants to invert a strictly positive function on some set, then every larger function also has no zeros, hence might as well also be inverted. Complements of prime convex ideals are concave multiplicative sets, as is S(1), the shadow of 1, consisting of all elements $s \geq 1$.
7. Maximal convex ideals are characterized by the property that the associated residue ring is a semi-field. By semi-field I mean that for each non-zero a, one has $ab \geq 1$ for some b. Such elements a (semi-units) are the analogues of units since they belong to no proper convex ideal.
8. Certain categorical constructions such as fibre sums, fibre products, direct and inverse limits are carried out in the various categories.
9. The set X of prime convex ideals is given the Zariski "topology",

and basic functorial and "topological" properties established. A structure sheaf of partially ordered rings is constructed by means of localizations. The stalks of the structure sheaf are the partially ordered rings one obtains by localizing with respect to complements of prime convex ideals. The ring of global sections is generally larger than the original ring. In fact, in (PORCK), the shadow of 1, $S(1) = \{s \geq 1\}$, consists of non-zero-divisors and the global sections of the structure sheaf is the ring one obtains by localizing with respect to $S(1)$. This is reasonable, since this localization exactly inverts all elements which belong to no prime convex ideals, that is, "functions nowhere zero on X". (In (POR), $S(1)$ can have zero divisors, a ring does not even inject into its global sections, in general, and the global sections cannot be described by a simple localization.)

10. A "universal bound" is obtained for roots of a monic polynomial with coefficients in a partially ordered ring. More generally, bounds are obtained for solutions of $f(x) \leq 0$, where f is a monic polynomial of even degree. As a corollary, one obtains a going-up theorem for prime convex ideals in what I call semi-integral extensions $A \subset B$. Namely, an element $x \in B$ is semi-integral over A if $f(x) \leq 0$ for some monic polynomial of even degree, with coefficients in A. Unfortunately, the going-up theorem requires a mild hypothesis on the partial orders on A and the extension B. So the theorem does not have the same applicability to, say, the real Nullstellensatz that the classical going-up theorem has to the Nullstellensatz over algebraically closed fields.

11. A theory of associated primes and isolated primary components is worked out, which is quite satisfactory for rings in (PORCK) satisfying the ascending chain condition for convex ideals. However, there is no general decomposition of convex ideals as intersections of primary convex ideals, even for Noetherian partially ordered rings. Here, as in the going-up theorem above, the gap between the abstract category (PORCK) and the specific study of finitely generated rings

over fields begins to widen.

12. Each subset X of a partially ordered ring A belongs to a smallest convex ideal $H(X)$ (the *hull* of X) and a smallest absolutely convex ideal $AH(X)$ (the *absolute hull* of X). If $I \subset A$ is any ideal, then $AH(I^n)AH(I^m) \subset AH(I^{n+m})$. The corresponding property for hulls does not seem to hold. Thus in (PORCK) there is a natural graded ring associated to the ideal I, namely $\bigoplus_{n \geq 0} AH(I^n)/AH(I^{n+1}) = G(A)$. Relations between A and $G(A)$ in (PORCK) parallel properties of ordinary associated graded rings in commutative algebra.

There are two fundamental and unavoidable reasons why partially ordered algebra is "harder" than commutative algebra and why real algebraic geometry is "harder" than algebraic geometry over algebraically closed fields. The first is that principal ideals are not convex in general. The smallest convex ideal containing a given element or set of elements is rather complicated. The second is that the statement that a given polynomial over an ordered field has roots in some ordered extension field, is a non-trivial statement, requiring specific verification. These two difficulties permeate the entire theory. The second especially is perhaps the main reason why the abstract theory of partially ordered rings is not as useful in real algebraic geometry as abstract commutative algebra is in classical algebraic geometry.

Here is a simple example which illustrates certain features of the theory. If $\mathbb{R}[X,Y,Z]$ is partially ordered as a ring of real valued functions on affine space, then the ideal (X^2+Y^2+1) has no zeros, but is not convex. (Any convex ideal containing X^2+Y^2+1 must also contain $X^2,Y^2,1$.) Slightly more subtle is the ideal (X^2+Y^2), which has zeros but for which the co-dimension of the zero set is too big. The smallest convex ideal containing X^2+Y^2 is the ideal (X^2,Y^2,XY), consisting of all functions which vanish to second order on the Z-axis. To see this, observe that the functions $(X+Y)^2$ and $(X-Y)^2$ are positive and $(X+Y)^2+(X-Y)^2 = 2(X^2+Y^2)$. So the smallest convex ideal containing X^2+Y^2 must also contain X^2,Y^2 and $(X+Y)^2$, $(X-Y)^2$. Conversely, if a sum of positive functions is in the ideal (X^2,Y^2,XY), the summands must vanish to second order on the Z-axis. One can even see that (X^2,Y^2,XY) is absolutely convex, so the residue ring $\mathbb{R}[X,Y,Z]/(X^2,Y^2,XY)$ is in (PORCK) and has nilpotent elements.

Roughly, the goal of the abstract affine theory is to interpret partially ordered rings A as rings of "functions" on some set (the prime or maximal convex ideals of A) modulo functions which "vanish to some order" on a subset. That is, some jet is zero on the subset. An algebraic set corresponds to zeros of some convex radical ideal $I = \sqrt{I}$. The jets associated to I are the (absolutely) convex J with $\sqrt{J} = I$.

One must contend not only with the rings A/I, A/J but with various partial orders on these rings. In geometric language, it is natural to think of A/I, $I = \sqrt{I}$, as a ring of functions on the zeros of I, and partially order accordingly. In the general case, it is natural to relate orders on A/J, $\sqrt{J} = I$, to more refined orders on A itself, obtained by regarding A as a ring of germs of functions defined on neighborhoods of the zeros of I. Neighborhoods in the space of prime or maximal convex ideals are defined by inequalities. In the right examples, neighborhoods are always Zariski dense, which is why A can be regarded as a ring of germs of functions defined "near" the zero set of an ideal. The use of inequalities thus enables one to define "strong" open sets and semi-algebraic sets in spaces of prime and maximal convex ideals. In good cases the structure sheaf soups up to these strong open sets, again using suitable localizations.

This discussion brings me back to the philosophy of rejecting point set topology altogether. Above I mentioned a Zariski "topology" on the set of convex prime ideals. Now, the open and closed sets in the Zariski topology on the prime ideal spectrum of a *Noetherian* ring are accessible by *finite* set arithmetic from sets defined by algebraic equalities and non-equalities. So, again, I regard viewing the Zariski topology as a classical topology to be an historical accident. In the partially ordered context, one has the added freedom of defining sets by equalities, non-equalities, and inequalities, but one should stick with finite methods and not refer to weak and strong topologies, nor allow infinite processes.

Continuing, it is clear that once an affine theory is established, one could globalize to define partially ordered preschemes. A modern viewpoint is that structures are conveniently organized by sheaves on spaces. But one still begins with "a topological space" and then imposes a structure sheaf of, say, smooth, analytic, or algebraic functions on open sets. In the algebraic case this seems unnatural because, although an algebraic function can be restricted to any open set, the *natural* domains of definition of algebraic functions are a more restricted class of sets. Also, in topology, analytic geometry and algebraic geometry, one generally comprehends the open sets in the underlying spaces in the first place in terms of inequalities or non-equalities involving functions of the structure sheaf. That is, one sees these functions before one sees the open sets.

Thus, it would seem preferable to dispense altogether with the original topological space. In the following paragraphs, I make some (vague) suggestions about alternative foundations. First, one might go backwards to Weil's method of constructing abstract varieties. Simply take some rings A_i (preferably finitely many, all Noetherian) and match up their local structure sheaves along open neighborhoods. These open neighborhoods in $\mathrm{Spec}(A_i)$ should also be finitely described by Boolean operations, equalities, non-equalities, and inequalities. Note the strong open sets in $\mathrm{Spec}(A_i)$ depend only on the ring A_i and its partial order, not on some preassigned topological space on which the ring A_i is imposed.

Naturally, in the global case, one wants an invariant definition, not dependent on a particular finite affine cover. A reasonable approach might be to follow Grothendieck and identify a partially ordered prescheme with a set-valued functor on some category of partially ordered rings. Another approach which looks mildly interesting is to axiomatize a notion of pseudo-ring, where sums and products of elements are sometimes but not always defined and ordinary ring axioms hold for all defined expressions. The model is provided by pairs (f,U) where U is open in some space and f is a suitable function on U.

The idea of working directly with "sheaves" in some axiomatic manner, bypassing an initial topological point set, is also the starting point of the theory of topoi. For the purposes of algebraic geometry, semi-algebraic geometry and algebraic topology, I think topos theory should be refined in two ways. First, rings should be built into the basic definitions, and secondly a purely finite theory should be developed. These two modifications are related.

One possible approach to a more finite topology is to simply look for substitutes for the axiom that "arbitrary" unions of open sets are open. Perhaps interesting alternatives do exist. However, I think that such an approach would suffer from lack of natural examples and with such an approach the appropriate notion of morphism would be obscure. Categorical products would probably also be awkward. On the other hand, in the theory of Grothendieck topoi, a very natural finiteness assumption does exist, namely the Noetherian assumption that every allowable cover $\{U_\alpha \to U\}$ has a finite subcover. Thus one should somehow blend set theory and the theory of Noetherian topoi. Here is where a structure ring or sheaf of rings is useful. If one has "rings of functions" $A(X)$ associated to sets X, which are used to define subsets of X in various ways, then one can associate to $X \times Y$ the ring $A(X) \otimes A(Y)$ ($\otimes$ over a suitable ground ring) and thus define various subsets in $X \times Y$. Morphisms and more general correspondences can be studied as suitable subsets of $X \times Y$. Once open sets are defined, the associated Noetherian Grothendieck topos, where covering families mean essentially finite covers, can be studied. Admittedly, in Chapter V, I consider old-fashioned topological sheaves. There, I study the prime convex ideal spectrum of arbitrary partially ordered rings. (Old habits die hard.) In Chapter VIII, however, I try to emphasize the significance of Noetherian topoi in semi-algebraic geometry.

In scheme theory, non-equalities in the structure rings define the open sets. In topology, the inequalities in the structure ring define the open sets. Note that if R is a real closed field, the "interval topology" is generally terrible from the standard viewpoint, as it tends to be totally disconnected, non-separable, non-locally compact, etc. But if one pursues the above finite approach to topology, all this pathology is irrelevant.

To repeat the main point: just because certain interesting functions, say polynomials, or more general algebraic functions, satisfy conditions such as continuity or differentiability, it does not follow that all continuous or differentiable functions are interesting objects of study. Over any real closed field, topology makes sense as the classification of semi-algebraic sets up to semi-algebraic isomorphism. At a point of an affine semi-algebraic set the germs of semi-algebraic functions can neatly be defined as those continuous germs which satisfy an algebraic relation with the coordinate functions. Alternatively, the semi-algebraic function germs are those germs with closed, bounded semi-algebraic graphs. It makes sense to say a germ is r-times continuously differentiable at a point of affine space, $1 \leq r \leq \infty$. The C^∞ semi-algebraic germs are called (germs of) Nash functions and correspond over any real closed field to formal power series solutions of algebraic equations. Thus various classification problems of differential topology have analogues over any real closed field.

In the second half of the book I restrict attention to finitely generated algebras over a totally ordered field. The starting point is the Artin-Schreier theory of ordered fields, which I think properly belongs to a more general theory of partially ordered rings. Most of the results can actually be found in recent and not so recent literature.

Here is a rough outline. Let $R[T_1 \cdots T_n]$ be the polynomial ring. Right from the beginning, two sorts of partial orders present themselves. On the arithmetic side, one can take as positive elements only the sums of squares, denoted $\mathfrak{P}_w$, or the polynomial combinations with only plus signs of squares and finitely many additional polynomials $g_1, \ldots, g_k$, denoted $\mathfrak{P}_w[g_1 \cdots g_g]$. A theorem of Artin assures this latter is admissible as a set of positives if and only if there is a point $x \in R^n$ with $g_i(x) > 0$, $1 \leq i \leq k$. (We assume R real closed.) On the geometric side, let $V \subset R^n$ be a Zariski dense set (that is, a polynomial vanishing on V vanishes everywhere). The set $\mathfrak{P}(V)$ of polynomials nowhere negative on V defines a partial order on $R[T_1 \cdots T_n]$.

Artin's solution of Hilbert's 17th problem can be interpreted as an algebraic relation between the orders $\mathfrak{P}_w$ and $\mathfrak{P}(R^n)$. Namely, if f is a nowhere negative polynomial, Artin proved that for suitable polynomials h, h_i, $h^2 f = \Sigma h_i^2$. In other words, as a rational function, f is a sum of squares.

Given a partial order $\mathfrak{P} \subset B$ generally, define the derived order $\mathfrak{P}_d$ to be the subset of those $x \in B$ which satisfy $px = q$, for some $p,q \in \mathfrak{P}$ with p not a zero divisor. Thus, Artin's theorem becomes $(\mathfrak{P}_w)_d = \mathfrak{P}(R^n)$.

The motivation for introducing the derived order is the following. The geometric notion of a function being nowhere negative is, to a large extent, a birational notion, depending only on behavior on a dense open set. Artin's theorem is quite reasonable, in fact, expected, when considered from this viewpoint. Similarly, an inequality such as $f^3 \geq 0$ does not abstractly yield $f \geq 0$. However, $f^3 \in \mathfrak{P}$ does give $f \in \mathfrak{P}_d$, at least if f is not a zero divisor, and certainly if we seek a function theoretic interpretation of our partially ordered rings, the deduction of $f \geq 0$ from $f^3 \geq 0$ is desirable.

Consider $\mathfrak{P} = \mathfrak{P}_w[g_1 \ldots g_k]$, $g_i \in R[T_1 \ldots T_n]$. Let $U\{g_1 \ldots g_k\} = \{x \in R^n | g_i(x) > 0,\ 1 \leq i \leq k\}$, an open set. A generalization of Artin's theorem reads $\mathfrak{P}_d = \mathfrak{P}(U)$. Of course, $\mathfrak{P}(U) = \mathfrak{P}(\overline{U})$, where $\overline{U}$ is the closure of U, but decidedly $\overline{U}$ may not include certain degenerate points y, where all $g_i(y) \geq 0$. Birationally, these degenerate points are lost from the semi-algebraic set $W = \{x \in R^n | g_i(x) \geq 0\}$.

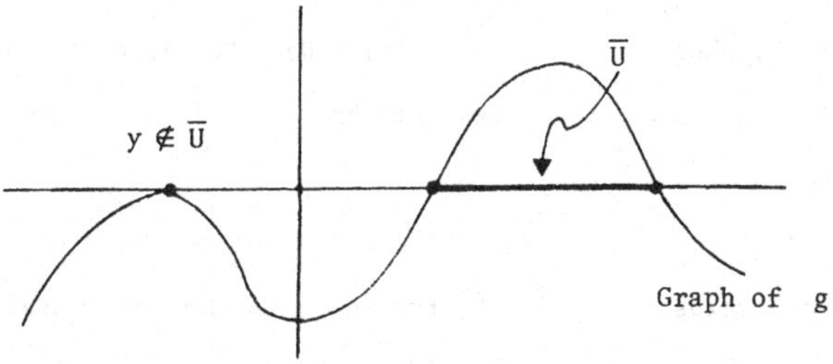

The degenerate points are also lost from the point of view of ideals. The real Nullstellensatz says that the maximal convex ideals of $R[T_1 \ldots T_n]$, relative to the weak order $\mathfrak{P}_w$, correspond precisely to the points $x \in R^n$ in the usual manner of evaluating functions at x. If I is any $\mathfrak{P}_w$-convex ideal, $\sqrt{I}$ consists of precisely the functions vanishing on the zeros of I. Suppose now $\mathfrak{P}_w$ is replaced by $\mathfrak{P} = \mathfrak{P}_w[g_1 \ldots g_k]$. The maximal $\mathfrak{P}$-convex ideals are the points $W = \{x \in R^n | g_i(x) \geq 0, 1 \leq i \leq k\}$. This retains an arithmetic character. However, the $\mathfrak{P}_d$ maximal convex ideals are just the points of $\overline{U}$, where $U = \{x | g_i(x) > 0, 1 \leq i \leq k\}$ as above. In both cases, the appropriate strong Nullstellensatz also holds. The set $\overline{U}$ is semi-algebraic. This is *not* trivial, but follows from the Tarski-Seidenberg theorem, which we discuss below.

Let $Q \subset A = R[T_1 \ldots T_n]$ satisfy $Q = \sqrt{Q}$, and Q convex for some order $\mathfrak{P} \subset A$. The general theory in this first volume guarantees that the finitely many associated primes of Q are also $\mathfrak{P}$-convex, $Q = \bigcap_{i=1}^{m} Q_i$. Also, if A/Q is given its residue partial order $\mathfrak{P}/Q$, the maximal convex ideals will correspond to the maximal $\mathfrak{P}$-convex ideals of A which contain Q. However, if one considers other orders such as $(\mathfrak{P}/Q)_d$, $\mathfrak{P}_d/Q$, $(\mathfrak{P}_d/Q)_d$, the maximal convex ideal spectrum may change--certain degenerate zeros of Q will not be convex. In fact, Q might not even be convex for $\mathfrak{P}_d$. (Remark that even if $\mathfrak{P} = \mathfrak{P}_d$, one does not necessarily have $\mathfrak{P}/Q = (\mathfrak{P}/Q)_d$.) For example, when $\mathfrak{P} = \mathfrak{P}_w[g_1 \ldots g_k]$ and Q is prime, the maximal convex ideal spectrum of $(A, \mathfrak{P})$ is $W = \{x \in R^n | g_i(x) \geq 0\}$ as before, and the maximal convex ideal spectrum of $(A/Q, \mathfrak{P}/Q)$ is the set of zeros of Q in W. If all $g_i \notin Q$, then Q will be $\mathfrak{P}_d$ convex. However, if some $g_j \in Q$, Q will not be $\mathfrak{P}_d$ convex unless Q has "enough zeros" in $\overline{U}$, where $U = \{x \in R^n | g_i(x) > 0\}$. Specifically, $\overline{U}$ must contain all the zeros of Q in some neighborhood of an algebraic simple zero of Q.

Continuing, if Q is $\mathfrak{P}_d$-convex, the maximal convex ideals of $(A/Q, \mathfrak{P}_d/Q)$ are the zeros of Q in $\overline{U}$, but the maximal convex ideals of $(A/Q, (\mathfrak{P}_d/Q)_d)$ are only those zeros x such that every neighborhood of x in $\overline{U}$ contains an entire neighborhood of simple points. If all $g_i \notin Q$, then $(\mathfrak{P}/Q)_d = (\mathfrak{P}_d/Q)_d$. If some $g_j \in Q$, obviously $\mathfrak{P}/Q$ coincides with the order $\mathfrak{P}_w[g_i | g_i \notin Q]/Q$, so the above discussion applies with fewer g_i. In any case, $(\mathfrak{P}/Q)_d$ consists of exactly the functions in A/Q nowhere negative on the maximal convex ideals of $(A/Q, (\mathfrak{P}/Q)_d)$. This is a nice generalization of Artin's theorem. The pictures below indicate how the sets above can differ.

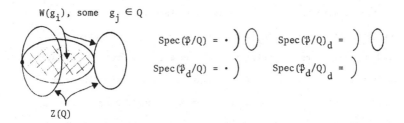

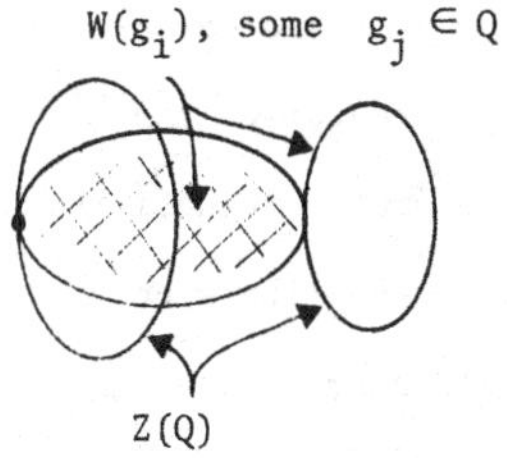

$\mathrm{Spec}(\mathfrak{P}/Q) = \cdot\ )\ ()$ $\qquad$ $\mathrm{Spec}(\mathfrak{P}/Q)_d = \ )\ ()$

$\mathrm{Spec}(\mathfrak{P}_d/Q) = \cdot\ )$ $\qquad$ $\mathrm{Spec}(\mathfrak{P}_d/Q)_d = \ )$

Another natural problem generalizing Artin's theorem is to find an algebraic characterization of functions non-negative on the set of all zeros of Q in W, that is, the maximal convex ideals of $(A/Q, \mathfrak{P}/Q)$. This is solved by a theorem of Stengle, which implies that such $f \in A/Q$ are precisely those for which an equation

$$(*) \qquad (f^{2n} + p)f \equiv q \pmod{Q}$$

holds, with $p, q \in \mathfrak{P}$. In the language of our structure sheaf of partially ordered rings on $\mathrm{Spec}(A/Q, \mathfrak{P}/Q)$, this condition translates to the statement that f is "positive" in the partially ordered ring of "sections over the basic open set $D(f)$". Similar results hold for $(A/Q, \mathfrak{P}_d/Q)$, characterizing the functions non-negative on all zeros of Q in $\bar{U}$.

The paragraphs above discuss "irreducible" affine semi-algebraic sets. A general closed, affine semi-algebraic set S is a union of finitely many sets, each defined by finitely many polynomial equalities and inequalities. By the affine coordinate ring of S, we mean the ring $A(S)$, obtained from the polynomial ring by dividing by the ideal $I(S)$ of functions which vanish on S, together with the partial order $\mathfrak{P}(S)$ consisting of functions nowhere negative on S. Our general theory allows us to identify S with the maximal convex ideal spectrum of $(A(S), \mathfrak{P}(S))$. Moreover, two general results allow us to reduce much of the study of arbitrary S to the irreducible case.

First, the associated primes of any absolutely convex ideal are always convex. Secondly, if $\mathfrak{P}_1$ and $\mathfrak{P}_2$ are two orders on a ring and Q is a prime ideal convex for $\mathfrak{P}_1 \cap \mathfrak{P}_2$, then Q is either $\mathfrak{P}_1$-convex or $\mathfrak{P}_2$-convex. (This last result was found by A. Klapper.) As a corollary, the union of two sets $W_1 = \{x | g_i(x) \geq 0, 1 \leq i \leq r\}$ and $W_2 = \{x | h_j(x) \geq 0, 1 \leq j \leq s\}$ can be identified with the maximal convex ideal spectrum of the order $\mathfrak{P}_w[g_i] \cap \mathfrak{P}_w[h_j]$, if indeed $\mathfrak{P}_w[g_i]$, $\mathfrak{P}_w[h_j]$ are orders. Otherwise, the ideals $I(W_1)$, $I(W_2)$ are non-trivial. In general, degenerate inequalities on one variety are handled by passing to non-degenerate inequalities on subvarieties. Finally, the theorems of Artin and Stengle characterizing non-negative functions on certain sets can be used to give necessary and sufficient conditions for $f \in A(S)$ to belong to $\mathfrak{P}(S)$. These conditions are purely algebraic formulas for f, like (*) above, expressed in terms of the original finite collections of polynomials which define S. Just as in algebra, where the ideal generated by a set of elements is more important than the specific basis, in our case the invariant notion of a polynomial being non-negative on a semi-algebraic set has more geometric significance than the particular defining equations and inequalities, and yet this invariant geometric notion is *algebraically* expressible in terms of the defining polynomials.

In the end, perhaps the following is the neatest characterization of the affine coordinate rings $(A(S), \mathfrak{P}(S))$ among all partially ordered rings $(A, \mathfrak{P})$. First, A should be a reduced algebra of finite type over R. Secondly, there should exist finitely many $\mathfrak{P}$-convex primes $P_i \subset A$ with $(0) = \cap P_i$ and orders $\mathfrak{P}_i \subset A_i = A/P_i$ which are derived orders of finite refinements of the weak order $\mathfrak{P}_i = (\mathfrak{P}_w[g_{ij}])_d$, such that $\mathfrak{P} = A \cap \Pi \mathfrak{P}_i$, where $A \to \Pi A_i$ is the natural inclusion. The P_i need not be distinct nor minimal, although the minimal primes do all occur. These basic building blocks $(A_i, \mathfrak{P}_i)$ are as discussed above. The maximal convex ideals will correspond to the points in the *closure* of the set of *simple* zeros x of P_i at which all $g_{ij}(x) > 0$. A set of primes $\{P_i\}$ which leads to such a formula for $(A(S), \mathfrak{P}(S))$ can be intrinsically described.

We also establish the basic results of dimension theory in our semi-algebraic category. That is, simple points do exist and the expected relations between transcendence degree and chains of convex prime ideals hold. For *any* real closed field we prove an implicit function theorem. The statement is analogous to the classical case of real numbers: if $f_1,\ldots,f_n$ are smooth algebraic functions defining a germ $f\colon (R^n,0) \to (R^n,0)$ with $df(0)$ non-singular, then f has a smooth algebraic inverse near 0. (This is the inverse function theorem. Implicit function theorems are routine corollaries.) Note that if one *has* an R-valued function germ, it makes sense to ask if it has derivatives. One should not get carried away and try to study "all differentiable functions." There are quite nice relations between formal algebraic derivations, abstract partially ordered rings, and the usual ε-δ definitions of derivatives. Once the implicit function theorem is available, it is routine to give Whitney type stratifications of semi-algebraic sets into non-singular manifold-like strata, with any real closed field as ground field.

The study of quotients A/Q, where $Q \neq \sqrt{Q}$ is more complicated. Suppose Q is $\mathfrak{P}$ absolutely convex $\mathfrak{P} = (\mathfrak{P}_w[g_1\cdots g_k])_d$. General theory yields the convexity of all associated primes of Q and also the absolute convexity of the isolated primary components. However, embedded primary components must be chosen carefully, before one can establish an absolutely convex primary decomposition of Q. Very crucial to the argument are (i) the restriction to finitely generated algebras over fields, (ii) the restriction to absolutely convex ideals (category (PORCK) rather than (POR)), (iii) the specific form of $\mathfrak{P}$ as the derived order of a finite extension of the weak order. Dropping any of these conditions leads to convex ideals which cannot be expressed as intersections of convex primary ideals, even for Noetherian ambient rings B.

A quick example might be in order. Consider $(X^2,XY) \subset R[X,Y]$, with isolated prime (X) and embedded prime (X,Y). The isolated primary component is also (X), and in pure algebra, one has a large choice for the embedded component, say (X^2,Y), $(X^2, Y-cX)$, (X^2,Y^2,XY), etc. Suppose now, however, that among the order relations are $0 \leq X$, $0 \leq Y$, $X \leq Y^n$. Any convex ideal containing Y^n must contain X. So one must choose as primary component belonging to (X,Y), one which contains X^2, but no power of Y less than Y^{n+1}. In fact, $(X^2,XY) = (X) \cap (X^2,Y^{n+1},XY)$ will work if $\mathfrak{P} = (\mathfrak{P}_w[X,Y,Y^n - X])_d$.

If one included infinitely many generators for the inequalities, say $0 \leq X,Y$ and $0 \leq Y^n - X$, all $n \geq 1$, then X becomes "infinitesimal". There would clearly be no primary decomposition of (X^2,XY), even though it is absolutely convex. Because, any ideal with radical (X,Y) contains a power of Y, hence X, if it were to be convex. This counterexample shows where the usual proof of primary decomposition breaks down. Basically, principal ideals, which are used in the classical proofs, are not generally convex, and putting one element in a convex ideal forces a lot of other elements to be included.

Order properties of A/Q, $Q \neq \sqrt{Q}$, are also complicated. As hinted at earlier, the results which seem most natural in this direction relate various partial orders on A/Q to orders on $A = R[X_1 \cdots X_n]$ obtained by regarding A as a ring of *germs* of functions on semi-algebraic neighborhoods of the zeros of Q. This infinitesimal study of functions, along with primary decomposition discussed above, is intimately connected with the relations between partial orders on rings and differential operators. I expect partially ordered rings with nilpotent elements to be useful in topological intersection theory, providing an alternative approach to transversality.

A nice application of partially ordered rings with nilpotent elements is the following characterization of geometric simple points. First note that in the semi-algebraic category, points may be simple which do not look simple to an algebraist. In Figure (a), the equation $y^3 + 2x^2y - x^4 = 0$

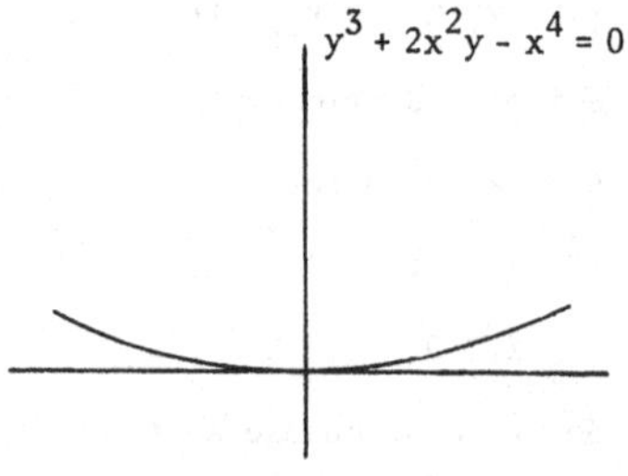

Figure (a)

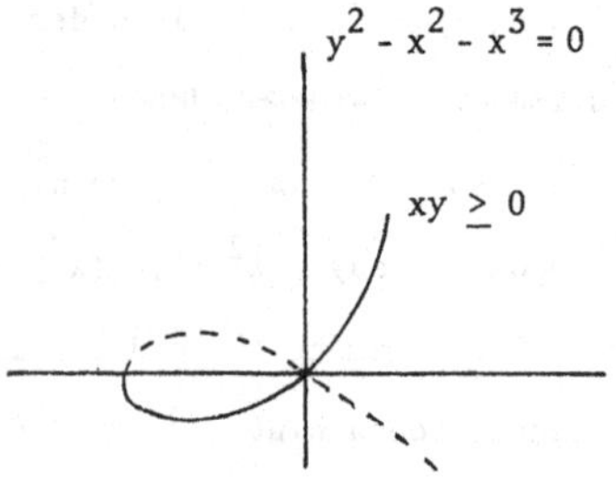

Figure (b)

has a formal power series solution (convergent, of course, over the real numbers) at the origin. By all rights, the origin should be a simple point. In Figure (b) the double point is made nice by restriction to the semi-algebraic domain $xy \geq 0$. Again, this branch in the first and third quadrants has a formal power series description. In general, suppose m is a maximal convex ideal of some affine coordinate ring $(B,\mathfrak{P})$ which we assume is an integral domain. We can localize, B_m. We can also localize the order $\mathfrak{P}$ at m, by taking $f \in \mathfrak{P}_m$ if f is non-negative on some neighborhood of m in the maximal convex ideal spectrum of $(B,\mathfrak{P})$. Here, $\mathfrak{P}_m$ will be an order if m is in the closure of suitable algebraic simple points. Our criterion is, then, that m is a geometric simple point if the associated graded ring

$$G(B_m) = \bigoplus_{n \geq 0} \frac{AH(m^n B_m)}{AH(m^{n+1} B_m)}$$

is a polynomial ring and the positive elements in the induced order are the polynomials which are non-negative in some neighborhood of the origin in the appropriate affine space. Here, the absolute hulls are taken with respect to the order $\mathfrak{P}_m \subset B_m$. Thus, for example, in Figure (a), $0 \leq y \leq 2x^2$ in the order $\mathfrak{P}_m$, so x and y are not linearly independent in $mB_m/AH(m^2B_m)$. In Figure (b), $0 \leq y - x \leq 2x^2$ in the order $\mathfrak{P}_m$.

Non-singular boundary points of a semi-algebraic set can be defined by modifying the order requirement on the associated graded ring to a statement about germs at the origin in an affine half space. In general, in partially ordered algebra we have these two types of localization available, first, by inverting elements of a ring in the usual way, and, secondly, by leaving the ring alone and refining the order.

In commutative algebra, it is fashionable to study *all* rings and claim rings with nilpotent elements arise naturally in geometry, say in specializations, intersections, and fibres of a morphism, as fibre products. But the formation of a fibre product requires *agreement* first on a suitable category. If, say, fibre products are to be taken in the category of all rings rather than in the category of rings without nilpotent elements, one must justify this by finding convincing illustrations of its geometric usefulness. In real algebraic geometry, there are already restrictions on the allowable rings without nilpotent elements, namely, non-trivial sums of squares cannot vanish. I have discussed two methods of introducing nilpotent elements, the categories (POR) and (PORCK). But, for the most part, the arguments given that (PORCK) is better have been based on purely algebraic properties. The above application of the graded ring construction (which exists in (PORCK)) to characterize simple points, is meant as geometric propaganda for (PORCK). But one would like many such applications, say to intersections and fibres of a morphism. It is clearly inadequate to simply say "take fibre products" since fibre products exist in both (POR) and (PORCK). What is required is close scrutiny of the geometry involved in these problems in order to decide which category is giving the "desired" answers. It is, of course, possible that even (PORCK) is not the "right" category for all purposes.

Given a ring which admits a partial order, the set of all partial orders is an interesting structure. The partial orders themselves are partially ordered by refinement, arbitrary intersections of partial orders are partial orders, and Zorn's lemma implies any partial order admits maximal refinements. In an integral domain, all maximal orders are total orders. Also, in an integral domain, and in certain other situations, the intersection of all maximal refinements of an order $\mathfrak{P}$ is exactly the derived order $\mathfrak{P}_d$. In general, the orders $\mathfrak{P}$ which satisfy $\mathfrak{P} = \mathfrak{P}_d$ correspond exactly to partial orders on the larger ring obtained by inverting all non-zero divisors. Thus, derived orders on the polynomial ring $R[T_1 \dots T_n]$ correspond to all partial orders on the field $R(T_1 \dots T_n)$.

These derived partial orders on $R[T_1 \dots T_n]$ can be interpreted function-theoretically. Define a family $\mathcal{V}$ of open subsets of R^n as follows. For each finite collection of polynomials $g_1, \dots, g_k$ let $V\{g_1, \dots, g_k\} = \{x \mid \text{each } g_i \text{ is locally non-negative at } x\}$. (Thus V is the interior of the closure of the set U where all g_i are strictly positive. It is a theorem that these V are semi-algebraic sets.) We allow only non-empty V, that is, $g_1, \dots, g_k$ such that $\mathfrak{P}_w[g_1 \dots g_k]$ is a partial order on $R[X_1 \ldots X_n]$. Then there is a natural, bijective, refinement preserving correspondence between derived orders on $R[T_1 \ldots T_n]$ and filters $\mathscr{F}$ in the family of sets $\mathscr{V}$. Total orders correspond to ultrafilters. Given a filter $\mathscr{F}$ the associated order $\mathfrak{P}(\mathscr{F})$ is defined by $f \in \mathfrak{P}(\mathscr{F})$ provided $f|_V \geq 0$, some $V \in \mathscr{F}$. The proof is a simple argument based on Artin's work.

In any event, there are tremendously many partial orders, even total orders, on the function field $R(X_1 \ldots X_n)$. In fact, these total orders are quite analogous to valuations, or places. Krull, in his original work on valuation theory, pointed out that given a totally ordered field E and a subfield F, one obtains immediately a valuation ring in E, consisting of elements finite relative to F. The maximal ideal consists of elements infinitesimally close to 0, relative to F, and the residue field Δ contains F, is naturally ordered, and is Archimedean over F. Krull proved a converse, which he attributed to earlier work of R. Baer. Namely, given a place on a field E with residue field an ordered field, then one can lift this situation to an ordering of E, yielding the original place by the construction above. Lang generalized Krull's results and established further relations between total orders and real places. In particular, he emphasized how the theory of real places can be applied to the study of real algebraic varieties.

I go a little further and develop carefully the notion of signed place. A signed place is a place with values in a totally ordered field in which $+\infty$ and $-\infty$ are distinguished in an arithmetically coherent fashion. This is clearly the thing to study in real algebra. A useful signed place extension theorem is proved, which has applications to certain versions of the Nullstellensatz and to dimension theory. Signed places are intimately related to total orders and subfields. Geometrically, they correspond to studying a variety infinitesimally near a subvariety, and they also provide a convenient language for dealing with behavior at infinity in affine space.

A somewhat valid objection to the place theoretic approach is that it has a non-constructive character (Zorn's lemma) at least as complicated as some reasoning about limits and continuity that I am trying to circumvent in the first place. However, I think this is more illusory than real. By the approach of Tarski and Seidenberg to some of these same problems, the algebraic foundations of the theory of algebraically closed and real closed fields can be established quite constructively, more or less by the methods of classical elimination theory. This approach yields rather trivial proofs of the basic results concerning varieties, once the initial logical inductive arguments are made, but has the disadvantage of obscuring the underlying geometry. More precisely, the Tarski-Seidenberg theorem asserts the existence of a decision procedure for answering any *elementary* (in the logical sense) question about affine semi-algebraic sets, and semi-algebraic functions. Thus, if one can test a statement in some larger real closed field, or even over the classical real numbers, the answer will be the same in the original real closed field as in this new field. In this sense transcendental methods will play the same role in semi-algebraic geometry over any real closed field as they have played in classical algebraic geometry over algebraically closed fields of characteristic zero. In some cases the shortest proof of a theorem may be transcendental, but algebraists should still be interested in conceptually nice, direct algebraic proofs.

I would like to establish certain basic properties of semi-algebraic sets, including the result that an algebraic image of a semi-algebraic set is semi-algebraic by another method, namely by exploiting properties of integral or semi-integral extensions. (Recall $x \in B$ is semi-integral over A if $f(x) \leq 0$ for some monic, even degree polynomial f with coefficients in A.) I have not been very successful in this, but I still believe it might be possible. Semi-integral elements are related to signed places or total orders pretty much the same way that integral elements are related to places. Namely, given $A \subset B$ with partial orders, and $x \in B$ semi-integral over A, then x is necessarily finite relative to A in any total order on B extending the partial orders. If A,B are affine coordinate rings of semi-algebraic sets X,Y and $A \rightarrow B$ is a (POR) morphism inducing $Y \rightarrow X$, then B semi-integral over A means the pull-backs of closed, bounded subsets of X are closed and bounded in Y. The rings A,B need not have the same transcendence degree over the ground field.

As a final remark on the birational interpretation of Artin's solution of Hilbert's 17th problem, consider a different category, that of smooth manifolds and smooth real valued functions. Then Paul Cohen has shown me that (i) there exist nowhere negative smooth functions on any manifold which are not finite sums of squares of smooth functions (in fact, the zero set can be a single point) and (ii) given any nowhere negative smooth f, there are h, g, both smooth and h not a zero divisor, such that $h^2f = g^2$. The zero divisors are, of course, the smooth functions which vanish on some open set. If M is the manifold, $\mathfrak{P}(M) \subset C^\infty(M)$ the nowhere negative functions, and $\mathfrak{P}_w \subset C^\infty(M)$ the sums of squares, this result again reads $(\mathfrak{P}_w)_d = \mathfrak{P}(M)$. Note that because of the existence of smooth bump functions, any partial order on $C^\infty(M)$ lies between $\mathfrak{P}_w$ and $\mathfrak{P}(M)$.

Following the development of the affine theory of semi-algebraic sets, it would be interesting to develop the basic results about projective coordinates in the real domain. Actually, there are some important additional possibilities for homogeneous methods. It is true enough that the zeros of a real homogeneous ideal is naturally a subset of projective space. However, for forms of odd degree the set of points where the form is non-negative is only sensible on the sphere, the double cover of projective space. Thus, from the point of view of semi-algebraic geometry, spherical homogeneous coordinates are better than projective coordinates. A third alternative is the hemisphere, since given any form, its values on the sphere are determined by the values on any hemisphere. Also, the usual manner of passing from an affine variety to a projective variety, thus including points at infinity, seems in the real domain to be most naturally expressed by passing from a plane to a hemisphere by radial projection. Finally, from a topologists point of view, the hemisphere is the natural domain for working with pairs of spaces.

One of my goals is to develop such homotopy theoretical concepts as K-theory, cohomology, cohomology operations, and techniques for computing homotopy classes of maps in the framework of partially ordered algebras. So far my ideas are more speculative than concrete, but some discussion might be in order in this introduction, to encourage further interest.

The idea of recovering properties of a space from a ring of functions is quite classical, of course. There are the well-known results that a compact Hausdorff space X is the space of maximal ideals of the ring C(X) of all continuous real valued functions on X, and that there is a natural correspondence between real vector bundles over X and finitely generated projective C(X) modules. There are also definitions of cohomology of commutative rings in the literature which give the Alexander-Spanier or Cech cohomology of X with real coefficients when applied to the ring C(X). Few of these results emphasize order properties of C(X). This is perhaps because the non-negative functions are exactly the squares in C(X), so it is easy to disguise order agruments by making use of obvious properties of squares. For example, a sum of squares in C(X) has no zeros in X, unless the summands have a common zero. Also, any ring map $C(Y) \to C(X)$, X, Y both compact Hausdorff spaces, is necessarily order preserving, and gives rise to $X \to Y$. For smaller rings, such as finitely generated rings of functions, the order properties are not as close to the algebra.

The definition of real K-theory seems to provide no problem. First, it is pretty clear that K-theory is more naturally associated to sheaves of rings, rather than to single rings. So the natural thing to consider is the Grothendieck group associated to sheaves of finitely generated projective modules over the structure sheaf of a semi-algebraic variety X. This works and the proof is already in the literature in slightly different form. Namely, all one really has to do is localize affine coordinate rings by inverting all polynomials with no zeros on the semi-algebraic set X. This gives enough continuous functions on X to approximate the idempotent operators on free C(X) modules by idempotent operators on free modules over this much smaller localized affine coordinate ring.

Inverting functions with no zeros on X also has the effect of reducing the ordinary maximal ideal spectrum exactly to the points of X, and this is perhaps regarded generally as the algebraic rationale for this localization. I regard the ordinary maximal ideals as irrelevant. Besides, the localization does not eliminate all unwanted prime ideals. In the partially ordered context, the maximal convex ideal spectrum and prime convex ideal spectrum are both regulated by the partial order on the ring, but are not further affected by the above localization. The localization does occur naturally in the construction of structure sheaves, however, and K-theory should involve the structure sheaf.

Similarly, there are examples of affine coordinate rings for real curves whose ideal class groups change after this localization, the class group after localization being closer to the geometry. If one defines an ideal class group in the partially ordered category by considering only absolutely convex ideals and interpreting principal ideals as those ideals which are minimal absolutely convex ideals containing some single element, then, again, the localization above does not change anything. In higher dimensions, one has a similar discussion for divisors, and an algebraic formulation of $H^1(\ ,\mathbb{Z}/2)$. For curves, of course, $H^1(\ ,\mathbb{Z}/2)$ and K-theory coincide.

Perhaps a harder invariant to recover neatly from a sheaf of partially ordered rings is the cohomology (with general coefficients) of a semi-algebraic variety. An approach which probably works is to simply write down some simplicial complex, or semi-simplicial complex, of the correct homotopy type. For example, Sullivan and De Ligne have suggested that given X defined by real polynomial equalities and inequalities, there is an algorithm for paving affine space with, say, rectangles which along with their faces intersect the solution set X in a contractible space, if at all. The nerve of the resulting cover of X is a simplicial complex of the correct homotopy type. Alternatively, one could probably construct a kind of semi-algebraic singular complex, or even a triangulation, using only constructive algebraic steps. In some sense, these approaches are not in the right spirit. On the other hand, they are related to the general problem of understanding how the "topological" type (by this I mean classification up to semi-algebraic isomorphism) or homotopy type of a real semi-algebraic set changes, with changes in the parameters, the coefficients of the polynomial equalities and inequalities which define the set. Presumably, such changes can occur only along semi-algebraic locii in the parameter space, *defined over* Q, and this seems to me to be the real reason that algebraic topology can be reduced to combinatorics and number theory.

More direct algebraic approaches to cohomology might be based on sheaf theory or on algebraic versions of Alexander-Spanier cohomology or de Rham cohomology. The Alexander-Spanier method is based on functions on multiple products of X with itself, modulo functions which "vanish near the diagonal". This is not so hard to formulate algebraically, using tensor products, augmentation ideals and completions, or inverse limits. In fact, this approach already exists in algebraic geometry, and no doubt provides an algebraic computation of the cohomology with complex coefficients of a complex affine variety. I believe the correct interpretation would also work for real semi-algebraic varieties, with any coefficients.

De Rham cohomology, on the other hand, is problematical. It is true enough that the complex cohomology of complex affine varieties whose singularities are not too bad, can be computed as the cohomology of the exterior algebra complex on the module of Kähler differentials over the affine coordinate ring. For example, the variety defined by $XY - 1 = 0$ has coordinate ring $\mathbb{C}[X,X^{-1}]$ and the cohomology is generated by the constant 0-form 1, and the closed but not exact 1-form dx/x. In the real domain, the variety $XY - 1 = 0$ is R^*, not $\mathbb{C}^*$, but the de Rham complex will produce the same answer which is now wrong. The point seems to be that in the complex domain, if an algebraic form has any integral at all, it has an algebraic integral, while in the real domain, forms like dx/x on R^* have integrals, but outside the world of algebra.

At first, I hoped to salvage de Rham cohomology for real semi-algebraic sets, perhaps by modifying the notion of module of differential forms in some appropriate way, or by justifying the formal adjunction of solutions of algebraic differential equations to the structure sheaves, whenever such solutions were formally consistent. This latter viewpoint would be somewhat similar to the two conceptions of, say, $\sqrt{2}$, first as a limit of rational numbers and secondly as a formal solution of $X^2 - 2 = 0$ in some ordered field extension of the rationals.

Some such interpretation of de Rham cohomology might work. However, I have come to take the philosophical position that it should not be expected to work and that it works as well as it does in the complex case only by luck or for some reason we do not yet understand. Certainly, the notion of algebraic differential equation seems close to the dividing line between algebra and geometry. But I believe that the mechanism of modeling natural phenomena on the continuous solutions of differential equations was never meant as more than an approximation of hopelessly complicated combinatorial situations. The purely mathematical introduction of differential equations into science reinforces the use of arbitrary analytic or even differentiable functions in geometry. But the invariants of algebraic topology, such as cohomology, are global combinatorial invariants, which should retain their significance long after the real numbers, at least most of them, are forgotten. Thus, there seems no reason to believe that de Rham cohomology, with its dependence on the completeness of the real numbers, can be interpreted naturally in a purely algebraic context over a real closed field.

Another piece of mathematics close to the boundary between algebra and topology is Morse theory and the theory of singularities in general. Morse theory originated in analysis and is generally regarded as a tool of differential topology, but I believe it has a purely algebraic significance, which shows up most clearly in the well-known applications of Morse theory to the topology of real or complex varieties. In these applications, one begins with a variety and some real algebraic function. Then one describes information about the critical points, say second order behavior, by algebraic computations. Then one says, the variety is a smooth manifold, and smooth manifolds with smooth real functions with certain critical point behavior have such and such local and global geometric properties. It seems to me that this appeal to analysis should be unnecessary. The critical point behavior of real algebraic functions on real semi-algebraic sets should be directly interpretable in abstract semi-algebraic geometry. In fact, this discussion reaches to the very heart of the distinction between algebraic and transcendental methods. In differential topology, one proves varieties equivalent by integrating vector fields, say, provided by Morse functions. But if the varieties are algebraic, then Nash's approximation theorem guarantees they are also algebraically equivalent, in a suitable way. (Their sheaves of Nash functions are isomorphic.) Thus one ought to investigate whether the transcendental methods are really essential or whether they serve merely as a crutch which has been used simply because the algebra is too hard.

Certainly, semi-algebraic and smooth algebraic classification problems are extremely difficult. A more modest goal would be to find an algebraic approach to cohomology or homotopy theory, based on an abstract Morse theory. As supporting evidence, I cite the result of Harrison and others that signature type homomorphisms from the Witt ring of quadratic forms over a field to the integers correspond to possible orderings of the field. Knebusch has applied quadratic forms to the algebraic study of components of real curves.

One of the most important techniques for studying real algebraic varieties is to regard them as the fixed points of an associated complex variety on which conjugation acts as an involution. The germ of this technique is obviously the equality $\mathbb{C} = \mathbb{R}[i]$. Now, among the fantastic results of Artin-Schreier theory are that if R is any real closed field, then $R[i]$ is algebraically closed, any algebraically closed field of characteristic zero can be written $R[i]$ for some real closed field R (although non-unique, in general, even up to isomorphism), and the only fields of finite codimension under their algebraic closure are the real closed fields. Certainly, no analytical or topological result about complex varieties can be proved without ultimately some appeal to properties of real numbers, precisely because the norm is defined in terms of the real numbers.

Nonetheless, many very nice results on real algebraic varieties are established in the literature by making use of the existence of a complexification. For example, (i) Whitney used the complexification to study stratifications of real algebraic varieties by dimension, (ii) properties of real algebraic curves are often deduced from classical results about complex curves, (iii) the usual proof that a real algebraic variety carries a fundamental $\mathbb{Z}/2$-homology class (there are an even number of sheets at the singular set) makes use of the fundamental $\mathbb{Z}$-homology class of the complexified variety and conjugation. Sullivan has generalized these last results. He shows that a point on a complex variety has a neighborhood which is a cone on a space of zero Euler characteristic and deduces that a point on a real variety has a neighborhood which is a cone on a space of even Euler characteristic.

More in the direction of algebraic topology, Atiyah has shown how real K-theory can be efficiently studied via complex vector bundles with conjugation, and Sullivan has suggested studying the homotopy theory of a real algebraic variety by means of the canonical involution on the étale homotopy type of its complexification.

I feel that this extensive use of the complexification may be attributable more to a lack of suitable algebraic foundations in the real case, rather than to any natural priority of the algebraically closed case. Also, semi-algebraic sets do not complexify, so a real theory is certainly necessary for their study. In any event, efficient proofs of the results described above, entirely within the realm of partially ordered algebra, offers a reasonable challenge to the theory. The study of stratifications, and basic properties of one dimensional semi-algebraic sets are in fact not difficult, using the machinery of this book.

I · Partially ordered rings

1.1. Definitions

By "ring" we always mean commutative ring with unit. All ring homomorphisms map unit elements to unit elements. We allow $1 = 0$, but obviously this holds only in the trivial ring (0).

By a *partially ordered ring*, we mean a ring A and a subset $\mathfrak{P} \subset A$ which satisfies

(i) $\mathfrak{P} \cap (-\mathfrak{P}) = \{0\}$

(ii) $\mathfrak{P} + \mathfrak{P} \subset \mathfrak{P}$ and $\mathfrak{P} \cdot \mathfrak{P} \subset \mathfrak{P}$

(iii) $a^2 \in \mathfrak{P}$ for all $a \in A$.

By a *morphism* of partially ordered rings, $f\colon (A, \mathfrak{P}_A) \to (B, \mathfrak{P}_B)$, we mean a ring homomorphism $f\colon A \to B$ with $f(\mathfrak{P}_A) \subset \mathfrak{P}_B$. We thus have a category of partially ordered rings which we denote (POR).

If $(A, \mathfrak{P}) \in$ (POR), we refer to $\mathfrak{P}$ as an *order* on A. We call elements of $\mathfrak{P}$ *positive* and elements of $\mathfrak{P}^+ = \mathfrak{P} - \{0\}$ *strictly positive*. In fact, the subset $\mathfrak{P} \subset A$ defines a partial order relation on A by $a \leq b$ if and only if $b - a \in \mathfrak{P}$. It is easy to redefine partially ordered rings and morphisms in terms of the relation $\leq$. For example, $f\colon A \to B$ is a morphism in (POR) if $a \leq a'$ implies $f(a) \leq f(a')$. The notation $b - a \in \mathfrak{P}$ is somewhat preferable to $a \leq b$, since the former makes it more clear which partial order relation on A is involved. Nonetheless, the latter is often notationally simpler, and we use it routinely.

If we add axioms to the definition of partially ordered rings, we can define subcategories of (POR). For example, one such axiom simply prohibits nilpotent elements.

(iv) If $a^2 = 0 \in A$, then $a = 0$.

This is a natural axiom for studying rings of real valued functions on a set. We denote the resulting subcategory (PORNN) $\subset$ (POR), partially ordered rings with no nilpotents.

A more subtle axiom is the following.

(v) If $(p_1 + p_2)x = 0$, $p_i \in \mathfrak{P}$, $x \in A$, then $p_1x = p_2x = 0$.

If $(A,\mathfrak{P}) \subset (POR)$, a set $C \subseteq A$ is *convex* if $p_1 + p_2 \in C$, $p_i \in \mathfrak{P}$ implies $p_1, p_2 \in C$. Thus axiom (v) says the annihilator of x is convex, all $x \in A$. We denote the resulting subcategory (PORCK) $\subset$ (POR), partially ordered rings with convex killers. Roughly, the motivation for axiom (v) is that it is a natural axiom for studying real valued functions on a set modulo those functions which vanish to some preassigned order on a subset.

Axioms (i) and (iv) imply (v). Namely, if $(p_1 + p_2)x = 0$, then $p_1x^2 + p_2x^2 = 0$, hence $p_1x^2 = p_2x^2 = 0$, by (i). But then $p_1^2x^2 = p_2^2x^2 = 0$ and by (iv), $p_1x = p_2x = 0$. Thus, (PORNN) $\subset$ (PORCK) $\subset$ (POR). Since the definition of morphism is unchanged, each subcategory is a full subcategory.

A third axiom which might be interesting is

(vi) If $(p_1 + p_2)x \in \mathfrak{P}$, $p_i \in \mathfrak{P}$, $x \in A$, then $p_1x, p_2x \in \mathfrak{P}$.

We refer to this subcategory as (PORPP), partially ordered rings with plenty of positives. It is easy to see that (i) and (vi) imply (v). Thus (PORPP) $\subset$ (PORCK).

Now, let A be a ring. If $\mathfrak{P}, \mathfrak{P}'$ are orders on A, with $\mathfrak{P} \subset \mathfrak{P}'$, we say that $\mathfrak{P}'$ *refines* $\mathfrak{P}$ (or that $\mathfrak{P}'$ is stronger than $\mathfrak{P}$ or $\mathfrak{P}$ weaker than $\mathfrak{P}'$). The set of orders on a fixed ring A is itself partially ordered by refinement. An arbitrary intersection of orders is again an order and the union of a chain of orders (totally ordered subset of the set of orders) is an order. Thus, any ring A which admits an order has a weakest order $\mathfrak{P}_w$, and, by Zorn's lemma, any order $\mathfrak{P}$ has maximal refinements. These remarks apply equally to the categories (POR), (PORNN), (PORCK) and (PORPP).

1.2. Existence of Orders

Let A be a ring.

Proposition 1.2.1.

(a) A admits an order $\mathfrak{P}$ if and only if the following condition holds:

$$\sum_{i=1}^{n} a_i^2 = 0 \qquad \text{implies} \qquad a_j^2 = 0, \qquad 1 \leq j \leq n.$$

(b) A admits an order $\mathfrak{P}$ with $(A,\mathfrak{P}) \in$ (PORNN) if and only if

$$\sum_{i=1}^{n} a_i^2 = 0 \qquad \text{implies} \qquad a_j = 0, \qquad 1 \leq j \leq n.$$

(c) A admits an order $\mathfrak{P}$ with $(A,\mathfrak{P}) \in$ (PORCK) if and only if

$$\left(\sum_{i=1}^{n} a_i^2\right)x = 0 \qquad \text{implies} \qquad a_j^2 x = 0, \qquad 1 \leq j \leq n. \qquad \square$$

The proof is a simple exercise. In all three categories (POR), (PORNN), (PORCK), the weakest order on a ring A is the same,

$$\mathfrak{P}_w = \left\{ \sum_{i=1}^{n} a_i^2 \,\middle|\, a_i \in A \right\},$$

consisting of just the sums of squares.

If A is a field, the three conditions (a), (b), (c) in the proposition are obviously equivalent, and, in fact, are equivalent to the assumption that -1 is not a sum of squares in A. This is the usual definition of a formally real field.

1.3. Extensions and Contractions of Orders

Let $f\colon A \to B$ be a ring homomorphism and let $\mathfrak{P}_A \subset A$ be an order on A. We ask if there is an order on B, $\mathfrak{P}_B \subset B$, such that $f(\mathfrak{P}_A) \subseteq \mathfrak{P}_B$, that is, such that f is a morphism in (POR). Clearly, the answer is yes if and only if the following condition holds:

$$\sum_{i=1}^{n} f(p_i)b_i^2 = 0, \quad p_i \in \mathfrak{P}_A, \quad b_i \in B, \quad \text{implies} \quad f(p_j)b_j^2 = 0, \quad 1 \leq j \leq n.$$

The weakest such order on B is then

$$f_*\mathfrak{P}_A = \{\sum_{i=1}^{n} f(p_i)b_i^2 \mid p_i \in \mathfrak{P}_A,\ b_i \in B\},$$

which we call the *extension* of $\mathfrak{P}_A$ to B.

More generally, if $f_\alpha\colon A_\alpha \to B$ is a family of ring homomorphisms and $\mathfrak{P}_\alpha \subset A_\alpha$ are orders on the A_α, there are always ideals $J \subset B$ such that the family of maps $\pi f_\alpha\colon A_\alpha \to B \to B/J$ satisfies a generalized extension condition, which can be formulated as follows:

$$\sum_{i=1}^{k}\Big(\prod_{j=1}^{n} f_{\alpha_{ij}}(p_{ij})\Big)b_i^2 \in J,\quad p_{ij} \in \mathfrak{P}_{\alpha_{ij}},\quad b_i \in B,\quad \text{implies}\quad \prod_{j=1}^{n} f_{\alpha_{ij}}(p_{ij})b_i^2 \in J,\quad 1 \le i \le k.$$

One could take $J = B$, for example. More to the point, the family of all ideals $\{J\}$ of B with this property is closed under arbitrary intersections, hence there is a smallest such ideal, $I \subset B$. Then on the quotient ring $\overline{B} = B/I$, one can impose an order $\overline{\mathfrak{P}}$ such that the maps $\pi f_\alpha\colon A_\alpha \to \overline{B}$ are simultaneously order preserving, $\pi f_\alpha(\mathfrak{P}_\alpha) \subset \overline{\mathfrak{P}}$, all α.

Next, suppose $f\colon A \to B$ is a ring homomorphism and $\mathfrak{P}_B \subset B$ is an order on B. One might ask if there is an order $\mathfrak{P}_A \subset A$ such that $f(\mathfrak{P}_A) \subset \mathfrak{P}_B$. Obviously, there is such if and only if A admits an order, namely, just take $\mathfrak{P}_A = \mathfrak{P}_w$, the weakest order on A.

This is not too satisfactory, but we next observe that the set of orders $\mathfrak{P}_A \subset A$ with $f(\mathfrak{P}_A) \subset \mathfrak{P}_B$ admits maximal elements, by Zorn's lemma. In general, however, there will not be a maximum such order which we could call $f^*\mathfrak{P}_B$.

If $f\colon A \to B$ is injective, there obviously is such a maximum, namely the order $\mathfrak{P}_B \cap A \subset A$ (more precisely, $f^{-1}(\mathfrak{P}_B) \subset A$). We call $\mathfrak{P}_B \cap A$ the *contraction* of $\mathfrak{P}_B$ to the subring A, and sometimes denote it $f^*\mathfrak{P}_B$.

If $f\colon A \to B$ is injective and $\mathfrak{P}_A$ is an order on A which extends to B, then $\mathfrak{P}_A \subset f^*(f_*\mathfrak{P}_A)$. Similarly, if $\mathfrak{P}_B$ is an order on B, then $f_*(f^*\mathfrak{P}_B) \subset \mathfrak{P}_B$. If $f\colon A \to B$, $g\colon B \to C$ are two homomorphisms and $\mathfrak{P}_A \subset A$ is an order, then $g_*(f_*\mathfrak{P}_A) = (g \circ f)_*\mathfrak{P}_A \subset C$, whenever $f_*\mathfrak{P}_A$ and either of $g_*(f_*\mathfrak{P}_A)$ or $(g\circ f)_*\mathfrak{P}_A$ are defined. Similarly, if $\mathfrak{P}_C \subset C$ is an order, $f^*(g^*\mathfrak{P}_C) = (g\circ f)^*\mathfrak{P}_C \subset A$, when defined.

Note that in any order $\mathfrak{P} \subset A$ we have $1 = 1^2 \in \mathfrak{P}$, hence $0 \le 1 \le 2 \le 3 \le \cdots$

If ever $n = 0 \in A$, $n \in \mathbb{Z}$, then $1 = 0$ and $A = (0)$. Thus $1 \neq 0 \in A$ implies the unique ring map $\mathbb{Z} \to A$ is injective and order preserving. The ring $\mathbb{Z}$ with its unique order is the initial object in the categories (POR), (PORNN), (PORCK) and (PORPP).

1.4. Simple Refinements of Orders

Let A be a ring, $\mathfrak{P}' \subset A$ an order, and $\mathfrak{P} \subset \mathfrak{P}'$ a subset closed under sums and products and containing all squares in A. Then $\mathfrak{P}$ is an order.

Suppose $a \in \mathfrak{P}' - \mathfrak{P}$ and consider the set

$$\mathfrak{P}[a] = \{p_1 + p_2 a | p_1, p_2 \in \mathfrak{P}\}.$$

Since $a^2 \in \mathfrak{P}$, we see that, in fact, $\mathfrak{P}[a]$ is an order refining $\mathfrak{P}$. Clearly, $\mathfrak{P}[a]$ is the weakest refinement of $\mathfrak{P}$ containing a. We will refer to $\mathfrak{P}[a]$ as a *simple refinement* of $\mathfrak{P}$.

Given an order $\mathfrak{P} \subset A$ and an element $a \notin \mathfrak{P}$, we ask when the set $\mathfrak{P}[a]$ is, in fact, an order. The answer is easy.

Proposition 1.4.1. If $(A,\mathfrak{P}) \in (\text{POR})$, $a \in A$, then $\mathfrak{P}[a] = \{p_1+p_2 a|p_1,p_2\in\mathfrak{P}\}$ is an order on A if and only if both the conditions below hold:

(i) $p_1 + p_2 a = 0$, $p_i \in \mathfrak{P}$, implies $p_1 = p_2 a = 0$.

(ii) $q_1 a+q_2 a = 0$, $q_i \in \mathfrak{P}$, implies $q_1 a = q_2 a = 0$.

Equivalent conditions are:

(i)' $\mathfrak{P}a \cap -\mathfrak{P} = \{0\}$

(ii)' $\mathfrak{P}a \cap -\mathfrak{P}a = \{0\}$. □

Given $(A,\mathfrak{P}) \in (\text{POR})$, we define certain subsets of A.

$$D\mathfrak{P} = \{a \in A|pa \in \mathfrak{P} \quad \text{for some} \quad p \in \mathfrak{P}^+\}$$

$$D\mathfrak{P}^+ = \{b \in A|qb \in \mathfrak{P}^+ \quad \text{for some} \quad q \in \mathfrak{P}^+\}.$$

$D\mathfrak{P}$ and $D\mathfrak{P}^+$ are called the *derived sets* of $\mathfrak{P}$ and $\mathfrak{P}^+$, respectively.

In general, $D\mathfrak{P}$ will not be an order on A and, in general, $D\mathfrak{P} - D\mathfrak{P}^+ \neq \{0\}$. However, there are important relations between the derived sets $D\mathfrak{P}$ and $D\mathfrak{P}^+$ and refinements of $\mathfrak{P}$.

Proposition 1.4.2. If $(A,\mathfrak{P}) \in$ (POR) and $-a \notin D\mathfrak{P}$, then $\mathfrak{P}[a]$ is an order on A.

Proposition 1.4.3.

(a) If $(A,\mathfrak{P}) \in$ (POR), then $\mathfrak{P} \cup (-D\mathfrak{P}^+) = A$ implies $\mathfrak{P}$ is a maximal order.

(b) If $(A,\mathfrak{P}) \in$ (POR) and $\mathfrak{P}$ is a maximal order, then $\mathfrak{P} \cup (-D\mathfrak{P}) = A$.

Remark. Note the similarity of these results with the *definition* of a total order on A, $\mathfrak{P} \cup (-\mathfrak{P}) = A$.

Proof of 1.4.2. We verify the two conditions of 1.4.1. Condition (i) is clearly equivalent to $-a \notin D\mathfrak{P}^+$. But $D\mathfrak{P}^+ \subset D\mathfrak{P}$. As for condition (ii), if $(q_1 + q_2)a = 0$, then $(q_1 + q_2)(-a) = 0$. Hence $-a \notin D\mathfrak{P}$ implies $q_1 = q_2 = 0$. □

Proof of 1.4.3(a). If $\mathfrak{P} \cup (-D\mathfrak{P}^+) = A$ and $a \notin \mathfrak{P}$, then $-a \in D\mathfrak{P}^+$. Thus, condition (i) of 1.4.1 does not hold for a, hence $\mathfrak{P}[a]$ is not an order, and $\mathfrak{P}$ is maximal. □

Proof of 1.4.3(b). If $\mathfrak{P}$ is maximal and $a \notin \mathfrak{P}$, then $\mathfrak{P}[a]$ is not an order. By 1.4.1, we must have $-a \in D\mathfrak{P}$, hence $\mathfrak{P} \cup (-D\mathfrak{P}) = A$. □

1.5. Remarks on the Categories (PORNN) and (PORCK)

We want to extend the discussions of sections 1.3 and 1.4 to (PORNN) and (PORCK). Section 1.3 applies without much modification to (PORNN). This is because the assumption $(A,\mathfrak{P}) \in$ (PORNN) is really a purely algebraic assumption about A, and does not involve $\mathfrak{P}$. The only change required concerns the souped up extension condition for $f_\alpha\colon A_\alpha \to B$. One must restrict to ideals $J \subset B$, with $J = \sqrt{J}$, the nil radical.

Section 1.4 also applies to (PORNN), but there are simplifications. For example,

Proposition 1.5.1.

(a) If $(A,\mathfrak{P}) \in$ (PORNN), $a \in A$, then $\mathfrak{P}[a]$ is an order if and only if $p_1 + p_2 a = 0$ implies $p_1 = p_2 a = 0$, that is, $-a \notin D\mathfrak{P}^+$.

(b) If $(A,\mathfrak{P}) \in$ (PORNN), then $\mathfrak{P}$ is a maximal order if and only if $\mathfrak{P} \cup (-D\mathfrak{P}^+) = A$.

Proof. The point is, condition (ii) of 1.4.1 is automatically satisfied in (PORNN). In fact, this condition is the defining axiom 1.1.(v) for (PORCK), and we showed in 1.1 that (PORNN) $\subset$ (PORCK). □

Despite the redundancy of condition (ii) of 1.4.1 in (PORCK), one does not quite have 1.5.1 in (PORCK). This is because the assertion $(A,\mathfrak{P}[a]) \in$ (PORCK involves more than just that $\mathfrak{P}[a]$ is an order on A. Specifically, define

$$D_{CK}(\mathfrak{P}^+) = \{a \in A \mid p_1 x = p_2 ax \neq 0, \text{ some } p_i \in \mathfrak{P},\ x \in A\}.$$

Thus $-a \notin D_{CK}(\mathfrak{P}^+)$ if and only if $(p_1 + p_2 a)x = 0$ implies $p_1 x = p_2 ax = 0$.

Proposition 1.5.2.

(a) If $(A,\mathfrak{P}) \in$ (PORCK) and $a \in A$, then $\mathfrak{P}[a]$ is an order with $(A,\mathfrak{P}[a]) \in$ (PORCK) if and only if $-a \notin D_{CK}(\mathfrak{P}^+)$.

(b) If $(A,\mathfrak{P}) \in$ (PORCK), then $\mathfrak{P}$ is a maximal (PORCK)· order on A if and only if $\mathfrak{P} \cup (-D_{CK}(\mathfrak{P}^+)) = A$. □

We leave the proof as an exercise. Also, we mention that if $(A,\mathfrak{P}) \in$ (PORNN), then $D_{CK}(\mathfrak{P}^+) = D\mathfrak{P}^+$, so 1.5.2 contains 1.5.1.

The extension conditions in 1.3 also require modification in (PORCK). Suppose $(A,\mathfrak{P}_A) \in$ (PORCK), $f\colon A \to B$ a ring homomorphism. Then there is an order $\mathfrak{P}_B$ on B with $f(\mathfrak{P}_A) \subseteq \mathfrak{P}_B$ and $(B,\mathfrak{P}_B) \in$ (PORCK) if and only if

$$\left(\sum_{i=1}^{n} f(p_i)b_i^2\right)x = 0,\ p_i \in \mathfrak{P}_A,\ b_i,\ x \in B \text{ implies } f(p_j)b_j^2 x = 0,\ 1 \leq j \leq n.$$

This is contained in a souped up condition concerning a family of maps $f_\alpha: A_\alpha \to B$, $(A_\alpha, \mathfrak{P}_\alpha) \in$ (PORCK). Namely, there is always a smallest ideal $I \subset B$ which satisfies:

$$\sum_{i=1}^{k} \prod_{j=1}^{n} \left(f_{\alpha_{ij}}(p_{\alpha_{ij}})\right) b_i^2 x = 0, \quad p_{\alpha_{ij}} \in \mathfrak{P}_{\alpha_{ij}}, \quad b_i, x \in B,$$

$$\text{implies} \quad \prod_{j=1}^{n} f_{\alpha_{ij}}(p_{\alpha_{ij}}) b_i^2 x = 0, \quad 1 \le i \le k.$$

The ring $\overline{B} = B/I$ admits an order $\overline{\mathfrak{P}}$ with $(\overline{B}, \overline{\mathfrak{P}}) \in$ (PORCK) such that all the compositions $\pi f_\alpha: (A_\alpha, \mathfrak{P}_\alpha) \to (\overline{B}, \overline{\mathfrak{P}})$ are (PORCK)-morphisms.

The concept of contraction of order to a subring is unchanged in (PORCK). If $(B, \mathfrak{P}_B) \in$ (PORCK) and $A \subset B$ is a subring, then $(A, \mathfrak{P}_A) \in$ (PORCK), where $\mathfrak{P}_A = A \cap \mathfrak{P}_B$.

1.6. Remarks on Integral Domains

If we restrict our attention to rings without zero divisors, that is, to integral domains, the theory of orders begins to look very much like the classical theory of formally real fields.

Proposition 1.6.1. Let $(A, \mathfrak{P}) \in$ (POR), A an integral domain. Then

(a) The derived set $D\mathfrak{P} \subset A$ is an order on A.

(b) Every maximal refinement of $\mathfrak{P}$ contains $D\mathfrak{P}$ and $D\mathfrak{P}$ is the intersection of the maximal refinements of $\mathfrak{P}$.

(c) Every maximal refinement of $\mathfrak{P}$ is a total order on A.

Proof.

(a) Suppose $p_1 a_1, p_2 a_2 \in \mathfrak{P}$ with $p_1, p_2 \in \mathfrak{P}^+$. Then $p_1 p_2 (a_1 a_2) \in \mathfrak{P}$ and $p_1 p_2 (a_1 + a_2) \in \mathfrak{P}$ and $p_1 p_2 \neq 0$. Thus $D\mathfrak{P}$ is closed under products and sums. Since $\mathfrak{P} \subset D\mathfrak{P}$, all squares belong to $D\mathfrak{P}$. Finally, if $a \in D\mathfrak{P} \cap (-D\mathfrak{P})$, then we would have $pa \le 0 \le qa$ with $p, q \in \mathfrak{P}^+$. Thus $qpa \le 0 \le pqa$, which implies $pqa = 0$. Again, since A is an integral domain, $a = 0$.

(b) Suppose $\mathfrak{P} \subset \mathfrak{P}'$. Then $D\mathfrak{P} \subset D\mathfrak{P}'$. If $\mathfrak{P}'$ is a maximal refinement of $\mathfrak{P}$, $\mathfrak{P}' = D\mathfrak{P}'$. For the second statement, suppose $a \notin D\mathfrak{P}$. By 1.4.1,

$\mathfrak{P}[-a]$ is an order, and can be refined to a maximal order $\mathfrak{P}'$. Thus $-a \in \mathfrak{P}'$, so $a \notin \mathfrak{P}'$.

(c) This is easy from 1.4.3(b). Namely, if $\mathfrak{P}'$ is a maximal order, $A = \mathfrak{P}' \cup (-D\mathfrak{P}') = \mathfrak{P}' \cup (-\mathfrak{P}')$, so $\mathfrak{P}'$ is total. □

We observed in 1.2 that a field E admits an order $\mathfrak{P} \subset E$, if and only if E is formally real. We claim for any such order $\mathfrak{P} = D\mathfrak{P}$. Namely, if $0 < a < b$, then $0 < \frac{1}{b} < \frac{1}{a}$, since $(\frac{1}{a} - \frac{1}{b}) = (b-a)/ab = ab(b-a)/(ab)^2 > 0$, and $\frac{1}{a} = (\frac{1}{a})^2 a > 0$. Thus if $pa = q$, $p,q > 0$, then $a = (\frac{1}{p})q > 0$.

It follows from $\mathfrak{P} = D\mathfrak{P}$ that any partial order $\mathfrak{P}$ on a field E is the intersection of the total orders which refine $\mathfrak{P}$. Thus any element $a \in E$ with $a \notin \mathfrak{P}$ and $-a \notin \mathfrak{P}$ can be made either positive or negative in some refinement. In particular, this applies to the weak order $\mathfrak{P}_w = \{\Sigma a_i^2 | a_i \in E\}$.

Let $(A,\mathfrak{P})$ be a partially ordered integral domain and let $i\colon A \to E$ be the inclusion of A in its field of fractions. The following is now an easy consequence of our definitions.

Proposition 1.6.2.

(a) The extended order $i_*\mathfrak{P} \subset E$ is defined. An element $a/b \in E$, $a,b \in A$, belongs to $i_*\mathfrak{P}$ if and only if $abp^2 \in \mathfrak{P}$ for some $p \in \mathfrak{P}$.

(b) $i^*i_*\mathfrak{P} = D\mathfrak{P} \subset A$.

(c) The functions i_*, i^* define inclusion preserving bijections between the set of orders $\mathfrak{P}$ on A which satisfy $\mathfrak{P} = D\mathfrak{P}$, and the set of all orders on E. In particular, total orders on A correspond bijectively to total orders on E. □

We will prove a more general result in a later chapter on localization.

1.7. Some Examples

(1) The ring $\mathbb{Z}$ of integers admits a unique order, as does its quotient field Q. If A admits an order and if every element or its negative is a sum of squares in A, then A admits a unique order, the weak order, which is a total order.

(2) Let $(A,\mathfrak{P}) \in$ (POR) and let S be a set. Then A^S, the ring of A-valued functions on S, can be ordered. Two natural orders are the weak order $\mathfrak{P}_w$, and the *affine order* $\mathfrak{P}_S = \{f\colon S \to A \mid f(s) \geq 0, \text{ all } s \in S\}$. These orders can be contracted to any subring of A^S, and the relation between the orders is an interesting problem in many instances. Examples include the ring of continuous real valued functions on a topological space X, the ring of C^∞-real functions on a C^∞-manifold M, and the ring $\mathbb{R}[X_1,\ldots,X_n]$ of polynomial functions on affine n-space $\mathbb{R}^{(n)}$.

More generally, let S be a set and let $(A_s,\mathfrak{P}_s)$, $s \in S$, be a family of partially ordered rings indexed by the elements of S. Then the ring of sections of the projection $\coprod_{s \in S} A_s \to S$ can be ordered in various ways, as can any subring. For example, X might be a topological space equipped with a sheaf of partially ordered rings. Then the ring of global sections of the sheaf becomes a partially ordered ring in various ways.

(3) Since the square root of a nowhere negative continuous real function is continuous on a space X, we see that the weak order and the affine order on C(X) coincide. That is, the positive functions in C(X) are exactly the squares. Thus the algebra structure of C(X) determines the order structure. Any ring homomorphism from C(X) to any ring which admits an order, e.g., C(Y), where Y is a topological space, is necessarily order preserving.

If X is reasonable, C(X) actually has only this one order $\mathfrak{P}$. Namely, suppose $f \notin \mathfrak{P}$, that is $f(x) < 0$ for at least one point $x \in X$. Let $p\colon X \to \mathbb{R}$ be a nonnegative continuous function with $p(x) > 0$ and $\rho \equiv 0$ outside some small neighborhood of x. Then $pf \in (-\mathfrak{P}^+)$ and $-f \in D\mathfrak{P}^+$. This proves that the order $\mathfrak{P}$ on C(X) admits no simple refinements.

(4) Let M be a C^∞-manifold, $C^\infty(M)$ the ring of C^∞ real functions on M. The affine order is $\mathfrak{P}_M = \{f \in C^\infty(M) \mid f(X) \geq 0 \text{ all } x \in M\}$. Just as in the case above, $\mathfrak{P}_M$ admits no refinements since there always exist C^∞-functions on M positive at a point and vanishing outside any small neighborhood of the point.

On the other hand, one cannot characterize $\mathfrak{P}_M$ as the squares in $C^\infty(M)$. For example, the function x^2+y^2 on $\mathbb{R}^2$ is not the square of a C^∞ function. One next might guess that $\mathfrak{P}_M = \mathfrak{P}_w$, the sums of squares. This is still not correct.

Note that the zero divisors in $C^\infty(M)$ are exactly the functions which vanish on some open set of M. Suppose $f \in C^\infty(M)$ satisfies $h^2 f = \sum_{i=1}^{m} g_i^2$, $h, g_i \in C^\infty(M)$ and h *not* a zero divisor. Then $f \in \mathfrak{P}_M$, as is easily checked. The set of such f defines a natural order $\mathfrak{P}_d$, refining $\mathfrak{P}_w$, and it turns out that $\mathfrak{P}_w \neq \mathfrak{P}_M$, but $\mathfrak{P}_d = \mathfrak{P}_M$.

I am indebted to Paul Cohen for proofs of these results. In fact, one can always write $h^2 f = g^2$ for suitable h,g, if $f(x) \geq 0$ for all $x \in M$. In the other direction, there are C^∞ nowhere negative functions on the real line, which vanish only at the origin, but which are not sums of squares of C^∞ functions.

(5) We consider the polynomial ring $\mathbb{R}[X_1 \ldots X_n]$, ordered as real functions on affine space $\mathbb{R}^{(n)}$. It is true that if $n = 1$ and $f(x) \geq 0$ for all $x \in \mathbb{R}$, then f is a sum of squares of polynomials. However, if $n \geq 2$, there are polynomials $f(X_1 X_2 \ldots X_n) \geq 0$, which are not sums of squares. Examples are due to Hilbert. However, if we pass to rational functions, then every positive polynomial is a sum of squares. This was a Hilbert problem, solved by Artin. We may rewrite the result as follows. If $f \in \mathbb{R}[X_1 \ldots X_n]$ and $f(x_1 \ldots x_n) \geq 0$ for all $(x_1 \ldots x_n) \in \mathbb{R}^{(n)}$, then for suitable $h, g_i \in \mathbb{R}[X_1 \ldots X_n]$, $h \neq 0$, $h^2 f = \sum_{i=1}^{m} g_i^2$.

$\mathbb{R}[X_1 \ldots X_n]$ is an integral domain, of course. If $\mathfrak{P}_w$ is the weak order and $\mathfrak{P}_{\mathbb{R}^{(n)}}$ is the order defined by positive functions on $\mathbb{R}^{(n)}$, then Artin's result becomes in our notation $\mathfrak{P}_{\mathbb{R}^{(n)}} \subseteq D\mathfrak{P}_w$. But also, $D\mathfrak{P}_w \subseteq \mathfrak{P}_{\mathbb{R}^{(n)}}$ since if $h^2 f = \sum_{i=1}^{m} g_i^2$ and if $f(x) < 0$ for some $x = (x_1 \ldots x_n) \in \mathbb{R}^{(n)}$, then h vanishes identically on some open neighborhood of x in $\mathbb{R}^{(n)}$. Since h is a polynomial, this implies $h = 0$. Thus $\mathfrak{P}_{\mathbb{R}^{(n)}} = D\mathfrak{P}_w$.

(6) Return to the general construction of Example (2), $(A,\mathfrak{P}) \in$ (POR),

S is a set. Suppose $B \subset A^S$ is a subring. We may as well assume B separates points of S, by identifying s and s' if $f(s) = f(s') \in A$, all $f \in B$.

Suppose $Y \subset S$ is a subset such that $f \in B$, $f|_Y = 0$ implies $f = 0$. We call such a Y · B-*Zariski dense*. Y is, in fact, dense in the weakest topology on S for which all sets $Z_g = \{s | g(s) = 0\}$ are closed, $g \in B$.

A B-Zariski dense Y determines an order on B, say $\mathfrak{P}_Y$, by $g \in \mathfrak{P}_Y$ if $g|_Y \geq 0$ (as a function $Y \to A$). Thus $\mathfrak{P}_S$ is the affine order on B, which we defined earlier. Clearly, if Y_1 and Y_2 are two such sets, then $Y_1 \subset Y_2$ implies $\mathfrak{P}_{Y_2} \subset \mathfrak{P}_{Y_1}$.

More generally, let $\mathcal{Y}$ be a family of B-Zariski dense subsets of S. Suppose $\mathcal{F} \subset \mathcal{Y}$ is a filter, that is, $Y_1, Y_2 \in \mathcal{F}$ implies $Y_1 \cap Y_2 \in \mathcal{F}$ and $Y_3 \in \mathcal{F}$, $Y_4 \supseteq Y_3$ implies $Y_4 \in \mathcal{F}$. Then $\mathfrak{P}_{\mathcal{F}} = \bigcup_{Y_i \in \mathcal{F}} \mathfrak{P}_{Y_i}$ is an order on B, and $g \in \mathfrak{P}_{\mathcal{F}}$ if $g|_{Y_i} \geq 0$ for some $Y_i \in \mathcal{F}$. Clearly, if $\mathcal{F}_1 \subset \mathcal{F}_2$ are filters in $\mathcal{Y}$, then $\mathfrak{P}_{\mathcal{F}_1} \subset \mathfrak{P}_{\mathcal{F}_2}$.

(7) Again suppose $(A, \mathfrak{P}) \in (POR)$, S a set, $B \subset A^S$ a subring of functions which separates points of S. Let $Z \subset S$ be a subset and let $I \subset B$ be the ideal of functions in B which vanish on Z. Then the ring B/I is,in a natural way, a subring of the ring of functions A^Z, and separates points.

(8) We consider as a special case of (7), $A = \mathbb{R}$, $S = \mathbb{R}^{(n)}$, affine n-space, and $B = \mathbb{R}[X_1 \ldots X_n]$, the polynomials in n-variables. Given $Z \subset \mathbb{R}^{(n)}$, $I \subset \mathbb{R}[X_1 \ldots X_n]$ the ideal of polynomials vanishing on Z, then $\mathbb{R}[X_1 \ldots X_n]/I$ is a ring of algebraic functions on some real algebraic set, specifically the Zariski closure of Z in $\mathbb{R}^{(n)}$. However, the affine order $\mathfrak{P}_Z$ on $\mathbb{R}[X_1 \ldots X_n]/I$ very definitely depends on Z itself, not just the Zariski closure.

(9) For example, let K be a finite simplicial complex, with vertices $v_1, \ldots, v_n$. Embed $|K|$ rectilinearly in $\mathbb{R}^{(n)}$ be sending each vertex v_i

to the ith unit basis vector $\vec{e}_i = (0\ldots 0,1,0\ldots 0)$, and let $Z = |K| \subset \mathbb{R}^{(n)}$. If the X_i are interpreted as coordinate functions on $\mathbb{R}^{(n)}$, then Z is contained in the hyperplane $\sum_{i=1}^{n} X_i = 1$. The Zariski closure of $Z = |K|$ is the union of all affine linear subspaces of this hyperplane through k-simplexes of $|K|$, $k \geq 0$. The ideal $I \subset \mathbb{R}[X_1 \ldots X_n]$ vanishing on $|K|$ is generated by $\sum_{i=1}^{n} X_i - 1$ and monomials $X_{i_o} \cdot \ldots \cdot X_{i_k}$, one for each set of vertices $(\vec{e}_{i_o}, \ldots, \vec{e}_{i_k})$ of $|K|$ which *do not* span a simplex of $|K|$. In the affine order on $\mathbb{R}[X_1 \ldots X_n]/I$, one has $0 \leq X_i \leq 1$, and the X_i become the baracentric coordinate functions on $|K|$.

(10) Suppose X is a compact Hausdorff space, $\{U_i\}_{1 \leq i \leq n}$ is a finite cover of X and $\varphi_i\colon X \to \mathbb{R}$ is a subordinate partition of unity. That is, $\sum_{i=1}^{n} \varphi_i(x) = 1$, $0 \leq \varphi_i(x) \leq 1$, all $x \in X$, and $\overline{\text{supp}(\varphi_i)} \subset U_i$. The functions $\varphi_1, \ldots, \varphi_n$ generate over $\mathbb{Z}$ (or Q or $\mathbb{R}$) a ring of real valued functions on X, and this ring is a quotient of $\mathbb{Z}[X_1 \ldots X_n]$ (or $Q[X_1 \ldots X_n]$ or $\mathbb{R}[X_1 \ldots X_n]$, depending on the coefficients). The ideal of relations is just the ideal of relations among the φ_i as functions on X. One has a natural partial order, on this quotient, with $0 \leq X_i \leq 1$. One might call such rings "partially ordered rings with a positive partition of unity".

(11) It is easy to find rings with maximal orders which are not total. For example, let $A = \prod_i A_i$ be a direct product of rings. Then any order $\mathfrak{P}$ on A is a product $\prod \mathfrak{P}_i$, where $\mathfrak{P}_i \subset A_i$ are orders. This is immediate from the equation

$$(a_i)(1_j) = (0 \ldots a_j \ldots 0)$$

where $1_j = (0 \ldots 1_j \ldots 0) = 1_j^2$. A product order $\prod \mathfrak{P}_i$ is clearly maximal if each $\mathfrak{P}_i \subset A_i$ is maximal, but $\prod \mathfrak{P}_i$ will essentially never be total on A.

II · Homomorphisms and convex ideals

2.1. Convex Ideals and Quotient Rings

Fix $(A,\mathfrak{P}) \in (POR)$. A subset $C \subset A$ is *convex* if $0 \leq p \leq q$, $q \in C$ implies $p \in C$. Equivalently, C is convex if $p_1 + p_2 \in C$, $p_i \in \mathfrak{P}$, implies $p_i \in C$, $i = 1,2$. Clearly, arbitrary unions and intersections of convex sets are convex. An ideal $I \subset A$ is said to be convex if it is convex as a subset.

Proposition 2.1.1. An ideal $I \subset A$ is the kernel of a morphism in (POR) if and only if I is convex.

Proof. First, if $I = \text{kernel}(f\colon (A,\mathfrak{P}) \to (A',\mathfrak{P}'))$, and $0 \leq p \leq q$ in A with $f(q) = 0$, then $0 \leq f(p) \leq f(q) = 0$ in A', hence $f(p) = 0$. Thus I is convex.

Conversely, suppose I is convex. Consider the projection $\pi\colon A \to A/I$. According to 1.3, we can impose an order on A/I so that π is order preserving if, whenever $\sum_{i=1}^{n} p_i a_i^2 \in I$, $p_i \in \mathfrak{P}$, $a_i \in A$, then $p_j a_j^2 \in I$, $1 \leq j \leq n$. But this is immediate from the second characterization of convexity above. □

The weakest order on A/I such that π is order preserving, namely

$$\pi_*(\mathfrak{P}) = \{p + I \mid p \in \mathfrak{P}\} \subset A/I$$

will be called the *induced order*, or *quotient order*, on A/I, and will be denoted $(A/I, \mathfrak{P}/I)$.

Let $f\colon (A,\mathfrak{P}) \to (A',\mathfrak{P}')$ be a morphism in (POR). Let $I \subset A$ be the convex ideal kernel (f). Then the morphism f factors as a composition

$$(A,\mathfrak{P}) \xrightarrow{f_1} (A/I, \mathfrak{P}/I) \xrightarrow{f_2} (A/I, \mathfrak{P}' \cap \mathfrak{P}/I) \xrightarrow{f_3} (A',\mathfrak{P}')$$

where f_1 is a quotient projection, f_2 is a refinement of order (that is, f_2 is the identity as a ring homomorphism), and f_3 is an inclusion of a subring with the contracted order.

Notice that because of the existence of refinements of order, a map can be an order preserving isomorphism of rings, without being an isomorphism in the category (POR). In general, if A is a ring and $\mathfrak{P} \subset \mathfrak{P}'$ are orders on A, then A will have fewer $\mathfrak{P}'$-convex ideals than $\mathfrak{P}$-convex ideals.

We now consider the operations on convex ideals of (i) intersection, (ii) forming nil radicals, (iii) sums, (iv) products, and (v) quotients.

Proposition 2.1.2. Fix $(A,\mathfrak{P}) \in$ (POR). Then

(a) An arbitrary intersection of convex ideals is convex.

(b) If $I \subset A$ is convex, then $\sqrt{I} = \{a \in A \mid a^n \in I,\ \text{some } n \geq 1\}$ is convex.

Proof. Statement (a) is trivial. For statement (b), suppose $0 \leq a \leq b \in \sqrt{I}$. Then $0 \leq a^n \leq b^n$ for all n, since $b^n - a^n = (b-a)(b^{n-1} + ab^{n-2} + \cdots + a^{n-1})$. If $b^n \in I$, then $a^n \in I$, hence $a \in \sqrt{I}$. □

It turns out that sums, products, and quotients of convex ideals need not be convex. Examples will be given in 2.8.

2.2. Convex Hulls

Let $(A,\mathfrak{P}) \in$ (POR), $X \subset A$ a subset. We denote by $H(X)$ the smallest convex ideal of A which contains X. Thus

$$H(X) = \bigcap_{\substack{I \text{ convex ideal} \\ X \subset I}} I .$$

It is possible that $H(X) = A$, of course. We refer to $H(X)$ as the *convex hull*, or simply *hull*, of X.

Here is a construction of $H(X)$. For any subset $Y \subset A$, let $J(Y)$ be the ordinary ideal of A generated by Y. Define

$$H_0(X) = J(X)$$

$$H_{n+1}(X) = J(H_n(X) \cup \{b \in A \mid 0 \leq b \leq a,\ a \in H_n(X)\})\ .$$

<u>Proposition 2.2.1.</u> $H(X) = \bigcup_{n \geq 0} H_n(X)$.

<u>Proof.</u> Clearly, $H_0(X) \subset H_1(X) \subset \cdots$ is an increasing sequence of ideals of A, hence $\bigcup_{n \geq 0} H_n(X)$ is an ideal. Suppose $0 \leq x \leq y$, $y \in H_m(X)$, some m. Then $x \in H_{m+1}(X)$. This proves that $\bigcup_{n \geq 0} H_n(X)$ is convex. Finally, for all m, $H_{m+1}(X)$ is obviously contained in any convex ideal which contains $H_m(X)$. Thus $\bigcup_{n \geq 0} H_n(X)$ is the smallest convex ideal containing X, as claimed. □

The following "compactness" result is useful.

<u>Proposition 2.2.2.</u> Let $X \subset A$ be any subset and let $y \in H(X)$. Then there is a finite subset of X, say $\{x_1 \cdots x_k\}$, such that $y \in H(\{x_1 \cdots x_k\})$.

<u>Proof.</u> Let $I = \{y \in A \mid y \in H(\{x_1 \cdots x_k\}),\ \text{some } x_i \in X\}$. Then (i) $X \subset I \subset H(X)$ and (ii) I is a convex subset of A. If I is an ideal, then necessarily $I = H(X)$.

Obviously, if $y \in I$, $a \in A$, then $ay \in I$. But also, if $X \subset Y \subset A$, $H(X) \subset H(Y)$, hence if $X_1, X_2 \subset A$, $H(X_1) \cup H(X_2) \subset H(X_1 \cup X_2)$. It follows easily that if $y_1, y_2 \in I$, then $y_1 + y_2 \in I$. □

The next result is also extremely useful. We characterize the nil radical of a convex hull.

<u>Proposition 2.2.3.</u> Let $Z \subset \mathfrak{P}$ be any set of positive elements. Then

$$\sqrt{H(Z)} = \{z \in A \mid 0 \leq z^{2s} \leq \sum_{i=1}^{k} p_i z_i \text{ for some integer } s > 0,\ z_i \in Z \text{ and } p_i \in \mathfrak{P}\}.$$

Although stated for positive subsets, 2.2.3 actually characterizes the nil radical of any convex hull. Namely, if $X \subset A$ is any subset, let $X^2 = \{x^2 \mid x \in X\} \subset \mathfrak{P}$. Then $X \subset \sqrt{H(X^2)}$, hence $\sqrt{H(X)} \subset \sqrt{H(X^2)}$, since $\sqrt{H(X^2)}$

is convex and coincides with its own nil radical. But also $H(X^2) \subset H(X)$, so $\sqrt{H(X)} = \sqrt{H(X^2)}$. Thus 2.2.3 has the following corollary.

Corollary 2.2.4. Let $X \subset A$ be any subset. Then

$$\sqrt{H(X)} = \{z \in A \mid 0 \leq z^{2s} \leq \sum_{i=1}^{k} p_i x_i^2, \text{ some } s > 0,\ p_i \in \mathfrak{P} \text{ and } x_i \in X\}.$$

Proof of 2.2.3. Let $I = \{y \in A \mid 0 \leq y^{2r} \leq \sum_{i=1}^{k} p_i z_i\}$. We want to prove $I = \sqrt{H(Z)}$. It suffices to prove (i) I is an ideal, (ii) I is convex, and (iii) $I = \sqrt{I}$. Because, obviously, $\sqrt{H(Z)} \supset I \supset Z$. Then (i) and (ii) imply $I \supset H(Z)$ and (iii) implies $I \supset \sqrt{H(Z)}$.

Now, (ii) and (iii) are trivial, so we will prove (i). It is clear that if $y \in I$, say $0 \leq y^{2r} \leq \sum_{i=1}^{k} p_i z_i$, and $a \in A$, then $0 \leq (ay)^{2r} \leq \sum_{i=1}^{k} (a^{2r} p_i) z_i$. Thus $ay \in I$.

Finally, we must show that I is closed under sums. Let $y_1, y_2 \in I$. Then for suitable $b \in A$ and any $s \geq 1$,

$$(y_1 + y_2)^{2s} - y_1^{2s} = by_2 ,$$

hence for any $r \geq 1$,

$$((y_1 + y_2)^{2s} - y_1^{2s})^{2r} = b^{2r} y_2^{2r} . \tag{*}$$

Expanding the left-hand side of (*) gives

$$\sum_{j=0}^{r} c_j (y_1+y_2)^{2s(2r-2j)} y_1^{2s(2j)} - \sum_{j=1}^{r} d_j (y_1+y_2)^{2s(2r-2j+1)} y_1^{2s(2j-1)}$$

where the c_j and d_j are (positive) binomial coefficients. In particular, $c_0 = 1$, so transfering the negative terms to the right-hand side of (*) gives

$$\begin{aligned} 0 &\leq (y_1+y_2)^{4sr} \\ &\leq \sum_{j=0}^{r} c_j (y_1+y_2)^{2s(2r-2j)} y_1^{2s(2j)} \end{aligned} \tag{**}$$

$$\leq b^{2r}y_2^{2r} + \sum_{j=1}^{r} d_j(y_1+y_2)^{2s(2r-2j+1)}y_1^{2s(2j-1)}.$$

But $y_1, y_2 \in I$, say $0 \leq y_1^{2s} \leq \sum_{j=1}^{k} p_j z_j$ and $0 \leq y_2^{2r} \leq \sum_{j=1}^{k'} p_j' z_j'$, $p_j, p_j' \in \mathfrak{P}$, $z_j, z_j' \in Z$. Replacing y_1^{2s}, y_2^{2r} in (**) by these larger terms results in an inequality which implies $y_1 + y_2 \in I$. □

2.3. Maximal Convex Ideals and Prime Convex Ideals

We fix $(A,\mathfrak{P}) \in (POR)$ throughout this section. A convex ideal $Q \subset A$ is a *prime* convex ideal, if $Q \neq A$ and whenever $ab \in Q$, either $a \in Q$ or $b \in Q$. Equivalently, Q is convex and A/Q is an integral domain, with $1 \neq 0$.

A convex ideal $Q \subset A$ is a *maximal* convex ideal if $Q \neq A$ and whenever $Q \subset Q'$, Q' a convex ideal, either $Q' = Q$ or $Q' = A$.

The concept prime convex ideal just means ordinary prime ideal which is convex. On the other hand, a maximal convex ideal may not be maximal in the family of ordinary ideals.

We first establish the existence of maximal convex ideals.

Proposition 2.3.1. Let $(A,\mathfrak{P}) \in (POR)$, $1 \neq 0$ in A. Let $I \subset A$ be a convex ideal, $1 \notin I$ (equivalently, $I \neq A$). Then I is contained in at least one maximal convex ideal.

Proof. The usual Zorn's lemma argument applies, since the family of all convex ideals containing I but not containing 1 is non-empty, partially ordered by inclusion, and satisfies the chain condition. □

In particular, since $I = (0)$ is always a convex ideal, we conclude that any non-zero $(A,\mathfrak{P}) \in (POR)$ has maximal convex ideals.

Next we characterize maximal convex ideals by the quotient ring they define. If $(A,\mathfrak{P}) \in (POR)$, we say $(A,\mathfrak{P})$ is a *semi-field* if $1 \neq 0$ in A and for all $a \in A$, $a \neq 0$, there exists $b \in A$ with $1 \leq ab$. Note that this definition depends on the order $\mathfrak{P} \subset A$, and not just the algebraic structure.

Proposition 2.3.2. If $(A,\mathfrak{P}) \in (POR)$, the following conditions are equivalent.

(i) $(A,\mathfrak{P})$ is a semi-field.

(ii) Every homomorphism of $(A,\mathfrak{P})$ in (POR) is zero or injective.

(iii) $(0) \subset A$ are the only convex ideals.

(iv) For all $a \in A$, $a \neq 0$, we have $1 \in H(a)$.

(v) For all $a \in A$, $a \neq 0$, we have $1 \leq pa^2$, some $p \in \mathfrak{P}$.

Proof. The implications (i) ⇒ (ii) ⇒ (iii) ⇒ (iv) and (v) ⇒ (i) are trivial. Corollary 2.2.4 gives (iv) ⇒ (v). □

In general, if $Q \subset A$ is a convex ideal, $Q \neq A$, then Q is a maximal convex ideal if and only if every homomorphism of $(A/Q, \mathfrak{P}/Q)$ in (POR) is zero or injective. This is more or less obvious, but in any event will be clarified in the next section, thus 2.3.2 has the following corollary.

Corollary 2.3.3. Let $(A,\mathfrak{P}) \in (POR)$, $Q \subset A$ a convex ideal, $Q \neq A$. Then Q is a maximal convex ideal if and only if $(A/Q, \mathfrak{P}/Q)$ is a semi-field. Equivalently, for all $a \in A$, $a \notin Q$, there should exist $b \in A$ and $q \in Q$ with $1 \leq ab + q$. □

Proposition 2.3.4. Partially ordered fields are semi-fields and semi-fields are integral domains.

Proof. The inequality $1 \leq ab$ implies $1 \leq a^2b^2$, so in any semi-field, $a \neq 0$ implies $a^2 \neq 0$. If now $ax = 0$, $a \neq 0$, we would get $0 \leq x^2 \leq a^2b^2x^2 = 0$, for some b, hence $x^2 = 0$ and thus $x = 0$.

Corollary 2.3.5. Maximal convex ideals are prime. □

We will give a second proof of 2.3.5., also based on Proposition 2.2.3. First, if $Q \subset A$ is a maximal convex ideal, then $Q = \sqrt{Q}$, since $\sqrt{Q}$ is convex. Suppose $ab \in Q$, $b \notin Q$. Then $1 \in H(Q,b^2)$. By 2.2.3, we will have

$$0 \leq 1 \leq \sum_{i=1}^{k} p_i q_i + pb^2$$

for suitable p_i, $p \in \mathfrak{P}$, $q_i \in Q$. Multiplying by a^2 gives

$$0 \leq a^2 \leq \sum_{i=1}^{k} (p_i a^2) q_i + pa^2 b^2 \in Q.$$

Since Q is convex, $a^2 \in Q$, and, since $Q = \sqrt{Q}$, $a \in Q$. Thus Q is prime. □

Finally, we relate nilpotent elements and prime convex ideals.

Proposition 2.3.6. Let $(A,\mathfrak{P}) \in (POR)$. Then

$$\sqrt{0} = \bigcap_{\substack{Q \subset A \text{ prime} \\ \text{convex ideal}}} Q.$$

Proof. Clearly $\sqrt{0} \subset Q$ if Q is prime. Conversely, suppose $a \notin \sqrt{0}$. Then the set of convex ideals not containing any power of a is non-empty, since (0) is such an ideal. Zorn's lemma applies to give a convex ideal, maximal in this set. We assert such an ideal, say Q, is prime. Otherwise, let $bc \in Q$, $b \notin Q$, $c \notin Q$. Then $a^n \in H(Q,b)$, $a^m \in H(Q,c)$ for some integers $n,m \geq 1$. By 2.2.4, there are integers $r,s \geq 1$, $q_i \in Q$, $p_i \in \mathfrak{P}$, $i = 1,2$, with

$$0 \leq a^{2r} \leq q_1 + p_1 b^2$$

$$0 \leq a^{2s} \leq q_2 + p_2 c^2.$$

Multiplying these equations gives

$$0 \leq a^{2(r+s)} \leq q \in Q,$$

which is a contradiction.

Corollary 2.3.7. Let $I \subset A$ be a convex ideal. Then

$$\sqrt{I} = \bigcap_{\substack{P \text{ convex prime} \\ I \subset P}} P.$$

Proof. We apply 2.3.6 to the partially ordered ring $(A/I, \mathfrak{P}/I)$, and use the proposition of the next section.

2.4. Relation Between Convex Ideals in $(A,\mathfrak{P})$ and $(A/I, \mathfrak{P}/I)$

Fix $(A,\mathfrak{P}) \in$ (POR), and let $I \subset A$ be a convex ideal.

Proposition 2.4.1.

(a) There is a natural, bijective, inclusion preserving correspondence between $\mathfrak{P}$-convex ideals J of A which contain I and $\mathfrak{P}/I$-convex ideals J/I of A/I.

(b) For any such $J \supset I$, there is a natural isomorphism in (POR),

$$(A/J, \mathfrak{P}/J) \xrightarrow{\sim} (A/I\big/J/I, \mathfrak{P}/I\big/J/I).$$

(c) The prime convex ideals of A which contain I correspond bijectively to the prime convex ideals of A/I and the maximal convex ideals of A which contain I correspond bijectively to the maximal convex ideals of A/I.

Proof. Given a $\mathfrak{P}$-convex $J \supset I$, the identity on A induces an order preserving morphism $(A/I, \mathfrak{P}/I) \to (A/J, \mathfrak{P}/J)$, with kernel J/I. Thus J/I is $\mathfrak{P}/I$-convex. Conversely, if $\pi\colon A \to A/I$ is projection, and $\bar{J} \subset A/I$ is $\mathfrak{P}/I$ convex, then $J = \pi^{-1}(\bar{J}) \subset A$ is $\mathfrak{P}$-convex. The remaining details are equally simple. □

2.5. Absolutely Convex Ideals

Although the category (PORCK) is a full subcategory of (POR), one sees fewer ideals as kernels of morphisms in (PORCK). The same is true of the category (PORNN), where obviously an ideal $I \subset A$, $(A,\mathfrak{P}) \in$ (PORNN), is the kernel of a (PORNN)-morphism if and only if $I = \sqrt{I}$ and I is $\mathfrak{P}$-convex. In this section we will investigate kernels of morphisms in (PORCK).

Let $(A,\mathfrak{P}) \in$ (POR), $I \subset A$ an ideal. We say that I is *absolutely convex* if $0 \leq a \leq b$ and $bx \in I$ implies $ax \in I$. Equivalently, I is absolutely convex if $(p+q)x \in I$, $p,q \in \mathfrak{P}$, implies $px \in I$ and $qx \in I$.

Proposition 2.5.1.

(a) Absolutely convex ideals are convex.

(b) Arbitrary intersections of absolutely convex ideals are absolutely convex.

(c) Any convex ideal I with $I = \sqrt{I}$ is absolutely convex. In particular, convex prime and maximal ideals are absolutely convex.

(d) If $\mathfrak{P} \subset \mathfrak{P}'$ is a refinement of order and $I \subset A$ is $\mathfrak{P}'$-absolutely convex, then I is $\mathfrak{P}$-absolutely convex.

(e) If $f\colon (A,\mathfrak{P}) \to (A',\mathfrak{P}')$ is a morphism in (POR) and $I' \subset A'$ is absolutely convex, then $I = f^{-1}(I') \subset A$ is absolutely convex.

Proof.

(a) Let $x = 1$, the unit in A.

(b) Obvious.

(c) If $0 \leq a \leq b$ and $bx \in I$, then $0 \leq ax^2 \leq bx^2$, hence $ax^2 \in I$, hence $(ax)^2 \in I$. Since $I = \sqrt{I}$, $ax \in I$.

(d), (e) Obvious. □

Any subset $X \subset A$ is contained in a smallest absolutely convex ideal, namely the intersection of all absolutely convex ideals containing X. We denote this smallest absolutely convex ideal $AH(X)$, the *absolute hull* of X.

It is easy to give a construction of $AH(X)$ as a countable union of an increasing chain of ideals of A, just as was done in 2.2 for $H(X)$. Namely, define

$$AH_o(X) = J(X)$$

$$AH_{n+1}(X) = J(AH_n(X) \cup \{ax \mid 0 \leq a \leq b,\ bx \in AH_n(X)\}),$$

where $J(Y)$ is the ordinary ideal generated by Y.

Proposition 2.5.2.

(a) $AH(X) = \bigcup_n AH_n(X)$.

(b) If $y \in AH(X)$, then there exists a finite subset of X, $\{x_1 \cdots x_k\}$, with $y \in AH(\{x_1 \cdots x_k\})$.

(c) $\sqrt{AH(X)} = \sqrt{H(X)}$.

(d) $AH(X)AH(Y) \subset AH(XY)$ where $XY = \{xy | x \in X, y \in Y\}$.

Proof:

(a), (b) Very similar to 2.2.1 and 2.2.2

(c) Clearly $H(X) \subset AH(X)$. Also, $\sqrt{H(X)}$ is absolutely convex by 2.5.1(c), hence $AH(X) \subset \sqrt{H(X)}$. Thus, $\sqrt{AH(X)} = \sqrt{H(X)}$.

(d) We prove by induction that $AH_n(X)AH_m(Y) \subset AH_{n+m}(XY)$. If $n = m = 0$, this is clear. If $bx \in AH_{n-1}(X)$, $0 \le a \le b$, and $y \in AH_m(Y)$, we must show $(ax)y \in AH_{n+m}(XY)$, since ax is a typical new generator of $AH_n(X)$. But, by induction, $(bx)y = b(xy) \in AH_{n+m-1}(XY)$ and $0 \le a \le b$, hence $a(xy) \in AH_{n+m}(XY)$. □

Proposition 2.5.2(d) allows us to construct graded rings naturally in (PORCK). Namely, if $(A,\mathfrak{P}) \in$ (PORCK) and $I \subset A$ is any ideal, we have $A \supset AH(I) \supset AH(I^2) \supset AH(I^3) \supset \cdots$. By 2.5.2(d), $AH(I^n)AH(I^m) \subset AH(I^{n+m})$. We thus have a graded ring associated to I,

$$GA = \bigoplus_{n \ge 0} AH(I^n)/AH(I^{n+1}) .$$

This construction does not seem natural in (POR) because the analogue of 2.5.2(d) for hulls, rather than absolute hulls, is not true.

Our next result is that absolutely convex ideals are closed under the quotient operation, $(I : J) = \{b \in A | bJ \subset I\}$.

Proposition 2.5.3. Let $(A,\mathfrak{P}) \in$ (POR), $I \subset A$ an ideal, $Y \subset A$ any subset, $(I : Y) = \{b \in A | bY \subset I\}$. Then I is absolutely convex if and only if $(I : Y)$ is absolutely convex for all subsets $Y \subset A$. Moreover, if I is absolutely convex, then $(I : Y) = (I : AH(Y))$.

Proof. Since $(I: \{1\}) = I$, one direction is trivial. Conversely, if I is absolutely convex, $0 \leq a \leq b$ and $bx \in (I : Y)$, then $(bx)y = b(xy) \in I$ for all $y \in Y$. Hence $a(xy) = (ax)y \in I$ for all $y \in Y$. Thus $ax \in (I : Y)$, which proves $(I : Y)$ is absolutely convex.

For the final statement, if $bY \subset I$, then $bJ(Y) \subset I$. Assume inductively that $b \cdot AH_n(Y) \subset I$. It suffices to prove $b \cdot AH_{n+1}(Y) \subset I$, and, in particular, if $0 \leq a \leq c$, $cx \in AH_n(Y)$, it suffices to prove $b(ax) \in I$. But, $b(cx) = c(bx) \in I$, hence $a(bx) = b(ax) \in I$. □

In 2.8 we will give examples of convex ideals which are not absolutely convex. We will also give examples of sums and products of absolutely convex ideals which are not absolutely convex.

We now establish the relation between absolutely convex ideals and the category (PORCK). The defining axiom 1.1.(v) for (PORCK) is exactly the condition (0) absolutely convex. That is, $(p_1 + p_2)x = 0$, $p_i \in \mathfrak{P}$ implies $p_1x = p_2x = 0$. In fact, we have

Proposition 2.5.4. If $(A,\mathfrak{P}) \in$ (POR), the following are equivalent.

(i) $(A,\mathfrak{P}) \in$ (PORCK).

(ii) $(0) \subset A$ is absolutely convex.

(iii) $(0 : a)$ is absolutely convex for all $a \in A$.

(iv) $(0 : a)$ is convex for all $a \in A$.

(v) $(0 : X)$ is absolutely convex for all subsets $X \subset A$.

(vi) $(0 : X)$ is convex for all subsets $X \subset A$.

Proof. We have just observed that (i) ⇔ (ii). Proposition 2.5.3 gives (ii) ⇒ (iii) and (iii) ⇒ (iv) is trivial. Now, (iv) ⇒ (ii) because if $0 \leq a \leq b$ and $bx = 0$, then $(0 : x)$ convex implies $a \in (0 : x)$, hence $ax = 0$. The equivalences (iii) ⇔ (v) and (iv) ⇔ (vi) follow from

$$(0 : X) = \bigcap_{a \in X} (0 : a) . \qquad \square$$

Remark: The proposition is true with (0) replaced by I, for any ideal $I \subset A$, if statement (i) is replaced by $(A/I, \mathfrak{P}/I) \in$ (PORCK). The proof is the same.

As a consequence of the results above, we state

Corollary 2.5.5.

(a) Let $(A,\mathfrak{P}) \in$ (PORCK) [respectively $(A,\mathfrak{P}) \in$ (PORNN)]. An ideal $I \subset A$ is the kernel of a (PORCK)-morphism [respectively (PORNN)-morphism] if and only if I is absolutely convex [respectively $I = \sqrt{I}$].

(b) If $(A,\mathfrak{P}) \in$ (PORCK) and $I \subset A$ is absolutely convex [respectively $(A,\mathfrak{P}) \in$ (PORNN), $I = \sqrt{I}$], then there is a natural bijection between absolutely convex ideals J of A containing I and absolutely convex ideals J/I of A/I [respectively ideals J of A, $J = \sqrt{J}$, containing I and ideals J/I of A/I, $\sqrt{J/I} = J/I$].

(c) The inclusion functors (PORNN) $\to$ (PORCK) $\to$ (POR) have adjoints (POR) $\to$ (PORCK) $\to$ (PORNN), defined by assigning to $(A,\mathfrak{P}) \in$ (POR) the partially ordered quotients $(A/AH(0), \mathfrak{P}/AH(0)) \in$ (PORCK) and $(A/\sqrt{0}, \mathfrak{P}/\sqrt{0}) \in$ (PORNN).

Proof. Whatever remains to be proved will be left as an exercise. □

2.6. Semi-Noetherian Rings

It should not be surprising that various finiteness conditions on partially ordered rings lead to structure theorems and results which cannot possibly be proved in general. Since a partially ordered ring consists of a ring A, together with an order $\mathfrak{P} \subset A$, natural finiteness conditions can involve either the $\mathfrak{P}$-convex ideal structure or the order $\mathfrak{P}$ itself, as an extension of the weak order $\mathfrak{P}_w \subset A$. We are ultimately interested in partially ordered structures on finitely generated extensions of real closed fields. (See Chapters 7 - 8.) Such rings are, of course, Noetherian in the classical sense, hence no finiteness assumptions are necessary on chains of convex ideals. However, even for this restricted class of rings, many classical results fail to generalize to the partially ordered context. For example, one cannot always decompose convex ideals in such a ring as intersections of primary convex ideals. One needs additional finiteness conditions on the order. When one examines the classical proofs of primary decomposition and other results which fail to generalize the point which causes the difficulty often involves the fact that in commutative algebra the multiples of a single element always form an ideal, while in partially ordered algebra the smallest convex ideal containing a given element may be quite large.

On the other hand, chain conditions on convex ideals have certain interesting consequences, regardless if the underlying ring A is Noetherian in the classical sense or if the order $\mathfrak{P}$ satisfies extra conditions. Sometimes these results are proved most easily in the classical context of Noetherian rings as applications of primary decomposition, rather than by exploiting the chain conditions directly. Thus it actually gives one some added insight into these classical results to investigate chain conditions in the partially ordered context, where certain techniques of proof are unavailable.

It is the purpose of this section to initiate this study. We work entirely in the category (PORCK) because of our frequent use of the quotient construction $(I : X)$ for ideals $I \subset A$ and subsets $X \subset A$. However, 2.6.1 through 2.6.4 have obvious (POR) versions as well.

Definition 2.6.1. A ring $(A,\mathfrak{P}) \in$ (PORCK) is *semi-Noetherian* if any of the three equivalent conditions below hold:

(i) The absolutely convex ideals of $(A,\mathfrak{P})$ satisfy the ascending chain condition.

(ii) Any non-empty collection of absolutely convex ideals of $(A,\mathfrak{P})$ contains a maximal element with respect to inclusion.

(iii) Every absolutely convex ideal I of $(A,\mathfrak{P})$ may be written $I = AH(x_1,\ldots,x_k)$ for some choice of finitely many elements $x_1,\ldots,x_k \in I$.

As simple applications of the definition, we state some standard results.

Proposition 2.6.2. If $(A,\mathfrak{P}) \in$ (PORCK) is semi-Noetherian, $I \subset A$

is absolutely convex and $A' = A/I$, then $(A',\mathfrak{P}')$ is semi-Noetherian for any refinement $\mathfrak{P}'$ of $\mathfrak{P}/I$.

Proof. This is immediate from 2.5.5 (b). □

Let us call an absolutely convex ideal I *irreducible* if $I = I_1 \cap I_2$, I_j absolutely convex, implies $I = I_1$ or $I = I_2$.

Proposition 2.6.3. If $(A,\mathfrak{P}) \in (\mathrm{PORCK})$ is semi-Noetherian, then every absolutely convex ideal is a finite intersection of irreducible absolutely convex ideals.

Proof. An absolutely convex ideal maximal among those not so expressible leads to an immediate contradiction. □

Proposition 2.6.4. If $(A,\mathfrak{P}) \in (\mathrm{PORCK})$ is semi-Noetherian and $I = \sqrt{I}$ is a radical convex ideal, then there is a unique expression $I = P_1 \cap \cdots \cap P_k$ where the P_j are prime convex ideals and $P_i \not\subset P_j$ if $i \neq j$. Moreover, the P_i are precisely the minimal prime ideals containing I, which are necessarily convex.

Proof. In fact, one only needs the ascending chain condition for radical convex ideals. The proof is just as in the classical case.

Uniqueness is clear since if a prime ideal P contains $P_1 \cap \cdots \cap P_k$, then P contains some P_i. Thus if $P_1 \cap \cdots \cap P_k = P_1' \cap \cdots \cap P_n'$, then $P_1 \supset P_i' \supset P_j$ for some i,j, hence $j = 1$ and $P_1 = P_i'$. The argument also shows that the P_i which occur are exactly the minimal primes containing I, which are thus necessarily convex. (This result will be generalized in section 3.9.)

Existence of the stated decomposition is established as follows. Each radical convex ideal is a finite intersection of irreducible radical convex ideals, by the chain condition on radical convex ideals. But if $I = \sqrt{I}$ is an irreducible radical convex ideal, then I is prime. Otherwise, let $ab \in I$, $a,b \notin I$. Let $\{P_\alpha\}$ be all the prime convex ideals containing I, let $\{P_{\alpha'}\}$ be those P_α containing a and let $\{P_{\alpha''}\}$ be those P_α containing b. Then $I = \bigcap_\alpha P_\alpha$ by 2.3.7. On the other hand, $\{P_\alpha\} = \{P_{\alpha'}\} \cup \{P_{\alpha''}\}$

since $ab \in I$, hence $I = I' \cap I''$ where $I' = \cap P_{\alpha'}$, $I'' = \cap P_{\alpha''}$. Since $a \in I'$, $b \in I''$, this contradicts the irreducibility of I. □

Directly from 2.6.3, absolutely convex ideals in a semi-Noetherian ring can be understood in terms of irreducible absolutely convex ideals. However, irreducible absolutely convex ideals need not be primary, even if the ring is Noetherian. Despite the lack of primary decomposition, the associated primes of an absolutely convex ideal are accessible in any semi-Noetherian ring.

<u>Proposition 2.6.5</u>. Suppose $(A,\mathfrak{P}) \in$ (PORCK) is semi-Noetherian, $I \subset A$ an absolutely convex ideal. Then the following sets of primes coincide:

(i) $\{P \mid P$ prime and $P = (I : x)$, some $x \in A\}$

(ii) $\{P \mid P$ prime and $(I : x)$ is P-primary, some $x \in A\}$

(iii) $\{P \mid P$ prime and $P = \sqrt{I : x}$, some $x \in A\}$

(iv) $\{P \mid P$ a minimal prime containing $(I : x)$, some $x \in A\}$.

Moreover, this set of *associated primes* of I is non-empty and finite.

<u>Proof</u>. Passing to $(A/I, \mathfrak{P}/I)$, we may assume $I = (0)$. Certainly, Set (i) $\subset$ Set (ii) $\subset$ Set (iii) $\subset$ Set (iv). Moreover, among all the absolutely convex ideals $(0 : y)$, $y \neq 0$, choose a maximal one, say $(0 : z)$. Then $(0 : z) = P$ is prime. For suppose $abz = 0$, $bz \neq 0$. Then $(0 : z) \subset (0 : bz)$, hence by maximality, $(0 : z) = (0 : bz)$ and $az = 0$. Thus Set (i) $\neq \emptyset$.

Next, suppose $P \in$ Set (iv), say P minimal over $(0 : x)$. Among the ideals $(0 : y)$ with $(0 : x) \subseteq (0 : y) \subseteq P$, let $(0 : z)$ be maximal. Thus if $c \notin P$, $(0 : z) = (0 : cz) \subseteq P$. In fact, if $bz \neq 0$, then $(0 : bz) \subseteq P$, hence $(0 : z) = (0 : bz)$. Otherwise, $bcz = 0$, some $c \notin P$, hence $b \in (0 : cz) = (0 : z)$, contradicting $bz \neq 0$. We now claim $(0 : z)$ is prime, hence $(0 : z) = P$. Namely, if $abz = 0$, $bz \neq 0$, then $(0 : z) = (0 : bz)$, hence $az = 0$. This proves Set (iv) $\subseteq$ Set (i).

Finally, we must prove there are only finitely many associated primes. Suppose the set of associated primes is $\{P_\alpha\} = \{(0 : x_\alpha)\}$ and suppose there are infinitely many. Let $AH(\{x_\alpha\}) = AH(x_{\alpha_1} \cdots x_{\alpha_k})$. If $x_\alpha \neq x_{\alpha_i}$, then $P_\alpha \neq P_{\alpha_i}$, but $P = (0 : x_\alpha) \supset \bigcap_{j=1}^{k} (0 : x_{\alpha_j}) = \bigcap_{j=1}^{k} P_{\alpha_j}$ by Proposition 2.5.3. Thus $P_\alpha \supsetneq P_{\alpha_i}$, some i. Choose an index among the α_i, call it α_0, such that for infinitely many β, $P_\beta \supsetneq P_{\alpha_0}$. Let $AH(\{x_\beta\}) = AH(x_{\beta_1} \cdots x_{\beta_n})$. If $\beta \neq \beta_j$ all j, then $P_\beta \supsetneq P_{\beta_i}$ some i, by the same argument. So infinitely many P_γ properly contain some P_{β_0}. Iterating this process leads to an infinite chain $P_{\alpha_0} \subsetneq P_{\beta_0} \subsetneq P_{\gamma_0} \subsetneq \ldots$, which is impossible. □

Proposition 2.6.6. Suppose $(A,\mathfrak{P}) \in$ (PORCK) is semi-Noetherian, $I,J \subset A$ absolutely convex ideals. Let P_j, $1 \leq j \leq k$, be the associated primes of I. Then $J \subset P_j$, some j, if and only if $(I : J) \neq I$. In particular, the set of zero divisors in A consists of all elements belonging to some associated prime of (0).

Proof. If $J \subset P_j = (I : x_j)$, then $x_j J \subset I$, hence $(I : J) \neq I$. Conversely, if $xJ \subset I$, $x \notin I$, then $J \subseteq (I : x)$ is contained in some maximal ideal of the form $(I : y)$, which, by the proof of 2.6.5, is an associated prime of I. □

Proposition 2.6.7. If $(A,\mathfrak{P}) \in$ (PORCK) is semi-Noetherian and $I \subset A$ is absolutely convex, then any prime P containing I contains an associated prime of I. In particular, the minimal associated primes of I, $\{P_i\}$, are exactly the minimal primes containing I, and $\sqrt{I} = \cap P_i$.

Proof. Suppose $(I : x) \not\subset P$, where $(I : x)$ is some associated prime of I. Let $xy \in I$, $y \notin P$. Then $(I : y) \subset P$ since $I \subset P$. By the proof of 2.6.5, a maximal ideal among the $(I : z)$ with $(I : y) \subset (I : z) \subset P$ will be prime, hence will be an associated prime of I contained in P. □

Although irreducible absolutely convex ideals of a semi-Noetherian ring are not necessarily primary, we can prove that an irreducible ideal has a unique maximal associated prime. In fact, for any absolutely convex ideal $I \subset A$ and prime ideal P, $I \subset P \subset A$, define

$$I(P) = \{x \in A \mid xs \in I, \text{ some } s \notin P\}.$$

(The ideal $I(P)$ is the kernel of the composite homomorphism $A \to A_{(P)} \to A_{(P)}/IA_{(P)}$, where $A_{(P)}$ is the localization of A at P. This point of view will be considered in the next chapter.) The following proposition lists some properties of the ideals $I(P)$. We assume $(A,\mathfrak{P}) \in$ (PORCK), $I \subset A$ absolutely convex.

Proposition 2.6.8.

(a) $I(P) = \bigcap_{s \notin P} (I : s)$, hence $I(P)$ is absolutely convex.

(b) $I \subseteq I(P) \subseteq P$.

(c) If P is a minimal prime containing I, then $I(P)$ is P-primary; in fact, $I(P)$ is the smallest P-primary ideal containing I.

Moreover, if $(A,\mathfrak{P})$ is semi-Noetherian and $\{P_j\}$ is the set of associated primes of I, then

(d) $P_1 \subset P_2$ if and only if $I(P_2) \subset I(P_1)$.

(e) $I = \cap I(P_j)$. In particular, if I is irreducible, it has a unique maximal associated prime.

(f) The associated primes of $I(P_j)$ are exactly the associated primes of I contained in P_j.

Proof. (a) and (b) are trivial. The result (c) is actually true in arbitrary commutative rings, for arbitrary ideals I. This point will be discussed again in the next chapter, so we postpone the proof of (c).

If $P_1 \subset P_2$, then certainly $I(P_2) \subset I(P_1)$. Conversely, suppose $I(P_2) \subset I(P_1)$, but $P_1 \not\subset P_2$. Let $y \in P_1 - P_2$, $P_1 = (I : x_1)$. Then $x_1 y \in I$, $y \notin P_2$ implies $x_1 \in I(P_2) \subset I(P_1)$. But this is ridiculous, since clearly $x \in I(P)$ is equivalent to $(I : x) \not\subset P$. This proves (d).

Also, we deduce (e) since if $x \in \cap I(P_j)$, then $(I : x) \not\subset P_j$, all j. But any proper ideal $(I : x)$ is contained in an associated prime of I, by the proof of 2.6.5. Thus $x \in I$. The second statement of (e) now follows from (d), since the intersection formula $I = \cap I(P_j)$ is non-trivial unless there is a unique maximal P_j.

Finally, we verify (f). If $(I : x_i) = P_i \subset P_j$, then arguments just as above show $(I(P_j) : x_i) = P_i$. Also, if $(I(P_j) : x) = Q_j$ is prime, then

$Q_j \subset P_j$. We must show Q_j is also an associated prime of I. Let $Q_j = AH(y_1 \cdots y_k)$. Then $xy_js_j \in I$ for some $s_j \notin P_j$. It is now easy to verify $(I : xs_1 \cdots s_k) = Q_j$, using 2.5.3. □

Remark. Note that 2.6.8(c), (d), (e) imply that if I has no embedded associated primes, then I is an irredundant intersection of convex primary ideals. A standard argument shows such a representation is unique.

2.7. Convex Ideals and Intersections of Orders

Let A be a ring, $\mathfrak{P}_1, \mathfrak{P}_2 \subset A$ two orders on A. Then $\mathfrak{P}_{12} = \mathfrak{P}_1 \cap \mathfrak{P}_2$ is an order on A. We are interested in relations between the convex ideals for $\mathfrak{P}_1$, $\mathfrak{P}_2$, and $\mathfrak{P}_{12}$.

For example, A could be a ring of functions on a set S, with values in an ordered ring, and $\mathfrak{P}_1$, $\mathfrak{P}_2$ could be affine orders defined by suitable subsets $S_1, S_2 \subset S$, as in Example (6) of 1.7. That is, for $j = 1,2$,

$$\mathfrak{P}_j = \{f \in A \mid f(s) \geq 0, \text{ all } s \in S_j\}.$$

Then $\mathfrak{P}_{12}$ is clearly the affine order corresponding to the subset $S_1 \cup S_2 \subset S$,

$$\mathfrak{P}_{12} = \{f \in A \mid f(s) \geq 0, \text{ all } s \in S_1 \cup S_2\}.$$

Points $s \in S_j$ define $\mathfrak{P}_j$-convex ideals, $j = 1,2$, namely, the ideal of functions $f \in A$ with $f(s) = 0$. Thus points of $S_1 \cup S_2$ define $\mathfrak{P}_{12}$ convex ideals.

More generally, for any A and $\mathfrak{P}_1$ and $\mathfrak{P}_2$, if $I \subset A$ is either $\mathfrak{P}_1$-convex or $\mathfrak{P}_2$-convex, certainly I is $\mathfrak{P}_{12}$-convex. If I_j is $\mathfrak{P}_j$-convex, $j = 1,2$, then $I_1 \cap I_2$ is $\mathfrak{P}_{12}$-convex. Both these remarks are equally obvious for absolutely convex ideals.

If $I \subset A$ is $\mathfrak{P}_{12}$ convex and $I = \sqrt{I}$ is radical, then we will show below that we can write $I = I_1 \cap I_2$, where $I_j = \sqrt{I_j}$ is a $\mathfrak{P}_j$-radical convex ideal, $j = 1,2$. For non-radical ideals, even primary ideals, such a decomposition is impossible in general. We will illustrate these phenomena with examples in 2.8.

The main positive result in this direction is the following very useful fact, discovered by Andrew Klapper.

Proposition 2.7.1 (Klapper). Suppose $P \subset A$ is a prime $\mathfrak{P}_{12}$-convex ideal, $\mathfrak{P}_{12} = \mathfrak{P}_1 \cap \mathfrak{P}_2$. Then P is either $\mathfrak{P}_1$-convex or $\mathfrak{P}_2$-convex.

Proof. Suppose P was neither $\mathfrak{P}_1$-convex nor $\mathfrak{P}_2$-convex. Choose $0 \leq x \leq y$ (rel $\mathfrak{P}_1$), $0 \leq u \leq v$ (rel $\mathfrak{P}_2$), with $y,v \in P$, $x,u \notin P$. Then

$$0 \leq x^2u^2 \leq y^2u^2 \leq y^2u^2 + x^2v^2 \quad (\text{rel } \mathfrak{P}_1)$$

and

$$0 \leq x^2u^2 \leq x^2v^2 \leq y^2u^2 + x^2v^2 \quad (\text{rel } \mathfrak{P}_2) .$$

Thus $0 \leq x^2u^2 \leq y^2u^2 + x^2u^2$ (rel $\mathfrak{P}_{12}$). But $y^2u^2 + x^2v^2 \in P$, hence $x^2u^2 \in P$, contradicting $x,u \notin P$. □

Corollary 2.7.2. Any maximal $\mathfrak{P}_{12}$-convex ideal is either a maximal $\mathfrak{P}_1$-convex ideal or a maximal $\mathfrak{P}_2$-convex ideal.

Proof. Maximal convex ideals are prime. □

Corollary 2.7.3. Any radical $\mathfrak{P}_{12}$-convex ideal I can be written $I = I_1 \cap I_2$, where I_j is a radical $\mathfrak{P}_j$-convex ideal, $j = 1,2$.

Proof. $I = \cap P_\alpha$, the intersection taken over all $\mathfrak{P}_{12}$-convex prime ideals containing I. Let $I_1 = \cap P_{\alpha'}$, $I_2 = \cap P_{\alpha''}$ where the $P_{\alpha'}$ are those P_α which are $\mathfrak{P}_1$-convex and the $P_{\alpha''}$ are those P_α which are $\mathfrak{P}_2$-convex. □

Corollary 2.7.4. If $\mathfrak{P}_1,\ldots,\mathfrak{P}_k \subset A$ are orders and $P \subset A$ is a $\mathfrak{P}_1 \cap \cdots \cap \mathfrak{P}_k$-convex prime ideal, then P is $\mathfrak{P}_j$-convex for some j, $1 \leq j \leq k$. If P is a maximal $\mathfrak{P}_1 \cap \cdots \cap \mathfrak{P}_k$-convex ideal, then P is a maximal $\mathfrak{P}_j$-convex ideal, some j. If I is a radical $\mathfrak{P}_1 \cap \cdots \cap \mathfrak{P}_k$-convex ideal, then $I = I_1 \cap \cdots \cap I_k$ where the I_j are radical $\mathfrak{P}_j$-convex ideals.

Proof. Each statement follows by induction from the results above. □

Proposition 2.7.1 has a very useful application to the following problem. Given $(A,\mathfrak{P}) \in (POR)$, $Q \subset A$ a convex ideal, to what extent can the order $\mathfrak{P}$ be refined, keeping Q convex? In general, this problem seems difficult. However, for prime ideals in integral domains we have the following.

Proposition 2.7.5. Let $(K, \mathfrak{P}) \in (POR)$ with K a field. Let $A \subset K$ be a subring and let $Q \subset A$ be a $\mathfrak{P} \cap A$-convex prime ideal. Then there exist total orders $\tilde{\mathfrak{P}} \subset K$, refining $\mathfrak{P}$, with Q still $\tilde{\mathfrak{P}} \cap A$-convex.

Proof. Suppose $x \in K$, $x,-x \notin \mathfrak{P}$. Then from 1.6, both $\mathfrak{P}[x]$ and $\mathfrak{P}[-x]$ are orders on K. Moreover, $\mathfrak{P} = \mathfrak{P}[x] \cap \mathfrak{P}[-x]$ since if $f = p_1 + q_1x = p_2 - q_2x$, $p_i, q_j \in \mathfrak{P}$, then $(q_2 + q_1)f = q_2p_1 + q_1p_2$. Either $q_1 = q_2 = 0$ and $f = p_1 = p_2$ or $f = (q_2p_1 + q_1p_2)/(q_2 + q_1)$. We now apply 2.7.1 to deduce that $Q \subset A$ is either $(\mathfrak{P}[x]) \cap A$-convex or $(\mathfrak{P}[-x] \cap A)$-convex. Zorn's lemma then completes the proof. □

Corollary 2.7.6. Let $(A,\mathfrak{P}) \in (POR)$, A an integral domain. Let $Q \subset A$ be a prime $\mathfrak{P}$-convex ideal. Then there exist total orders $\tilde{\mathfrak{P}} \subset A$. refining $\mathfrak{P}$, with Q still $\tilde{\mathfrak{P}}$-convex if and only if Q is $D\mathfrak{P}$-convex.

Proof. Properties of derived orders $D\mathfrak{P}$ were studied in 1.6. In particular, $D\mathfrak{P}$ is the intersection of all total orders on A refining $\mathfrak{P}$, so necessity of the condition is clear. Conversely, the derived orders $D\mathfrak{P} \subset A$ are exactly the orders contracted from orders on the field of fractions of A, hence sufficiency follows from 2.7.5. □

The next result generalizes 2.7.1, but the first step in the proof is the argument used in 2.7.1.

Proposition 2.7.7. Let A be a ring, $\mathfrak{P}_1$, $\mathfrak{P}_2$, $\mathfrak{P}_{12} = \mathfrak{P}_1 \cap \mathfrak{P}_2$ orders on A. Let $Q \subset A$ be a primary $\mathfrak{P}_{12}$-convex ideal [resp. primary $\mathfrak{P}_{12}$-absolutely convex ideal]. Let $P = \sqrt{Q}$ be the (prime) radical. Then if P is not $\mathfrak{P}_2$-convex, Q is $\mathfrak{P}_1$-convex [resp. $\mathfrak{P}_1$ absolutely convex].

Proof. Suppose $0 \leq x \leq y$ (rel $\mathfrak{P}_1$), $y \in Q$. We need to show $x \in Q$. Since P is not $\mathfrak{P}_2$-convex, choose $0 \leq u \leq v$ (rel $\mathfrak{P}_2$), $v \in P$, $u \notin P$. Raising to a large even power, we may assume $v \in Q$ and $0 \leq u,v$ (real $\mathfrak{P}_1$). The argument for 2.7.1 shows $x^2u^2 \in Q$ and since Q is primary, $u^2 \notin P$, we conclude $x^2 \in Q$.

We now observe

$$p = x^2 + 2ux + v^2 = (x+u)^2 + (v^2 - u^2) \in \mathfrak{P}_{12}$$

since the first term is obviously in $\mathfrak{P}_1$ and the second term obviously in $\mathfrak{P}_2$. Similarly,

$$q = (y-x)^2 + 2u(y-x) + v^2 = (y-x+u)^2 + (v^2 - u^2) \in \mathfrak{P}_{12}.$$

Now $p+q = x^2 + (y-x)^2 + 2uy + 2v^2$. Since $x^2, y, v \in Q$, we have $p+q \in Q$, and since Q is $\mathfrak{P}_{12}$-convex, we deduce $p,q \in Q$. Thus $p - x^2 - v^2 = 2ux \in Q$. Again, since Q is primary, we conclude $x \in Q$.

The proposition for absolutely convex ideals is a slight extension of the argument. Assume $0 \leq x \leq y$ (rel $\mathfrak{P}_1$) and $yz \in Q$, some $z \in A$. We must show $xz \in Q$. Choose u,v as above. We have $0 \leq x^2u^2 \leq y^2u^2 + x^2v^2$ (rel $\mathfrak{P}_{12}$), as in the proof of 2.7.1, and $(y^2u^2 + x^2v^2)z \in Q$. Hence $x^2u^2z \in Q$ and $x^2z \in Q$. If $p,q \in \mathfrak{P}_{12}$ are also as above, then $(p+q)z \in Q$ as is easily checked. Thus $pz, qz \in Q$, $pz - x^2z - v^2z = 2uxz \in Q$, and finally $xz \in Q$ as desired. □

If $Q \subset A$ is a primary $\mathfrak{P}_{12}$-convex ideal and $\sqrt{Q} = P$ is *both* $\mathfrak{P}_1$ and $\mathfrak{P}_2$-convex, it seems difficult in general to relate Q to $\mathfrak{P}_1$ and $\mathfrak{P}_2$-convex ideals. In 2.8 we give an example of such a Q which cannot be written as an intersection $Q = Q_1 \cap Q_2$, where Q_j is $\mathfrak{P}_j$-convex.

We give one final generalization of 2.7.1. Again, the proof is very similar.

Proposition 2.7.8. Let A be a ring, $I_\alpha \subset A$ a finite collection of ideals with $\cap I_\alpha = (0)$, $\mathfrak{P}_\alpha \subset A_\alpha = A/I_\alpha$ orders, and set $\mathfrak{P} = A \cap \Pi \mathfrak{P}_\alpha$ with respect to the obvious inclusion $A \to \Pi A_\alpha$. Then a prime ideal $Q \subset A$ is $\mathfrak{P}$-convex if and only if for some α, $I_\alpha \subset Q$ and Q/I_α is $\mathfrak{P}$-convex.

Proof: We certainly know some $I_\alpha \subset Q$. If for all these, Q/I_α failed to be $\mathfrak{P}_\alpha$-convex, we get inequalities

$$0 \leq x_\alpha^2 \leq y_\alpha^2 \ (\text{rel } \mathfrak{P}_\alpha \subset A_\alpha), \qquad y_\alpha \in Q, \quad x_\alpha \notin Q.$$

Multiplying by a suitable square in $I_\beta - Q$, we may assume $x_\alpha, y_\alpha \in I_\beta$ for all I_β with $I_\beta \not\subset Q$. Then, just as in the proof of 2.7.1, we get for all γ

$$0 \leq \prod_\alpha x_\alpha^2 \leq \sum_{\alpha'} \Big(\prod_{\alpha \neq \alpha'} x_\alpha^2 \Big) y_{\alpha'}^2, \ (\text{rel } \mathfrak{P}_\gamma).$$

(Note for those γ with $I_\gamma \not\subset Q$, both sides vanish.) Since $\mathfrak{P} = A \cap \Pi \mathfrak{P}_\gamma$, we have the same inequality in $(A, \mathfrak{P})$. The right-hand side belongs to Q, so if Q is $\mathfrak{P}$-convex, we have a contradiction. This proves the "only if" part of the proposition. The "if" part is trivial. □

2.8. Some Examples

(1) The ring $\mathbb{Z}$ has a unique order. The only convex ideals are (0) and $\mathbb{Z}$, and $\mathbb{Z}$ is a semi-field.

(2) Partially order the ring $\mathbb{Z}[T]$ as a ring of real valued functions on the line. Let $f(T)$ be an irreducible polynomial which has real roots. Then $(f) \subset \mathbb{Z}[T]$ is a convex prime ideal, since, if $0 \leq p \leq q$ and $q \in (f)$, p must vanish at all roots of f, hence $p \in (f)$.

In fact, (f) is a maximal convex ideal. The only ideals of $\mathbb{Z}[T]$ properly containing (f) will also contain some integer $m \in \mathbb{Z}$, $m \neq 0$. If such an ideal is convex, it must contain 1. Thus $\mathbb{Z}[T]/(f)$, with the quotient order, provides another example of a semi-field which is not a field.

(3) Let $(A, \mathfrak{P}) \in (POR)$. If $C \subset A$ is a subset, let $C^{\frac{1}{2}} = \{a \in A \mid a^2 \in C\}$. If C is convex, then $C^{\frac{1}{2}}$ is convex, since $0 \leq a \leq b$ implies $0 \leq a^2 \leq b^2$. In fact, $0 \leq a \leq b$ implies $0 \leq a^2x^2 \leq b^2x^2$, all $x \in A$, hence if C is convex, then $C^{\frac{1}{2}}$ is absolutely convex.

Now suppose $I \subset A$ is a convex ideal. Then $I^{\frac{1}{2}}$ is an ideal, hence an absolutely convex ideal. Certainly, if $y \in I^{\frac{1}{2}}$, then $ay \in I^{\frac{1}{2}}$, all $a \in A$. Also, if $y_1, y_2 \in I^{\frac{1}{2}}$, then $(y_1 + y_2)^2 + (y_1 - y_2)^2 = 2y_1^2 + 2y_2^2 \in I$, hence $(y_1 + y_2) \in I^{\frac{1}{2}}$, because I is convex. For example, in any partially ordered ring $(0)^{\frac{1}{2}} = \{a \in A | a^2 = 0\}$ is an absolutely convex ideal.

We can say slightly more about $I^{\frac{1}{2}}$. If $y_1, y_2 \in I^{\frac{1}{2}}$, then $(y_1 + y_2)^2 - y_1^2 - y_2^2 = 2y_1y_2 \in I$. If either I is absolutely convex or $2 \in A$ is invertible, then $y_1, y_2 \in I^{\frac{1}{2}}$ implies $y_1y_2 \in I$.

(4) Let $(A,\mathfrak{P}) \in$ (POR). Define

$$T = \{b \in A | nb = 0, \text{ some integer } n \neq 0\}$$

$$K = \{c \in A | sc = 0, \text{ some } s \in A, \ 1 \leq s\}.$$

We have $T \subset K$, and both T and K are convex ideals, as is easily checked. We regard the convex ideals T and K as somewhat pathological. Note that if $(A,\mathfrak{P}) \in$ (PORCK), necessarily $T = K = (0)$.

(5) Order $\mathbb{Z}[X]$ as a ring of real valued functions on the line. Consider the ideal $(2X, X^2)$. Suppose $0 \leq f(X) \leq 2nX + X^2g(X)$, some $n \in \mathbb{Z}$, $f(X), g(X) \in \mathbb{Z}(X)$. Letting X approach 0 shows that $n = 0$ and at the same time that $f(X)$ is divisible by X^2. Thus $f(X) \in (2X, X^2)$ and $(2X, X^2)$ is convex. On the other hand, $(2X, X^2)$ is not absolutely convex since $0 \leq 1 \leq 2$, $2X \in (2X, X^2)$, but $X \notin (2X, X^2)$.

Consider the quotient ring $\mathbb{Z}[X]/(2X, X^2)$. Elements are uniquely expressible as $n + \varepsilon X$, $n \in \mathbb{Z}$, $\varepsilon = 0$ or 1. Since $(n + \varepsilon X)^2 = n^2 + 2n\varepsilon X + \varepsilon^2 X^2 \equiv n^2$, the quotient ring $\mathbb{Z}[X]/(2X, X^2)$ has a unique non-zero convex ideal, consisting of the two elements $\{0, X\}$. We have $\{0, X\} = T = K$ (as defined in Example (4)) and $\{0, X\} = (0)^{\frac{1}{2}}$ (as defined in Example (3)).

(6) We continue with the example in (5), $(2X, X^2) \subset \mathbb{Z}[X]$. The ideal $(2X, X^2)^2 = (4X^2, 2X^3, X^4) \subset \mathbb{Z}[X]$ is not convex since $0 \leq X^2 \leq 4X^2$ in any order. Thus, products of convex ideals need not be convex. We have $H((2X, X^2)^2) = (X^2)$, which is, in fact, absolutely convex. Namely, if

$0 \leq p(X) \leq q(X)$ and $q(X)f(X) \in (X^2)$, then qf vanishes to second order at the origin. Either q vanishes to second order, both q and f vanish, or f vanishes to second order. In any case, pf will also vanish to second order at the origin, hence $pf \in (X^2)$.

In the quotient ring $\mathbb{Z}[x]/(2X,X^2)$, consider the ideal $((0): (X))$. Since $2X \equiv 0$, $2 \in ((0): (X))$, but $1 \notin ((0): (X))$. Thus quotients of convex ideals need not be convex.

(7) Order $\mathbb{Z}[X,Y]$ as a ring of real valued functions on the plane. A typical non-convex ideal is the principal ideal $(X^2 + Y^2)$.

Certainly $X^2,Y^2 \in H(X^2 + Y^2)$. But also $0 \leq 2X^2 + 2Y^2 + XY \leq 4X^2 + 4Y^2$, hence $XY \in H(x^2 + y^2)$. In fact, $H(X^2 + Y^2) = AH(X^2 + Y^2) = (X^2,Y^2,XY)$, the absolutely convex ideal of functions vanishing to second order at the origin.

The argument in Example (6) shows that $(X^2),(Y^2)$ are also absolutely convex ideals of $\mathbb{Z}[X,Y]$. But $H((X^2) + (Y^2)) = (X^2,Y^2,XY)$, hence sums of absolutely convex ideals need not be convex.

(8) Order the ring $\mathbb{R}[X,Y,Z]/(Z^2-X-Y)$ as a ring of real valued functions on the piece of the surface $z^2 = x+y$ lying in the first octant of $\mathbb{R}^{(3)}$, that is, $0 \leq x,y,z$. Evaluation at $(0,0,0)$ defines an order preserving morphism $\mathbb{R}[X,Y,Z]/(Z^2-X-Y) \to \mathbb{R}$, with kernel (X,Y,Z). Thus (X,Y,Z) is a maximal convex ideal.

Consider the square $(X,Y,Z)^2 = (X^2,Y^2,Z^2,XY,XZ,YZ)$. Since $Z^2 = X + Y$ and $0 \leq X$, $0 \leq Y$, we see that $(X,Y,Z)^2$ is not convex. In fact, $H((X,Y,Z)^2)$ (X,Y,Z^2). This example shows that products of absolutely convex ideals need not be convex.

The ideal (X,Y,Z^2) is, in a real sense, exactly the ideal of functions on our piece of surface S which vanish to second order at the origin. For example, if c: $[0,\varepsilon] \to S$ is any smooth curve, $c(0) = (0,0,0)$, then

$$\frac{d}{dt} (X \cdot c)\Big|_{t=0} = \frac{d}{dt} (Y \cdot c)\Big|_{t=0} = 0 .$$

If we regard $\mathbb{R}[X,Y,Z]/(Z^2-X-Y)$ as a ring of real valued functions on the entire surface $z^2 = x + y$ in $\mathbb{R}^{(3)}$, then the functions X,Y are no longer positive. There are curves on the surface through the origin along which X and Y do not vanish to second order at $(0,0,0)$, say, $c(t) = (t,-t,0)$. In fact, with this order $(X,Y,Z)^2$ is convex. This discussion illustrates how the convex ideal structure of a ring of real valued functions on a space can reflect some of the geometry of the space.

(9) The zero set Z of the equation $1 = y^2 + x^2(x^2+1)$ in $\mathbb{R}^{(2)}$ is topologically a circle. Order the ring $\mathbb{R}[X,Y]/(1-Y^2-X^2(X^2+1)) = A$ as a ring of functions on this zero set. Now, A is a Dedekind domain, hence the ideal class group $C(A)$ is isomorphic to the (reduced) Grothendieck group $K_o(A)$. This example has been studied by Evans [27], in connection with the problem of relating $K_o(A)$ to the real K-theory of the zero set Z. Evans showed that the ideal $(Y-1,X)^2 \subset A$ is not principal, hence $C(A)$ contains elements not of order 2. Thus, the natural map $K_o(A) \to KO(Z)$ is not injective.

However, suppose in defining $C(A)$, we replace ideals by convex ideals and principal ideals by ideals of the form $H(a)$, $a \in A$. Then $H((Y-1),X)^2)$ $= H(Y-1)$, which is "principal". Namely, $(Y-1)(Y+1) = Y^2-1 \equiv X^2(X^2+1) \in A$, hence $X^2 \in H(Y-1)$. Thus $(Y-1,X)^2 \subset H(Y-1)$. Conversely, the same equality $Y^2-1 \equiv X^2(X^2+1)$ shows $Y^2-1 \in (Y-1,X)^2$. Also $(Y-1)^2 \in (Y-1,X)^2$, hence $(Y-1)((Y+1)-(Y-1)) = 2(Y-1) \in (Y-1,x)^2$. We conclude $H(Y-1) = H((Y-1,X)^2)$, as claimed.

Actually, we prefer a (PORCK) version of $C(A)$, that is, absolutely convex ideals should be used in the definition. The relation with real K-theory is not really so relevant. The correct (PORCK) analogue of $C(A)$ should really correspond to a "real divisor class group" of Z, which in turn should be $H^1(Z, \mathbb{Z}/2)$. The group $KO(Z)$ should still be approached through projective modules. Evans' method was to replace A by a localization A_S, where $S \subset A$ is the multiplicative set of functions with no zeros on Z. We will see that this localization occurs quite naturally in the category (PORCK).

(10) Let K be a finite simplicial complex, with vertices $v_o,\dots,v_n$ and associated barycentric coordinate functions $X_o,\dots,X_n$. Let $A_{\mathbb{Z}}(K)$

be the partially ordered ring of real valued functions on K generated over $\mathbb{Z}$ by the functions $X_o,\ldots,X_n$. Thus, $A_{\mathbb{Z}}(K) = \mathbb{Z}[X_o\ldots X_n]/I(K)$, where $I(K)$ is the ideal generated by polynomials $\sum_{i=o}^{n} X_i - 1$ and $X_{i_o}\cdot\ldots\cdot X_{i_r}$, for all $i_o < \cdots < i_r$ such that $(v_{i_o},\ldots,v_{i_r})$ do not span an r-simplex of K. (See 1.7, Example (9).) In particular, if K is an n-simplex Δ^n, then

$$A_{\mathbb{Z}}(\Delta^n) = \mathbb{Z}[X_o\ldots X_n]/\left(\sum_{i=0}^{n} X_i - 1\right).$$

Thus $A_{\mathbb{Z}}(K)$ is an integral domain if and only if K is a simplex. In $A_{\mathbb{Z}}(K)$ we have the order relations $0 \leq X_i \leq 1$.

We have the following assertions as exercises.

(i) The non-zero (POR)-morphisms $A_{\mathbb{Z}}(K) \to \mathbb{Z} = A_{\mathbb{Z}}(\Delta^o)$ correspond exactly to the vertices $v_i\colon \Delta^o \to K$ of K.

(ii) The (POR)-automorphism group of $A_{\mathbb{Z}}(\Delta^r)$ is exactly the symmetry group S_{r+1} which acts on Δ^r by permuting vertices.

(iii) The surjective (POR)-morphisms $A_{\mathbb{Z}}(K) \to A_{\mathbb{Z}}(\Delta^r)$ correspond exactly to the inclusions of r-simplices in K.

(iv) If K, L are two finite complexes,

$$\mathrm{Hom}_{(POR)}(A_{\mathbb{Z}}(K),A_{\mathbb{Z}}(L)) \cong \mathrm{Hom}_{(Simpl.)}(L,K).$$

We thus obtain a faithful embedding of the category of finite simplicial complexes in the dual category of (POR).

(11) In this example we study the ring of polynomials $Q[X]$, Q the rational numbers. It is convenient to think of $Q[X]$ as a ring of functions, say on the real line. For algebraic purposes it is slightly preferable to think of $Q[X]$ as a ring of functions on $\overline{Q}$, the real algebraic numbers.

Before beginning the discussion, we indicate two directions of generalization. First, Q could be replaced by any totally ordered field F, with real closure $\overline{F}$. The discussion below will apply to $F[X]$, provided F is dense in $\overline{F}$, in the interval topology. (This is a tricky point. In all examples, no element of $\overline{F}$ will be infinitesimally small relative to F.

That is, all intervals *around* 0 in $\overline{F}$ will contain non-zero elements of F. However, there are examples where some intervals of $\overline{F}$ contain no elements of F.) Secondly, the number of variables could be increased. This would require much more work. In fact, the study of polynomial rings in several variables over an ordered field is, in some sense, the final goal of the whole theory. Polynomials in one variable is essentially zero-dimensional algebraic geometry, which is trivial from a geometric point of view.

All ideals in Q[X] are principal. We will (i) characterize the convex ideals (f(X)), and (ii) characterize certain orders on the quotient rings Q[X]/(f(X)), $f \neq 0$. Since this is only an example, proofs will not be given. We fix the weak order $\mathfrak{P}_w \subset Q[X]$ throughout this example. It is known that $\mathfrak{P}_w$, the sums of squares, coincides with the affine order, the polynomials which assume no negative values on the real line.

Assertion 1. An ideal $(f(X)) \subset Q[X]$ is convex if and only if every irreducible factor of f(X) has real roots.

Assertion 2. All convex ideals $(f(X)) \subset Q[X]$ are absolutely convex.

Suppose (f(X)) is convex. Factor f, say $f = h_1^{k_1} \cdot \ldots \cdot h_r^{k_r}$, where the h_i are distinct irreducible polynomials, with real roots. We denote the real roots of h_i by α_{ij}. By the Chinese Remainder Theorem, we have an isomorphism of rings

$$\frac{Q[X]}{(f)} \approx \prod_{i=1}^{r} \frac{Q[X]}{(h_i^{k_i})} .$$

From 1.7, Example (11), we know that any order on a direct product of rings is a direct product of orders. Thus we are reduced to studying $Q[X]/(h_i^{k_i})$, h_i irreducible.

If $g(X) \in Q[X]$, write $g = g_0 + g_1 h_i + \cdots + g_m h_i^m$, where degree $(g_j) <$ degree (h_i). Let j, $0 \leq j \leq m$, be least such that $g_j \neq 0$.

Assertion 3. We have $g \in \mathfrak{P}_w/(h_i^{k_i}) \subset Q[X]/(h_i^{k_i})$, the weak order, if and only if $j \geq k_i$ (whence $g \equiv 0$) or $j < k_i$ is even and $g_i(\alpha_{ij}) > 0$ for all real roots α_{ij} of h_i.

Assertion 3 can be interpreted geometrically. Let $Z_i = \{\alpha_{ij}\}$ be the zero set of h_i. Regard $Q[X]$ as a ring of *germs* of functions, defined on neighborhoods of Z_i. The ideal $(h_i^{k_i})$ is the ideal of germs which vanish to k_i-th order on Z_i. Assertion 3 says that a function $g \notin (h_i^{k_i})$ is positive in the weak order on $Q[X]/(h_i^{k_i})$ if and only if the germ of g is locally non-negative near Z_i.

The geometric interpretation of the weak order suggests other orders on $Q[X]/(h_i^{k_i})$. Each root α_{ij} has two sides on the real line, which we denote by $\alpha_{ij}^+, \alpha_{ij}^-$. Any non-empty subset $S \subset \{\alpha_{ij}^+, \alpha_{ij}^-\}$ defines an order $\mathfrak{P}_S \subset Q[X]/(h_i^{k_i})$. A function $g \notin (h_i^{k_i})$ belongs to $\mathfrak{P}_S$ if the germ of g is locally non-negative on those sides of those roots α_{ij} picked out by the subset $S \subset \{\alpha_{ij}^+, \alpha_{ij}^-\}$.

Assertion 4. The orders $\mathfrak{P}_S \subset Q[X]/(h_i^{k_i})$ are all quotients of orders on $Q[X]$. The maximal orders on $Q[X]/(h_i^{k_i})$ are precisely the orders defined by single elements of $\{\alpha_{ij}^+, \alpha_{ij}^-\}$, that is, by one side of a single root of h_i. The maximal orders on $Q[X]/(h_i^{k_i})$ are thus total orders. All the orders $\mathfrak{P}_S$ are (PORCK)-orders.

(12) If we refine the order on $Q[X]$ in Example (11), the notion of convex ideal changes. For example, let $Y \subset \overline{Q}$ be any infinite subset of algebraic numbers. (Infinite is equivalent to Zariski dense.) As in 1.7, Example (6), define $\mathfrak{P}_Y \subset Q[X]$ by $f(X) \in \mathfrak{P}_Y$ if $f(y) \geq 0$, all $y \in Y$. Let $\overline{Y}$ be the usual topological closure of Y. Then an ideal $(h(X)) \subset Q[X]$ is $\mathfrak{P}_Y$ convex if and only if every irreducible factor of $h(X)$ has at least one real root in $\overline{Y}$.

(13) In several examples above we have studied polynomial rings with orders defined in terms of values of functions on subsets of affine space. There is a more intrinsic, algebraic approach to these same orders.

Let $(A,\mathfrak{P}) \in (POR)$ and let $\{I_\alpha\}$ be a collection of $\mathfrak{P}$-convex ideals, with $\bigcap_\alpha I_\alpha = (0)$. Then the natural map $A \to \prod_\alpha A/I_\alpha$ is injective, and we can define a refinement of $\mathfrak{P}$ by $\overline{\mathfrak{P}} = A \bigcap \prod_\alpha (\mathfrak{P}/I_\alpha)$. We are in the situation of

1.7, Example (2), where A is now interpreted as a "ring of functions" on the index set $\{\alpha\}$, with value at α in the partially ordered ring $(A/I_\alpha, \mathfrak{P}/I_\alpha)$.

The order $\overline{\mathfrak{P}} \subset A$ can also be characterized as follows.

(i) All I_α are $\overline{\mathfrak{P}}$-convex

(ii) $\mathfrak{P}/I_\alpha = \overline{\mathfrak{P}}/I_\alpha \subset A/I_\alpha$

(iii) $\overline{\mathfrak{P}}$ is the union of all orders on A satisfying (i) and (ii).

As a specific example, let $A = \mathbb{R}[X_1 \cdots X_n]$, with the weak order $\mathfrak{P}_w$. Each $y \in \mathbb{R}^{(n)}$ defines an evaluation homomorphism $A \to \mathbb{R}$, with kernel I_y a maximal $\mathfrak{P}_w$-convex ideal. (In fact, the I_y, $y \in \mathbb{R}^{(n)}$, are exactly the maximal $\mathfrak{P}_w$-convex ideals of A.) Certainly, $\bigcap_{y \in \mathbb{R}^{(n)}} I_y = (0)$, and the quotients A/I_y are all isomorphic to $\mathbb{R}$, which has a unique order. The order $\overline{\mathfrak{P}} \subset A$ is thus the affine order, of polynomials nowhere negative on $\mathbb{R}^{(n)}$.

(14) Reconsider the convex but not absolutely convex ideal $(2X, X^2) \subset \mathbb{Z}[X]$ of Example (5). The associated primes are (X) and $(2,X)$. However, $(2,X)$ is not convex. This illustrates an advantage of the category (PORCK), where associated primes are always convex.

(15) Consider $\mathbb{R}[X,Y]$ and the family of orders $\mathfrak{P}_n = \mathfrak{P}_w[X, Y, 1-Y, Y^n - X]$, $n \geq 1$. We have $\mathfrak{P}_n \subset \mathfrak{P}_{n+1}$ since $Y^n - X = Y^{n+1} - X + Y^n(1-Y)$. Let $\mathfrak{P} = \cup\, \mathfrak{P}_n$.

The ideal (X^2, XY) is $\mathfrak{P}_n$-absolutely convex for all n. To see this, write $(X^2, XY) = (X) \cap (X^2, XY, Y^m)$. It is relatively easy to check that (X) and (X^2, XY, Y^m) are $\mathfrak{P}_n$-absolutely convex if $m > n$. Thus (X^2, XY) is $\mathfrak{P}$-absolutely convex, since a contradiction of this assertion would involve only finitely many elements, hence would already be a contradiction in some $\mathfrak{P}_n$.

The associated primes of (X^2, XY) are (X) and (X,Y). We have just observed that (X^2, XY) has an absolutely convex primary decomposition for each of the orders $\mathfrak{P}_n$. However, (X^2, XY) has no convex primary decomposition for $\mathfrak{P}$, since, first, the primary component corresponding to the minimal prime (X) is necessarily (X) itself in any decomposition, and secondly, any $\mathfrak{P}$-convex ideal with radical (X,Y) contains some Y^m, hence X, since $0 \leq X \leq Y^m$ (rel $\mathfrak{P}$).

The ideal (X^2, XY) is actually irreducible among $\mathfrak{P}$-convex ideals. If $(X^2, XY) = I_1 \cap I_2$, then $(X) = \sqrt{(X^2, XY)} = \sqrt{I_1} \cap \sqrt{I_2}$. If $\sqrt{I_1} = (X)$, but $(X^2, XY) \neq I_1$, then clearly $I_1 = (X)$. We must then have $(X) \subsetneq \sqrt{I_2}$, since $X \notin I_1 \cap I_2$. It follows that $\sqrt{I_2}$ must be a finite intersection of maximal convex ideals $(X, Y - c_i)$, corresponding to points on the Y-axis. But now Proposition 2.6.8(c), (e) would imply that $(X^2, XY) = I_1 \cap I_2$ has a primary convex decomposition, which we know is impossible by the above discussion.

(16) (Andrew Klapper) We give an example of a ring A, two orders $\mathfrak{P}_j \subset A$, $j = 1,2$, prime $\mathfrak{P}_j$-convex ideals P_j, such that $I = P_1 \cap P_2$ is neither $\mathfrak{P}_1$-convex nor $\mathfrak{P}_2$-convex. Note $I = \sqrt{I}$, so Proposition 2.7.1 does not extend to radical ideals.

Let $A = \mathbb{R}[X] \times \mathbb{R}[Y]$. Let $\mathfrak{P}_1 = \mathfrak{P}_w[X-1] \times \mathfrak{P}_w$, $\mathfrak{P}_2 = \mathfrak{P}_w \times \mathfrak{P}_w[Y-1]$. Then $\mathfrak{P}_1 \cap \mathfrak{P}_2 = \mathfrak{P}_w \times \mathfrak{P}_w$, the weak order on A. Let $P_1 = \mathbb{R}[X] \times (Y)$, $P_2 = (X) \times \mathbb{R}[Y]$. Then $P_1 \cap P_2 = (X) \times (Y)$. Clearly P_1 is $\mathfrak{P}_1$-convex (but not $\mathfrak{P}_2$-convex) and P_2 is $\mathfrak{P}_2$-convex (but not $\mathfrak{P}_1$-convex). But $P_1 \cap P_2$ is not $\mathfrak{P}_1$-convex since otherwise $(X,Y) \in P_1 \cap P_2$ would imply $(1,Y) \in P_1 \cap P_2$. Similarly, $P_1 \cap P_2$ is not $\mathfrak{P}_2$-convex.

Geometrically, the ring A can be rewritten as a quotient of a polynomial ring in four indeterminates, $\mathbb{R}[X,Y,E,F]$. Namely, one divides by the relations $E(E-1) = 0$, $F(F-1) = 0$, $E + F = 1$, $XF = YE = 0$. The maximal convex ideals for the weak order then correspond to the points on the two lines in $\mathbb{R}^4$ defined by $E = 1$, $F = 0$, $Y = 0$ and $E = 0$, $F = 1$, $X = 0$. The variable X parametrizes the first line, while Y parametrizes the second. In this notation $\mathfrak{P}_1 = \mathfrak{P}_w[X-E]$, $\mathfrak{P}_2 = \mathfrak{P}_w[Y-F]$, $P_1 = (E,X,Y)$, $P_2 = (F,X,Y)$.

(17) We give an example of two orders $\mathfrak{P}_1, \mathfrak{P}_2$ on $\mathbb{R}[X,Y]$, primary ideals Q_1, Q_2 with $\sqrt{Q_1} = \sqrt{Q_2} = (X,Y)$ such that Q_1 is $\mathfrak{P}_1$-absolutely convex, Q_2 is $\mathfrak{P}_2$-absolutely convex, but $Q = Q_1 \cap Q_2$ is neither $\mathfrak{P}_1$-convex nor $\mathfrak{P}_2$-convex. Since Q is primary with $\sqrt{Q} = (X,Y)$, we see

that Proposition 2.7.1 does not extend to primary ideals.

We let $\mathfrak{P}_1 = \mathfrak{P}_w[X,Y,Y^2-X]$, $\mathfrak{P}_2 = \mathfrak{P}_w[X,Y,X^2-Y]$. Let $Q_1 = (X,Y^2)$, $Q_2 = (X^2,Y)$. It is not difficult to check that Q_1 is $\mathfrak{P}_1$-absolutely convex. In fact, Q_1 is $\mathfrak{P}_1'$-absolutely convex, where $\mathfrak{P}_1' = \{f \in \mathbb{R}[X,Y] \mid f(x,y) \geq 0$ if $0 \leq x,y,y^2-x\}$. Similarly, Q_2 is $\mathfrak{P}_2'$-absolutely convex, where $\mathfrak{P}_2' = \{f \in \mathbb{R}[X,Y] \mid f(x,y) \geq 0$ if $0 \leq x,y,x^2-y\}$. But $Q_1 \cap Q_2 = (X^2,Y^2,XY)$ is not $\mathfrak{P}_1$-convex since $0 \leq X \leq Y^2$ (rel $\mathfrak{P}_1$) and is not $\mathfrak{P}_2$-convex since $0 \leq Y \leq X^2$ (rel $\mathfrak{P}_2$).

(18) The example in (17) still leaves open the hope that if Q is a $\mathfrak{P}_1 \cap \mathfrak{P}_2$ convex primary ideal in some ring A with $P = \sqrt{Q}$ both $\mathfrak{P}_1$-convex and $\mathfrak{P}_2$-convex, then Q might always be written $Q = Q_1 \cap Q_2$ where Q_j is $\mathfrak{P}_j$-convex. However, we now give an example where such a decomposition is impossible.

Let $A = \mathbb{R}[X,Y]$, $\mathfrak{P}_1 = \mathfrak{P}_w[Y,X-Y]$, $\mathfrak{P}_2 = \mathfrak{P}_w[Y,-X-Y]$. Let $Q = (X,Y^2)$. Clearly Q is not $\mathfrak{P}_1$-convex since $0 \leq Y \leq X$ (rel $\mathfrak{P}_1$) and Q is not $\mathfrak{P}_2$-convex since $0 \leq Y \leq -X$ (rel $\mathfrak{P}_2$). However, Q is $\mathfrak{P}_1 \cap \mathfrak{P}_2$-convex. In fact, let $S = \{(x,y) \in \mathbb{R}^2 \mid 0 \leq y,\ y^2 \leq x^2\}$, and let $\mathfrak{P}(S) = \{f \in \mathbb{R}[X,Y] \mid f(x,y) \geq 0,\ (x,y) \in S\}$. Then $\mathfrak{P}_1 \cap \mathfrak{P}_2 \subset \mathfrak{P}(S)$ and we will show Q is $\mathfrak{P}(S)$-convex. (In fact, Q is $\mathfrak{P}(S)$-absolutely convex.)

Suppose $0 \leq f \leq g$ as functions on S and $g \in (X,Y^2) = Q$. Equivalently, $g(0,0) = 0$ and $(\partial g/\partial y)(0,0) = 0$. We must show $f(0,0) = 0$ and $(\partial f/\partial y)(0,0) = 0$. It is obvious that $f(0,0) = 0$.

We consider directional derivatives of f and g with respect to the vectors $\vec{v} = \vec{i} + \vec{j}$ and $\vec{w} = -\vec{i} + \vec{j}$, where $\vec{i}, \vec{j}$ are the usual unit vectors along the x and y axes, respectively. Since $0 \leq f \leq g$ on S, $\vec{v}$ and $\vec{w}$ lie in S, and both f and g vanish at $(0,0)$, we must have

$$0 \leq \frac{\partial f}{\partial \vec{v}}(0,0) \leq \frac{\partial g}{\partial \vec{v}}(0,0)$$

and

$$0 \leq \frac{\partial f}{\partial \vec{w}}(0,0) \leq \frac{\partial g}{\partial \vec{w}}(0,0) .$$

But then (suppressing evaluation at (0,0))

$$0 \leq \frac{\partial f}{\partial \vec{v}} = \frac{\partial f}{\partial x} + \frac{\partial f}{\partial y} \leq \frac{\partial g}{\partial x} + \frac{\partial g}{\partial y} = \frac{\partial g}{\partial x}$$

and

$$0 \leq \frac{\partial f}{\partial \vec{w}} = \frac{-\partial f}{\partial x} + \frac{\partial f}{\partial y} \leq \frac{-\partial g}{\partial x} + \frac{\partial g}{\partial y} = \frac{-\partial g}{\partial x}$$

which implies by adding that $2 \frac{\partial f}{\partial y}(0,0) = 0$, as desired.

The ideal $Q = (X,Y^2) \subset \mathbb{R}[X,Y]$ is irreducible among *all* ideals of $\mathbb{R}[X,Y]$. Thus there is certainly no decomposition $Q = Q_1 \cap Q_2$ where Q_j is $\mathfrak{P}_j$-convex.

III · Localization

3.1. Partial Orders on Localized Rings

Let A be a ring and let $T \subset A$ be a multiplicative set, that is, a subset closed under products and containing the unit $1 \in A$. We let A_T denote the ring obtained by "inverting elements of T". That is,

$$A_T = \{a/t \mid a \in A,\ t \in T\}/\sim$$

where $a_1/t_1 \sim a_2/t_2$ if there is $t \in T$ with $t(a_1t_2 - a_2t_1) = 0$. Sums and products are defined by

$$\frac{a_1}{t_1} + \frac{a_2}{t_2} = \frac{a_1t_2 + a_2t_1}{t_1t_2}$$

$$\frac{a_1}{t_1} \cdot \frac{a_2}{t_2} = \frac{a_1a_2}{t_1t_2}.$$

There is a canonical ring homomorphism $i_T\colon A \to A_T$, $i_T(a) = a/1$, such that $i_T(t)$ is invertible for all $t \in T$ and, in fact, (A_T, i_T) is the universal morphism in the category of rings with this property.

We ask if A has a partial order, say $(A,\mathfrak{P}) \in (\mathrm{POR})$, is there a natural partial order $\mathfrak{P}_T$ on A_T such that, first, $i_T\colon (A,\mathfrak{P}) \to (A_T,\mathfrak{P}_T)$ is order preserving and, secondly, the triple $(A_T, \mathfrak{P}_T)$, i_T is universal in some sense?

Proposition 3.1.1. Define $\mathfrak{P}_T \subset A_T$ by

$$\mathfrak{P}_T = \{a/t \mid ats^2 \in \mathfrak{P},\ \text{some}\ s \in T\}.$$

Then

(a) $\mathfrak{P}_T$ is an order on A_T.

(b) $i_T\colon (A,\mathfrak{P}) \to (A_T, \mathfrak{P}_T)$ is order preserving.

(c) Given a morphism $f\colon\ (A,\mathfrak{P}) \to (A',\mathfrak{P}')$ in (POR) such that $f(t)$ is invertible in A' for all $t \in T$, then there is a unique morphism in (POR), $g\colon\ (A_T,\mathfrak{P}_T) \to (A',\mathfrak{P}')$, such that

$$\begin{array}{ccc} (A,\mathfrak{P}) & \xrightarrow{i_T} & (A_T,\mathfrak{P}_T) \\ & {\scriptstyle f}\searrow \quad \swarrow{\scriptstyle g} & \\ & (A',\mathfrak{P}') & \end{array}$$

commutes.

(d) $\mathfrak{P}_T = (i_T)_*\mathfrak{P}$, the weakest order on A_T such that i_T is order preserving.

<u>Proof</u>. We first show that our definition $a/t \in \mathfrak{P}_T$ does not depend on the choice of representative, a/t, of the element of A_T. If $a/t \sim a'/t'$ then $at't'' = a'tt''$ for some $t'' \in T$. We thus have, for any $s \in T$,

$$(at)(t')^2(t'')^2s^2 = (a't')t^2(t'')^2s^2 .$$

Thus, if $ats^2 \in \mathfrak{P}$, then $a't'(stt'')^2 \in \mathfrak{P}$, hence $a'/t' \in \mathfrak{P}_T$.

(a) The equations in A

$$(a_1t_2 + a_2t_1)t_1t_2\,s_1^2\,s_2^2 = a_1t_1\,s_1^2\,t_2^2 + a_2t_2\,s_2^2\,t_1^2$$

$$a_1a_2t_1t_2\,s_1^2\,s_2^2 = (a_1t_1s_1^2)(a_2t_2s_2^2)$$

show that if a_1/t_1, $a_2t_2 \in \mathfrak{P}_T$, then $a_1/t_1 + a_2/t_2$, $(a_1/t_1)(a_2/t_2) \in \mathfrak{P}_T$. Also, since $(a/t)^2 = a^2/t^2$ and $a^2t^2 = a^2t^2\cdot 1^2 \in \mathfrak{P}$, we see that $\mathfrak{P}_T$ contains all squares.

Finally, suppose $a/t \in \mathfrak{P}_T$ and $-a/t \in \mathfrak{P}_T$. Say, $ats_-^2 \le 0 \le ats_+^2$. Then $ats_-^2s_+^2 \le 0 \le ats_+^2s_-^2$, hence $at(s_-s_+)^2 = 0$ and $a/t = 0 \in A_T$. Thus $\mathfrak{P}_T \subset A_T$ is an order.

(b) is trivial.

(c) We already know that there is a unique such morphism $g\colon\ A_T \to A'$ in the category of rings. Namely, $g(a/t) = f(a)f(t)^{-1} \in A'$. We need to check that g is order preserving.

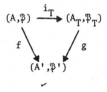

Lemma 3.1.2. Let $(A',\mathfrak{P}') \in (POR)$, and let $0 \leq x \leq y$ in A', with x,y invertible. Then $0 \leq y^{-1} \leq x^{-1}$.

The lemma implies our result, since if $a/t \in \mathfrak{P}_T$, then for some $s \in T$, $ats^2 \in \mathfrak{P}$. Hence $f(a)f(t)f(s)^2 \in \mathfrak{P}'$ and $f(a)f(t)f(s)^2(f(t)^{-1}f(s)^{-1})^2 = g(a/t) \in \mathfrak{P}'$.

The lemma itself is easy, since $y^{-1} = y(y^{-1})^2$ and $y^{-1} - x^{-1} = (x-y)(x^{-1}y^{-1})$. □

3.2. Sufficiency of Positive Multiplicative Sets

In general, if $T \subset A$ is a multiplicative set, there are many elements other than elements of T which are invertible in A_T. In fact, if

$$\tilde{T} = \{a \in A \mid ab \in T, \text{ some } b \in A\},$$

then $\tilde{T}$ is exactly the set of elements of A invertible in A_T. First, if $ab \in T$, then $b/ab = a^{-1} \in A_T$ and, conversely, if $(a/1)(x/t) = 1/1 \in A_T$, then $axs = ts \in T$ for some $s \in T$. Note that $\tilde{T}$ is a multiplicative set.

Proposition 3.2.1. Let $(A,\mathfrak{P}) \in (POR)$, $T \subset A$ a multiplicative set $\tilde{T} \subset A$ as above. If $T' \subset A$ is any multiplicative set with $T \subset T' \subset \tilde{T}$, there is a natural isomorphism in the category (POR)

$$i_{T,T'}\colon\ (A_T,\mathfrak{P}_T) \xrightarrow{\sim} (A_{T'},\mathfrak{P}_{T'}),$$

defined by

$$i_{T,T'}(a/t) = a/t.$$

Proof. It is routine that $i_{T,T'}$ is a ring isomorphism. It is also clear that $i_{T,T'}(\mathfrak{P}_T) \subset \mathfrak{P}_{T'}$. It is not quite so clear that $i^{-1}_{T,T'}$ is a morphism in (POR). However, since elements of T' are invertible in A_T, the universal property of $(A_{T'},\mathfrak{P}_{T'})$ *in (POR)* implies that $i^{-1}_{T,T'}\colon (A_{T'},\mathfrak{P}_{T'}) \to (A_T,\mathfrak{P}_T)$ is a morphism in (POR). (This could also be proved directly, of course.) □

Turning the argument around, we see that there are multiplicative sets *smaller* than T with the same localization in (POR).

Proposition 3.2.2. Let $T \subset A$ be a multiplicative set $(A,\mathfrak{P}) \in$ (POR). Define

$$T^+ = \{t \mid t \in T \cap \mathfrak{P}\}$$

$$T^2 = \{t^2 \mid t \in T\}.$$

Then $T^2 \subset T^+ \subset T$ are multiplicative sets and the natural maps

$$(A_{T^2}, \mathfrak{P}_{T^2}) \to (A_{T^+}, \mathfrak{P}_{T^+}) \to (A_T, \mathfrak{P}_T)$$

are isomorphisms in (POR).

Proof. The first assertion is trivial and the second follows from Proposition 3.2.1 and the observation that $T^2 \subset T^+ \subset T \subset (\widetilde{T^2})$. □

Thus, when localizing in (POR), we can get by with positive multiplicative sets.

3.3. Refinements of an Order Induced by Certain Localizations

Suppose $i_T\colon A \to A_T$ is injective, $(A,\mathfrak{P}) \in$ (POR). Then there is a natural refinement of $\mathfrak{P}$, namely, the contraction of $\mathfrak{P}_T$ to A, that is,

$$\mathfrak{P} \subset i_T^*(i_T)_*\mathfrak{P} = \mathfrak{P}_T \cap A = \{a \in A \mid as^2 \in \mathfrak{P}, \text{ some } s \in T\}.$$

It is easy to see that $\mathfrak{P}_T \cap A = \mathfrak{P}_{T^+} \cap A = \mathfrak{P}_{T^2} \cap A$, say from Proposition 3.2.2 of the preceding section. If A is an integral domain and $T = A - \{0\}$, then $T^+ = \mathfrak{P}^+$, the strictly positive elements of A. In this case

$$\mathfrak{P}_{T^+} \cap A = \{a \in A \mid pa \in \mathfrak{P} \text{ some } p \in \mathfrak{P}^+\}.$$

This is exactly the derived set $D\mathfrak{P}$ of $\mathfrak{P}$ discussed in 1.6 for integral domains.

More generally, if $(A,\mathfrak{P}) \in (POR)$ is arbitrary and T is the multiplicative set of non-zero divisors of A, then $i_T\colon A \to A_T$ is injective. We conclude that any order $\mathfrak{P}$ has a natural refinement $\mathfrak{P}_d = \mathfrak{P}_T \cap A$, where $T \subset A$ is the set of non-zero divisors. We will call $\mathfrak{P}_d$ the *derived order* of $\mathfrak{P}$ (hopefully remembering that the derived *set* $D\mathfrak{P}$ coincides with $\mathfrak{P}_d$ only in special cases).

Proposition 3.3.1. If $(A,\mathfrak{P}) \in (POR)$, then $(\mathfrak{P}_d)_d = \mathfrak{P}_d$.

Proof. If $T \subset A$ is the set of non-zero divisors, $i_T\colon A \to A_T$ the canonical map, one has (as a special case of a more general principle) that $(i_T)_* i_T^*(i_T)_*\mathfrak{P} = (i_T)_*\mathfrak{P} \subset A_T$. Now intersect with A. □

Remark. Note that if A is an integral domain $(A,\mathfrak{P}) \in (POR)$, then the following "simpler" definition of an order $\mathfrak{P}_{(0)}$ on the field of fractions $A_{(0)}$ of A *does not* work:

$$\mathfrak{P}_{(0)} = \{\tfrac{a}{b} \in A_{(0)} \,|\, ab \in \mathfrak{P}\}.$$

The condition doesn't always say the same thing for the *equal* elements a/b and as/bs of $A_{(0)}$ unless the order $\mathfrak{P}$ is sufficiently strong, say $\mathfrak{P} = \mathfrak{P}_d$.

3.4. Convex Ideals in $(A,\mathfrak{P})$ and $(A_T, \mathfrak{P}_T)$

The morphism $i_T\colon (A,\mathfrak{P}) \to (A_T, \mathfrak{P}_T)$, $T \subset A$ a multiplicative set, enables us to define correspondences between ideals of A and A_T. Namely, to $I \subset A$, assign the ideal $I_T = IA_T \subset A_T$, and to $\bar{J} \subset A_T$, assign the ideal $i_T^{-1}(\bar{J}) \subset A$.

If $(A,\mathfrak{P}) \in (POR)$ and $\bar{J} \subset A_T$ is $\mathfrak{P}_T$ convex, then $i_T^{-1}(\bar{J}) \subset A$ is $\mathfrak{P}$ convex, since i_T is a (POR)-morphism. The other correspondence also preserves convex ideals.

Proposition 3.4.1. If $I \subset A$ is $\mathfrak{P}$-convex, then $I_T \subset A_T$ is $\mathfrak{P}_T$-convex. Moreover, there is a natural isomorphism in (POR), induced by $i_T\colon A \to A_T$,

$$((A/I)_{\pi(T)}, (\mathfrak{P}/I)_{\pi(T)}) \xrightarrow{\sim} (A_T/I_T, \mathfrak{P}_T/I_T),$$

where π: $(A,\mathfrak{P}) \to (A/I, \mathfrak{P}/I)$ is the quotient projection.

Proof. One knows from commutative algebra that i_T: $A \to A_T$ induces a ring isomorphism $(A/I)_{\pi(T)} \simeq A_T/IA_T$. That is, "localization commutes with forming quotients." (Namely, A_T/IA_T has the right "universal property" to be $(A/I)_{\pi(T)}$.) Consider the commutative diagram

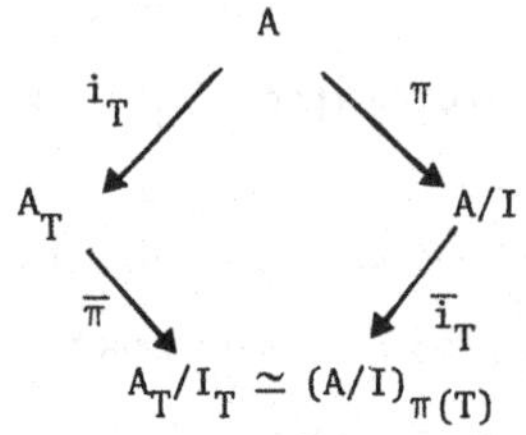

Following the arrows clockwise from A, we see that the order $\mathfrak{P} \subset A$ can be extended, first to $\mathfrak{P}/I \subset A/I$, then to $(\mathfrak{P}/I)_{\pi(T)} \subset (A/I)_{\pi(T)}$. Thus the order $\mathfrak{P}$ can also be extended going counterclockwise from A. It follows that $\mathfrak{P}_T \subset A_T$ can be extended under $\bar{\pi}$, which can only happen if kernel $(\bar{\pi})$ $= I_T$ is $\mathfrak{P}_T$-convex.

The equality of orders $(\mathfrak{P}/I)_{\pi(T)} = \mathfrak{P}_T/I_T$ follows since both are the weakest extensions of $\mathfrak{P} \subset A$, under either $\bar{i}_T \cdot \pi$ or $\bar{\pi} \cdot i_T$.

Remark. The result that I_T is $\mathfrak{P}_T$-convex could also be proved by direct computation.

Proposition 3.4.2.

(a) If $\bar{J} \subset A_T$ is an ideal, then $\bar{J} = i_T^{-1}(\bar{J})A_T$. Thus the correspondence $\bar{J} \mapsto i_T^{-1}(\bar{J})$ is injective from ideals of A_T to ideals of A. If $\bar{J}$ is a primary ideal of A_T, then $i_T^{-1}(\bar{J})$ is a primary ideal of A.

(b) If $Q \subset A$ is a primary ideal and $Q \cap T = \emptyset$, then $QA_T \subset A_T$ is a primary ideal, $\sqrt{Q} \cap T = \emptyset$, $\sqrt{Q}\, A_T = \sqrt{QA_T}$, and $Q = i_T^{-1}(QA_T)$. Thus the correspondence $Q \mapsto QA_T$ is a bijection between all primary ideals of A disjoint from T and all primary ideals of A_T. The correspondence commutes with the nil-radical operation and preserves inclusions, hence also provides a bijective correspondence between prime and maximal ideals of A_T and prime and maximal ideals of A, disjoint from T. (If $I \cap T \neq \emptyset$, any ideal $I \subset A$, then $IA_T = A_T$.)

Proof. This is a result in commutative algebra and doesn't really involve (POR). Of course, we get a result in (POR) as a corollary by replacing the word "ideal" in the proposition by "convex ideal." This is justified by Proposition 3.4.1 above. □

3.5. Concave Multiplicative Sets

Let $(A,\mathfrak{P}) \in$ (POR). A subset $S \subset A$ is *concave* if $0 \leq p \leq q$, $p \in S$ implies $q \in S$. equivalently, S is concave if and only if $A - S$ is convex. Arbitrary unions and intersections of concave sets are concave.

We are interested in concave multiplicative sets. Given any subset $Y \subset A$, we denote by $S(Y)$ the smallest concave multiplicative set containing Y. Thus, $S(Y) = \bigcap_\alpha S_\alpha$ where the S_α run over all concave multiplicative sets containing Y. We call S(Y) the *concave shadow* (or, simply, *shadow*) of Y.

We construct S(Y) explicitly as follows. Let

$$S_o(Y) = \{y_1 \cdot \ldots \cdot y_r | y_j \in Y,\ r \geq 0\} .$$

Then $S_o(Y)$ is the smallest multiplicative set containing Y. ($1 \in S_o(Y)$ is the empty product.) Let

$$S_{n+1}(Y) = S_o(S_n(Y) \cup \{q | 0 \leq p \leq q,\ p \in S_n(Y)\}) .$$

Then each $S_m(Y)$ is a multiplicative set, $S_m(Y) \subset S_{m+1}(Y)$ and $S(Y) = \bigcup_n S_n(Y)$ is the concave shadow of Y.

Proposition 3.5.1. Let $T \subset A$ be a multiplicative set $(A,\mathfrak{P}) \in$ (POR). Let $f\colon (A,\mathfrak{P}) \to (A',\mathfrak{P}')$ be a morphism in (POR) such that wherever $0 \leq t \leq s$ in A and $t \in T$, then f(s) is invertible in A'. Then there is a unique morphism $g\colon (A_{S(T)}, \mathfrak{P}_{S(T)}) \to (A',\mathfrak{P}')$ such that the diagram

$$\begin{array}{ccc} (A, \mathfrak{P}) & \xrightarrow{i_{S(T)}} & (A_{S(T)}, \mathfrak{P}_{S(T)}) \\ & f \searrow \quad \swarrow g & \\ & (A', \mathfrak{P}') & \end{array}$$

commutes.

Proof. First we prove a Lemma.

Lemma 3.5.2. $(S(T))^2 \subseteq S(T^2) \subseteq S(T^+) \subseteq S(T)$.

Proof. Only the first inclusion is not obvious. But $S(T) = \bigcup_n S_n(T)$ and $S_0(T) = T$, so $(S_0(T))^2 = T^2 = S_0(T^2)$. Suppose, inductively, $(S_m(T))^2 \subseteq S_m(T^2)$. Any element of $S_{m+1}(T)$ is a product of elements either in $S_m(T)$ or larger than positive elements of $S_m(T)$. It follows that $(S_{m+1}(T))^2 \subseteq S_{m+1}(T^2)$. Thus $(S(T))^2 \subseteq S(T^2)$. □

Returning to the proposition, from 3.2 we have canonical isomorphisms in (POR) (we suppress the symbols for the orders)

$$A_{(S(T))^2} \cong A_{S(T^2)} \cong A_{S(T^+)} \cong A_{S(T)} .$$

We now replace $(A_{S(T)}, \mathfrak{P}_{S(T)})$ by $(A_{S(T^+)}, \mathfrak{P}_{S(T^+)})$ and look for a suitable $g\colon (A_{S(T^+)}, \mathfrak{P}_{S(T^+)}) \to (A', \mathfrak{P}')$. But given a *positive* multiplicative set T^+,

$$S(T^+) = \{s \mid 0 \le t \le s,\ t \in T^+\},$$

as one verifies easily by showing the set on the right to be a concave multiplicative set. Thus, the hypothesis of the proposition clearly gives $g\colon (A_{S(T^+)}, \mathfrak{P}_{S(T^+)}) \to (A', \mathfrak{P}')$, as desired. □

3.6. The Shadow of 1

Let $(A, \mathfrak{P}) \in$ (POR). As an important example of a concave multiplicative set, we study $S(1) = \{b \in A \mid 1 \le b\}$, the shadow of 1.

First, an element $y \in A$ is invertible in $A_{S(1)}$ if and only if

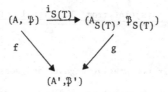

$1 \leq xy$ for some $x \in A$; that is, if and only if $1 \in H(y)$. In a commutative ring, any element not contained in some ordinary maximal ideal is a unit. If $(A,\mathfrak{P}) \in (POR)$, an element $y \in A$ is not contained in any maximal convex ideal if and only if $1 \in H(y)$. Thus, passing to $A_{S(1)}$ has the effect of making all such y units.

Secondly, the natural map $i_{S(1)}$: $A \to A_{S(1)}$ need not be injective. In fact,

$$\text{kernel}(i_{S(1)}) = \{a \in A \mid ab = 0, \text{ some } b, 1 \leq b\}.$$

If $1 \leq b$, then $0 \leq a^2 \leq a^2b^2$, so $a \in \text{kernel}(i_{S(1)})$ implies $a^2 = 0$. Consequently, $2a_1a_2 = 0$ for any $a_1, a_2 \in \text{kernel}(i_{S(1)})$, which one sees by expanding $(a_1 + a_2)^2$ and $(a_1 - a_2)^2$. We regard this kernel as somewhat pathological. It is the ideal K of 2.8, Example (4).

We now give an alternate characterization of semi-fields.

Proposition 3.6.1. Let $(A,\mathfrak{P}) \in (POR)$. Then $(A,\mathfrak{P})$ is a semi-field if and only if $i_{S(1)}$: $A \to A_{S(1)}$ is injective and $A_{S(1)}$ is a field. In such a case, $A_{S(1)}$ is the field of fractions $A_{(0)}$ of A.

Proof. Semi-fields were defined in 2.3. If $(A,\mathfrak{P})$ is a semi-field, A is an integral domain, so $i_{S(1)}$: $A \to A_{S(1)}$ is injective by the second paragraph above. Moreover, directly from the definition of semi-field and the first paragraph above, all non-zero elements of A are invertible in $A_{S(1)}$. Thus, $A_{S(1)} = A_{(0)}$, a field.

Conversely, if $i_{S(1)} = A \to A_{S(1)}$ is injective and $A_{S(1)}$ is a field, then all non-zero elements of A are invertible in $A_{S(1)}$. By the first paragraph above, this says exactly that $(A,\mathfrak{P})$ is a semi-field. □

We point out that any convex proper ideal of A is disjoint from $S(1)$. Thus, at least the primary convex ideals of $(A,\mathfrak{P})$ correspond bijectively with primary convex ideals of $(A_{S(1)}, \mathfrak{P}_{S(1)})$.

If $(A,\mathfrak{P}) \in (POR)$, we define the *Jacobson radical* of $(A,\mathfrak{P})$, $R(A,\mathfrak{P})$, to be the intersection of all maximal convex ideals

$$R(A,\mathfrak{P}) = \bigcap_{\substack{Q = \text{maximal} \\ \text{convex ideal}}} Q .$$

We have the following analog of a classical result.

<u>Proposition 3.6.2</u>. Let $I \subset A$ be a convex ideal. Then the following are equivalent.

(i) $I \subset R(A,\mathfrak{P})$

(ii) $1 + (a/b)$ is invertible in $A_{S(1)}$ for all $a \in I$, $b \in S(1)$.

(iii) For all $a \in I$, $b \in S(1)$, there is an $x \in A$ with $1 \leq x(b+a)$.

<u>Proof</u>. (ii) and (iii) are clearly equivalent. We will deduce the equivalence of (i) and (ii) from the classical result in commutative algebra. Namely, one first checks that

$$R(A,\mathfrak{P}) = R(A_{S(1)}, \mathfrak{P}_{S(1)}) \cap A$$

and

$$R(A_{S(1)}, \mathfrak{P}_{S(1)}) = R(A,\mathfrak{P})A_{S(1)},$$

using the correspondence between maximal convex ideals of $(A,\mathfrak{P})$ and $(A_{S(1)}, \mathfrak{P}_{S(1)})$. Thus $I \subset R(A,\mathfrak{P})$ if and only if $IA_{S(1)} \subset R(A_{S(1)}, \mathfrak{P}_{S(1)})$.

Secondly, one checks that $(A_{S(1)}, \mathfrak{P}_{S(1)}) = \{(A_{S(1)})_{S(1)}, (\mathfrak{P}_{S(1)})_{S(1)}\}$ in (POR). In particular, those elements of $A_{S(1)}$ not contained in any maximal convex ideal are already invertible in $A_{S(1)}$.

Now, if $IA_{S(1)}$ is contained in all maximal convex ideals of $A_{S(1)}$, then the elements $1 + (a/b)$, $a/b \in IA_{S(1)}$, are not in any maximal convex ideal, hence are invertible. Thus (i) $\Rightarrow$ (ii).

Conversely, if all $1 + (a/b)$ are invertible, $a \in I$, $b \in S(1)$, but $a \notin Q$, for some maximal $\mathfrak{P}$-convex ideal $Q \subset A$, then $1 \in H(Q,a)$. By Corollary 2.2.4, $1 \leq q + pa^2 = b$, for suitable $q \in Q$, $p \in \mathfrak{P}$. Computing in $A_{S(1)}$, $1 = b/b = q/b + pa^2/b$ and $1 - (pa^2/b) = q/b$ is invertible in $A_{S(1)}$, which is impossible since $q/b \in QA_{S(1)}$, a proper convex ideal. Thus (ii) $\Rightarrow$ (i). □

3.7. Localization at a Prime Convex Ideal

Another important concave multiplicative set S is the complement of a prime convex ideal Q, that is, $S = A - Q$. We denote by $(A_{(Q)}, \mathfrak{P}_{(Q)})$ the localized partially ordered ring $(A_{A-Q}, \mathfrak{P}_{A-Q})$. $A_{(Q)}$ is a local ring, whose unique maximal ideal $QA_{(Q)}$ is in fact convex.

More generally, if $\{Q_\alpha\}$ is a set of convex prime ideals, then $A - \bigcup_\alpha Q_\alpha = \bigcap_\alpha (A - Q_\alpha)$ is a concave multiplicative set. The maximal convex ideals of the corresponding localization $\left(A_{(\{Q_\alpha\})}, \mathfrak{P}_{(\{Q_\alpha\})}\right)$ are the $Q_\alpha A_{(\{Q_\alpha\})}$, (at least assuming there are no non-trivial inclusions $Q_\alpha \subset Q_\beta$).

If A is an integral domain, then for all multiplicative sets $T \subset A$, A_T is a subring of $A_{(0)}$, the field of fractions of A. The following is also an analog of a standard result in commutative algebra.

Proposition 3.7.1. If $(A,\mathfrak{P}) \in (POR)$, A an integral domain, then, in $A_{(0)}$,

$$\bigcap_{\substack{Q\,=\,\text{maximal}\\ \text{convex ideal}}} (A_{(Q)}, \mathfrak{P}_{(Q)}) = (A_{S(1)}, \mathfrak{P}_{S(1)}) .$$

Proof. $(A_{S(1)}, \mathfrak{P}_{S(1)}) \subset (A_{(Q)}, \mathfrak{P}_{(Q)})$ since $S(1) \subset A - Q$. Conversely, if $x/b \in A_{(0)}$ belongs to all $A_{(Q)} \subset A_{(0)}$, we can write for each Q, $x/b = x_Q/b_Q$, $b_Q \notin Q$. Then $1 \in H(\{b_Q\})$. We may as well assume $b_Q \in \mathfrak{P}$ (since $x_Q/b_Q = x_Q b_Q/b_Q^2$), hence we can write

$$1 \leq \sum_{i=1}^{k} p_{Q_i} b_{Q_i}$$

for suitable $p_{Q_i} \in \mathfrak{P}$ and $\{Q_i\}_{1 \leq i \leq k}$, a finite subset of the maximal convex ideals. But now, since $xb_{Q_i} = x_{Q_i} b$, we have $x/b = \sum_{i=1}^{k} p_{Q_i} x_{Q_i} \Big/ \sum_{i=1}^{k} p_{Q_i} b_{Q_i}$, which shows that $x/b \in A_{S(1)}$.

Finally, suppose $1 \leq b$, $0 \leq x/b$ in $(A_{(Q)}, \mathfrak{P}_{(Q)})$ for all maximal convex ideals Q. Then $0 \leq xb\,b_Q^2$ in $(A,\mathfrak{P})$, for suitable $b_Q \notin Q$. Again, $1 \in H(\{b_Q\})$, hence for suitable $p_{Q_i} \in \mathfrak{P}$,

$$1 \leq \sum_{i=1}^{k} p_{Q_i} b_{Q_i}^2 = b' \in S(1).$$

But now $0 \leq xbb'$ in $(A,\mathfrak{P})$, hence $0 \leq xb(b')^2$ and $0 \leq x/b$ in $(A_{S(1)},\mathfrak{P}_{S(1)})$.

3.8. Localization in (PORCK)

The results in this chapter extend to the category (PORCK), as we verify in this section. In fact, in (PORCK) certain desired results become true, that are not true in (POR).

Proposition 3.8.1. Let $(A,\mathfrak{P}) \in$ (PORCK), $T \subset A$ a multiplicative set. Then $(A_T,\mathfrak{P}_T) \in$ (PORCK).

Proof. Suppose $(a_1/b_1 + a_2/b_2)(x/b) = 0 \in A_T$, $a_i/b_i \in \mathfrak{P}_T$. Replacing a_i/b_i by $a_ib_is_i^2/(b_is_i)^2$ we may as well assume $a_i \in \mathfrak{P}$. Also, we may assume $b_1 = b_2$, by a similar trick. The hypothesis then implies that there is a $t \in T$ with $(a_1 + a_2)xt = 0 \in A$. Now, $(A,\mathfrak{P}) \in$ (PORCK) implies $a_1xt = a_2xt = 0$. Thus, $(a_i/b_i)(x/b) = 0 \in A_T$ and $(A_T, \mathfrak{P}_T) \in$ (PORCK). □

Proposition 3.8.2. Let $(A,\mathfrak{P}) \in$ (POR), $T \subset A$ a multiplicative set, $I \subset A$ an absolutely convex ideal. Then IA_T is an absolutely convex ideal of A_T.

Proof. Actually, we don't have to prove this, since we know that IA_T is convex and that in (POR) there is an isomorphism

$$\left((A/I)_{T/I}, (\mathfrak{P}/I)_{T/I}\right) \overset{\sim}{\rightarrow} (A_T/IA_T, \mathfrak{P}_T/IA_T) .$$

By Proposition 3.8.1, the left-hand side is in (PORCK), hence so is the right-hand side. □

Proposition 3.8.3. Let $(A,\mathfrak{P}) \in$ (PORCK). Then the concave multiplicative set $S(1) = \{b \in A | 1 \leq b\}$ consists of non-zero divisors. Thus $i_{S(1)}: A \rightarrow A_{S(1)}$ is always injective in (PORCK) and the order $\mathfrak{P}$ admits a canonical (PORCK) refinenent $\mathfrak{P}_{S(1)} = \mathfrak{P}_{S(1)} \cap A$.

Proof. If $bx = 0$, $0 \leq 1 \leq b$, then $1 \cdot x = x = 0$. □

Proposition 3.8.4. Let $(A,\mathfrak{P}) \in$ (PORCK), $S \subset S(1)$ a multiplicative set and $I \subset A$ an absolutely convex ideal. Then $I = IA_S \cap A$. Thus, there are natural bijective correspondences between the absolutely convex ideals of $(A,\mathfrak{P})$, $(A, \mathfrak{P}_S \cap A)$ and $(A_S, \mathfrak{P}_S)$.

Proof. In general, $IA_S \cap A = \{a \in A | sa \in I, \text{ some } s \in S\}$. But if I is absolutely convex, $sa \in I$, $0 \leq 1 \leq s$, then $1a = a \in I$. This proves $IA_S = I$.

The second statement follows, since by 3.8.2 IA_S is $\mathfrak{P}_S$ absolutely convex, hence $I = IA_S \cap A$ is $\mathfrak{P}_S \cap A$ absolutely convex. □

Lemma 3.8.5. Let $(A,\mathfrak{P}) \in$ (POR), $I \subset A$ absolutely convex with $\sqrt{I} = Q$ a maximal convex ideal. Then I is a primary ideal.

Proof. Let $ab \in I$, $a \notin \sqrt{I} = Q$. We must prove $b \in I$. Replacing a by a^2, we assume $a \in \mathfrak{P}$. Now, $1 \in H(Q,a)$, hence by 2.2.3, $0 \leq 1 \leq q + pa$, some $q \in Q$, $p \in \mathfrak{P}$. Raising to some large power, we get $0 \leq 1 \leq q' + p'a$, $q' \in I$. But now, $(q' + p'a)b \in I$ and since I is absolutely convex, $1 \cdot b = b \in I$. □

We can apply Lemma 3.8.5 and Proposition 3.4.2 to define symbolic powers in (POR). Let $(A,\mathfrak{P}) \in$ (POR), $Q \subset A$ a prime convex ideal. Then $(A_{(Q)}, \mathfrak{P}_{(Q)}) \in$ (POR) and $QA_{(Q)}$ is a maximal convex ideal. We have the absolutely convex ideal $AH(Q^n A_{(Q)}) \subset A_{(Q)}$ and certainly $\sqrt{AH(Q^n A_{(Q)})} = QA_{(Q)}$.

Definition 3.8.6. $Q^{(n)} = AH(Q^n A_{(Q)}) \cap A$. Then, $Q^{(n)} \subset A$ is a $\mathfrak{P}$-absolutely convex, primary ideal, with $\sqrt{Q^{(n)}} = Q$.

3.9. Applications of Localization, I - Some Properties of Convex Prime Ideals

In this section we first give an alternate proof of the fundamental result that maximal convex ideals are prime (Corollary 2.3.5). The argument is due to Andrew Klapper and is more elementary than the proof given in Chapter II, where we appealed to the characterization of the radical of a convex hull (Proposition 2.2.3). We then use localization to give a quick proof that the nil radical of a convex ideal is an intersection of convex prime ideals (Corollary 2.3.7). Finally, we use localization to study minimal primes.

We begin by pointing out that the existence of maximal convex ideals in any $(A,\mathfrak{P})$ is a trivial Zorn's lemma argument (Proposition 2.3.1). Next, if $I \subset A$ is convex, then the construction of the residue ring $(A/I, \mathfrak{P}/I)$ is elementary, as is the correspondence between $\mathfrak{P}/I$-convex ideals of A/I and $\mathfrak{P}$-convex ideals of A containing I. Thus, if $Q \subset A$ is maximal convex, then $(A/Q, \mathfrak{P}/Q)$ has no nonzero proper convex ideals.

Proposition 3.9.1. If $(A,\mathfrak{P}) \in (POR)$ and $Q \subset A$ is a maximal $\mathfrak{P}$-convex ideal, then A/Q is an integral domain. Thus Q is prime.

Proof. (Klapper) By the paragraph above, we are reduced to showing that if $(A,\mathfrak{P})$ has no nonzero proper convex ideals, then A is an integral domain. First notice that if $0 \neq p \in \mathfrak{P}$, then $(0:p) \subset A$ is convex, hence $(0:p) = (0)$. Namely, if $0 \leq x \leq y$, then $0 \leq xp \leq yp$, so if $yp = 0$, also $xp = 0$.

Now, suppose $ab = 0$, $a,b \neq 0$. Then a and b are not units. The principal ideals (a), (b) cannot be convex, hence we can find elements $p_1,p_2,q_1,q_2 \in \mathfrak{P}$ with $p_1 + p_2 \in (a)$, $q_1 + q_2 \in (b)$, but $p_i \notin (a)$, $q_j \notin (b)$. Multiplying, $(p_1 + p_2)(q_1 + q_2) = 0$, hence $p_i q_j = 0$. But this contradicts $(0:p) = 0$, $p \in \mathfrak{P}$. □

Our next observation is that if $(A,\mathfrak{P}) \in (POR)$, $S \subset A$ a multiplicative set, then the construction of $(A_S,\mathfrak{P}_S) \in (POR)$ in 3.1 is elementary, as are the results in 3.4 concerning the correspondence between prime $\mathfrak{P}_S$-convex ideals of A_S and prime $\mathfrak{P}$-convex ideals of A disjoint from S, and the permutability of localization and residue ring constructions. By the result just proved, we know there exist prime $\mathfrak{P}_S$-convex ideals of A_S, and this enables us to shorten certain arguments. For example, we can reprove 2.3.7.

Proposition 3.9.2. Let $(A,\mathfrak{P}) \in (POR)$, $I \subset A$ a convex ideal. Then $\sqrt{I} = \cap P_\alpha$, the intersection taken over all prime convex ideals $P_\alpha \supset I$.

Proof. Passing to $(A/I, \mathfrak{P}/I)$, we may assume $I = (0)$. The non-trivial step is showing that if $f^n \neq 0$ all n, then for some prime convex ideal P, $f \notin P$. Consider the localization $(A_f, \mathfrak{P}_f)$. Any prime $\mathfrak{P}_f$-convex ideal of A_f corresponds to a prime $\mathfrak{P}$-convex ideal of A disjoint from $\{f^n\}_{n\geq 0}$, as desired. □

The use of localization also aids in the study of minimal convex primes. If $(A,\mathfrak{P}) \in$ (POR) and I is a convex ideal, consider the family of multiplicative sets $S \subset A$ with $S \cap I = \emptyset$. For example, $\{1\}$ and $S(1)$, the shadow of 1, are such sets. Zorn's lemma easily implies that any such S is contained in a maximal S_0. Consider $(A_{S_0}, \mathfrak{P}_{S_0})$. Any prime $\mathfrak{P}_{S_0}$-convex ideal containing IA_{S_0} gives a prime $\mathfrak{P}$-convex ideal P_0, with $I \subset P_0 \subset A - S_0$ (such primes do exist since $A_{S_0}/IA_{S_0} \neq 0$). Thus $S_0 \subset A - P_0$ and $I \cap (A - P_0) = \emptyset$. By maximality of S_0, $S_0 = A - P_0$. Since our original S could have been $A - P$ for any prime $P \supset I$, we have proved the following (POR) extension of a familiar result in commutative algebra.

Proposition 3.9.3. If $(A,\mathfrak{P}) \in$ (POR) and $I \subset A$ is a $\mathfrak{P}$-convex ideal, then the minimal prime ideals of A containing I are always $\mathfrak{P}$-convex. Any prime ideal containing I contains a minimal such prime.

3.10. Applications of Localization, II - Zero Divisors

Our next application of localization will be to zero divisors. The results are most satisfactory if attention is restricted to the category (PORCK).

If A is any ring, let D denote the set of zero divisors of A and let N denote the non-zero divisors. A standard result in commutative algebra is that D is a union of prime ideals. More precisely, D is the union of those prime ideals which are minimal over ideals of the form $(0 : a)$, $a \in A$. Since $(0 : 1) = (0)$, the minimal primes of A are contained in D.

The proof of these assertions is as follows. Certainly D is the union of the ideals $(0 : a)$, so if we prove that whenever P is a prime minimal over $(0:a)$, then $P \subset D$, we are finished. But $S = A - P$ is a maximal multiplicative set disjoint from $(0:a)$. If $P \not\subset D$, choose $x \in P - D$ and observe that $\bigcup_{n \geq 0} S x^n$ is a multiplicative set strictly larger than S, but disjoint from $(0:a)$. This contradiction proves $P \subset D$.

Using the results that if $(A,\mathfrak{P}) \in$ (POR), any minimal prime is convex and that if $(A,\mathfrak{P}) \in$ (PORCK), all ideals $(0:a)$ are convex, hence so are minimal primes over $(0:a)$, we see that we have proved the following.

Proposition 3.10.1.

(a) If $(A,\mathfrak{P}) \in$ (POR) and P is a minimal convex prime ideal, then $P \subset D$.

(b) If $(A,\mathfrak{P}) \in$ (PORCK), then D is the union of all convex prime ideals minimal over ideals $(0:a)$, $a \in A$. □

For the sake of argument, we give two alternate proofs that in (PORCK), D is a union of convex prime ideals. These arguments are analogous to standard arguments for commutative rings.

First, if $(A,\mathfrak{P}) \in$ (PORCK), then the subset $D \subset A$ is convex. In fact, if $0 \leq a \leq b$ and $bc = 0$, then $ac = 0$, so if b is a zero divisor, then so is a. Now consider the family of absolutely convex ideals $I \subset A$ such that $I \subset D$. For example, (0) is such an ideal. By Zorn's lemma, let I be a maximal such ideal. Then $I = \sqrt{I}$. We claim I is prime. Otherwise, let $ab \in I$, $a,b \notin I$. Choose $x \in AH(I,a)$, $x \notin D$ and $y \in AH(I,b)$, $y \notin D$. By 2.2.4, we can write $0 \leq x^{2n} \leq q' + ar'$, $0 \leq y^{2n} \leq q'' + br''$ for suitable $n \geq 1$, $q',q'' \in I$, $r',r'' \in A$. Then $0 \leq (xy)^{2n} \leq q = (q' + ar')(q'' + br'') \in I$. But $I \subset D$ and D is a convex set, so $xy \in D$ and hence either $x \in D$ or $y \in D$.

The second proof also exploits the fact that D is convex if $(A,\mathfrak{P}) \in$ (PORCK). This is, of course, equivalent to the statement that $N = A - D$ is concave. N is also a saturated multiplicative set, in the sense that if $st \in N$, then $s \in N$ and $t \in N$.

We digress a moment to establish a lemma of independent interest. This lemma clarifies in some abstract sense how suitable localizations of partially ordered rings will have a "good" maximal ideal spectrum.

Lemma 3.10.2. Let $(A,\mathfrak{P}) \in (POR)$, $S \subset A$ a concave multiplicative set. Then any maximal ideal in A_S is $\mathfrak{P}_S$-convex.

Proof. If $Q \subset A_S$ is maximal, we know $Q = PA_S$ for some ordinary prime ideal $P \subset A$, maximal among those disjoint from S. Let $0 \leq q \leq p \in P$. We will show $q \in P$, hence P and Q are convex.

If $q \notin P$, the ideal $P + (q)$ intersects S. If $s = p' + a'q$, $p' \in P$, $a' \in A$, then $s^2 = p'' + a''q^2$ for some $p'' \in P$ and $a'' = (a')^2 \in A$. But $0 \leq s^2 = p'' + a''q^2 \leq p'' + a''p^2$ and since S is concave, $p'' + a''p^2 \in P \cap S$, a contradiction. □

Remark. As a corollary of 3.10.2 we deduce that all semi-units in A_S are actually units, since the semi-units belong to no maximal convex ideal, hence, by 3.10.2, to no ordinary maximal ideal. This result could also be verified by direct calculation.

We now complete our second alternate proof that if $(A,\mathfrak{P}) \in (PORCK)$, D is a union of convex prime ideals. By the above discussion, this follows immediately from the next lemma, by taking $S = N = A - D$.

Lemma 3.10.3. If $(A,\mathfrak{P}) \in (POR)$, then $S \subset A$ is a saturated, concave, multiplicative set if and only if $A - S$ is a union of convex prime ideals.

Proof. The if statement is trivial. Conversely, if S is saturated and concave, consider A_S. By the Remark above, the set of non-units in A_S is the union of the maximal convex ideals, hence the set of elements of A which are non-units in A_S is the union of those prime convex ideals which are disjoint from S. But the assumption that S is saturated simply says that the set of elements of A which are non-units in A_S is exactly $A - S$. □

3.11. Applications of Localization, III - Minimal Primes, Isolated Sets of Primes, and Associated Invariants.

Suppose A is a ring, $P \subset A$ a minimal prime ideal. A standard result in commutative algebra is that associated to P there is a *minimum*

P-primary ideal, $Q = \text{kernel}\ (A \to A_{(P)})$. If the ideal (0) has a primary decomposition, then P is necessarily an associated prime of (0) and Q is necessarily the associated primary component.

If $(A,\mathfrak{P}) \in (\text{POR})$, then any minimal prime P is convex, as is $Q = \text{kernel}\ (A \to A_{(P)})$. This remark establishes part (c) of Proposition 2.6.8, which was postponed in Chapter II.

In the interest of completeness, we prove the classical result. First if Q' is P-primary and $x \in Q = \text{kernel}\ (A \to A_{(P)})$, then $xs = 0$ for some $s \notin P$. Thus $x \in Q'$, and $Q \subset Q'$. Secondly, if $xy \in Q$, $u \notin P$, then $xys = 0$, some $s \notin P$. Since also $sy \notin P$, we conclude $x \in Q$. Finally, we must show $P \subset \sqrt{Q}$. Let $S = A - P$. Then S is a maximal multiplicative set disjoint from (0). But if $y \in P$ and $y^n \notin Q$ all n, then $\bigcup_{n \geq 0} Sy^n$ is a multiplicative set disjoint from (0), contradiction.

If A is a ring and $(0) = \cap Q_i$ is a finite intersection of primary ideals, then for any multiplicative set S, $(0) = \cap Q_i A_S \subset A_S$. Of course, this intersection reduces to the intersection of those $Q_j A_S$ with $Q_j \cap S = \emptyset$, since otherwise $Q_j A_S = A_S$. We then have $\text{kernel}\ (A \to A_S) = \bigcap_{Q_j \cap S = \emptyset} Q_j$. If P is a minimal prime of A and $S = A - P$, then only the primary component Q associated to P can be disjoint from S.

We state these results in our partially ordered context in the following

<u>Proposition 3.11.1</u>. If $(A,\mathfrak{P}) \in (\text{POR})$, [resp. (PORCK)], $I \subset A$ is a convex ideal [resp. absolutely convex ideal] and P is a minimal convex prime over I, then $Q = \text{kernel}\ (A \to A_{(P)}/I A_{(P)})$ is the smallest P-primary convex [resp. absoltely convex] ideal containing I. Moreover, if I has a decomposition as a finite intersection of primary ideals, then Q is the primary component belonging to P. □

<u>Remark</u>. We can apply 3.11.1 to construct convex primary ideals associated to certain sums and products. If $(A,\mathfrak{P}) \in (\text{POR})$ and $P \subset A$ is a fixed convex prime ideal, consider the family of convex ideals I such that $\sqrt{I} = P$. This family is closed under the operations $H(I_1 + I_2)$ and $H(I_1 I_2)$, (and also, of course, $I_1 \cap I_2$). Given any convex I with $\sqrt{I} = P$, Proposition

3.11.1 gives the minimal P-primary convex ideal containing I, specifically, kernel $(A \to A_{(P)}/I\,A_{(P)})$, which we abbreviate $I(P)$. For example, we could take $I = H(I_1 + I_2)$ or $H(I_1 I_2)$, if $\sqrt{I_j} = P$. This construction includes an analogue in (POR) of the *symbolic powers* of P, by taking $P^{(n)} =$ kernel $(A \to A_{(P)}/H(P^n)A_{(P)})$. As an exercise in the use of the explicit construction of a hull, $H(X) = \bigcup_{i \geq 0} H_i(X)$, given in 2.2, we state the following: If $I = H(P^{(n)}P^{(m)})$, then $I(P) = P^{(n+m)}$.

There is a completely analogous construction for absolutely convex ideals, using absolute hulls, $AH(I_1 + I_2)$, $AH(I_1 I_2)$, etc.

As final applications of localization, we discuss the semi-Noetherian case. If $(A,\mathfrak{P}) \in$ (PORCK) is semi-Noetherian, then so is $(A_S, \mathfrak{P}_S)$ for any multiplicative set $S \subset A$, by 3.4.2(a).

Proposition 3.11.2. Suppose $(A,\mathfrak{P}) \in$ (PORCK) is semi-Noetherian, $S \subset A$ is a multiplicative set, $I \subset A$ an absolutely convex ideal disjoint from S. Then the associated primes of $I A_S$ in A_S are exactly those primes PA_S where $P \subset A$ is an associated prime of I disjoint from S.

Proof. Since $(A_S/IA_S, \mathfrak{P}_S/IA_S) = ((A/I)_S, (\mathfrak{P}/I)_S)$, we may assume $I = (0)$. If $P = (0 : x) \subset A$ is prime and disjoint from S, then it is easy to check $PA_S = (0 : x)A_S = (0 : x/1) \subset A_S$. Thus PA_S is an associated prime of (0) in A_S. Conversely, suppose $PA_S = (0 : x/s) \subset A_S$. Let $P = AH(y_1 \ldots y_k)$, with $xy_i s_i = 0 \in A$, for suitable $s_i \in S$. Then $(0 : xs_1 \ldots s_k) = P \subset A$, hence P is an associated prime of (0) in A. □

If $(A,\mathfrak{P})$ is semi-Noetherian, $I \subset A$ absolutely convex, $P \subset A$ an associated prime of I, then the ideal $I(P) =$ kernel $(A \to A_{(P)}/IA_{(P)})$ is exactly the ideal $\{x \in A \mid xy \in I \text{ some } y \notin P\}$ considered in 2.6. According to 2.6.8(e), $I = \cap\, I(P_j)$, the intersection taken over the finitely many associated primes of I. If now $S \subset A$ is a multiplicative set, we have by 3.11.2 and 2.6.8, $IA_S = \cap\, IA_S(P_i A_S)$, the intersection taken over those associated primes P_i of I disjoint from S. Easily, $IA_S(P_i A_S) = I(P_i)A_S$ and $I(P_i) =$ kernel $(A \to A_S/I(P_i)A_S)$. Thus we have proved the following.

Proposition 3.11.3. If $(A,\mathfrak{P}) \in$ (PORCK) is semi-Noetherian, $I \subset A$ absolutely convex, $S \subset A$ a multiplicative set, then kernel $(A \to A_S/IA_S) = \cap I(P_i)$, the intersection taken over those associated primes P_i of I disjoint from S. □

Remark. As a special case of 3.11.3, let P be an associated prime of I and choose $f_i \in A$ with $f_i \in P_i - P$ for each associated prime P_i of I not contained in P. Let $f = \Pi f_i$ and let $S = \{f^n\}_{n \geq 0}$. Then the only associated primes of I disjoint from S are those contained in P. Thus $I(P) = \text{kernel } (A \to A_f/IA_f) = \bigcup_{n\geq 0} (I : f^n)$. Since $(I : f) \subset (I : f^2) \subset \ldots$ is an increasing chain of absolutely convex ideals, we must have $I(P) = (I : f^n)$, for suitably large n.

In the classical Noetherian case the ideals $I(P)$ are interpreted as the intersection of the primary components of I belonging to the set Σ_P of associated primes contained in P. This is a special case of the notion of an *isolated set* Σ of associated primes, meaning Σ contains all associated primes smaller than any one of its elements. The most general such Σ is a union of Σ_P. Thus for any isolated set of associated primes Σ, the intersection of all primary components of I belonging to primes of Σ is invariant and is characterized as $\bigcap_{P\in\Sigma} I(P)$. In our semi-Noetherian case we do not have primary decomposition in general, but the results above and in 2.6 show how many of the classical results involving isolated sets of associated primes do extend.

3.12. Operators on the Set of Orders on a Ring

Let $(A,\mathfrak{P}) \in$ (POR). We are interested in operators Δ which assign to $\mathfrak{P}$ a refined order $\Delta\mathfrak{P}$. In particular, functorial operators and idempotent operators $(\Delta^2\mathfrak{P} = \Delta\mathfrak{P})$ are natural objects of study.

The motivation is ultimately this: We seek to interpret partially ordered rings as "rings of functions" in some generalized sense. Certain formulas involving a "function" $f \in A$ ought to imply f positive. Thus if f^3 is positive and $(A,\mathfrak{P})$ is a ring of suitably valued functions on a set, then f should be positive. If p and $(1+p)f$ are positive, so should be f. But if the order $\mathfrak{P}$ is too small, say a weak order $\mathfrak{P}_w$ or a finitely generated refinement $\mathfrak{P}_w[g_1 \cdots g_k]$, such function theoretic results will not follow by simple algebraic manipulation. Thus we seek natural algebraic *extensions* of $\mathfrak{P}$, which more closely capture the behavior of functions.

We present four such operators here. The first three are related to localization.

<u>The operator</u> $\mathfrak{P}_d$. If $(A,\mathfrak{P}) \in$ (POR), (PORCK), or (PORNN), and $N \subset A$ is the multiplicative set of non-zero divisors, then we have seen that i_N: $(A,\mathfrak{P}) \to (A_N,\mathfrak{P}_N)$ is injective and induces a natural refinement $\mathfrak{P}_d = A \cap \mathfrak{P}_N$ of $\mathfrak{P}$. Specifically, $f \in \mathfrak{P}_d$ if $pf = q$, for some $p,q \in \mathfrak{P}$, p *not* a zero divisor. We have $(A,\mathfrak{P}_d) \in$ (POR), (PORCK), or (PORNN), whichever category held $(A,\mathfrak{P})$ originally. The operator $\mathfrak{P}_d$ is idempotent, $(\mathfrak{P}_d)_d = \mathfrak{P}_d$.

<u>The operator</u> $\mathfrak{P}_s$. Let $(A,\mathfrak{P}) \in$ (PORCK) or (PORNN). Then $S(1)$, the shadow of 1, is a multiplicative set, contained in the non-zero divisors. We thus have an injection $i_{S(1)}$: $(A,\mathfrak{P}) \to (A_{S(1)}, \mathfrak{P}_{S(1)})$, and we denote by $\mathfrak{P}_s$ the refinement of $\mathfrak{P}$ given by $A \cap \mathfrak{P}_{S(1)}$. Thus $f \in \mathfrak{P}_s$ if $(1+p)f = q$, $p,q \in \mathfrak{P}$.

We have $(A,\mathfrak{P}_s) \in$ (PORCK) or (PORNN), accordingly. In 3.8 we saw that the absolutely convex ideals of $(A,\mathfrak{P})$ and $(A,\mathfrak{P}_s)$ are identical. We also verified that $\mathfrak{P}_s$ is idempotent, $\mathfrak{P}_s = (\mathfrak{P}_s)_s$. Certainly $\mathfrak{P}_s \subset \mathfrak{P}_d$. The operator $\mathfrak{P}_s$ is a functorial operator in (PORCK) or (PORNN).

Given $f \in \mathfrak{P}_d$, we can write formally (or in A_N) $f = q/p$, $p,q \in \mathfrak{P}$. From the point of view of functions, this gives information about f, at least off the zero set of p. On the other hand, if $f \in \mathfrak{P}_s$, then $f = \frac{q}{1+p}$ gives a globally defined formula for the function f. Thus functions in $\mathfrak{P}_s$ are "non-negative" in rather a strong way.

Suppose A is a ring, and $P_i \subset A$ is a finite collection of primes with $\cap P_i = (0)$. Then A has no nilpotent elements and A has only finitely many minimal primes, say P_α, which are included among the P_i. Moreover, $\cap P_\alpha = (0)$ and $\cup P_\alpha$ is exactly the set of zero divisors of A.

Proposition 3.12.1. In the situation of the paragraph above, assume either an order $\mathfrak{P} \subset A$ with all P_i convex, and set $(A_i, \mathfrak{P}_i) = (A/P_i, \mathfrak{P}/P_i)$, or assume orders $\mathfrak{P}_i \subset A_i$ and set $\mathfrak{P} = A \cap \Pi\, \mathfrak{P}_i$. Consider the inclusion $A \to \Pi A_\alpha$. Then $\mathfrak{P}_d = A \cap \Pi(\mathfrak{P}_\alpha)_d$.

Proof: There are two points here. First given $(A, \mathfrak{P})$, then $\mathfrak{P}_d$ can be computed by passing to the integral domains $(A/P_\alpha, \mathfrak{P}/P_\alpha)$. Secondly, if there are more P_i than P_α and $\mathfrak{P} \subset A$ is defined using all the $(A_i, \mathfrak{P}_i)$, then $\mathfrak{P}/P_\alpha$ will generally be a much weaker order than $\mathfrak{P}_\alpha$ on A_α. Still $\mathfrak{P}_d$ will be strong enough to agree with $A \cap \Pi(\mathfrak{P}_\alpha)_d$.

The inclusion $\mathfrak{P}_d \subset \Pi(\mathfrak{P}_\alpha)_d$ is clear since an equation $pf = q$, $p, q \in \mathfrak{P}$, p not a zero divisor, can be reduced modulo P_α, since $p \notin P_\alpha$. Conversely, suppose given $f \in A$ with $h_\alpha^2 f = q_\alpha (\text{mod } P_\alpha)$, where $h_\alpha \notin P_\alpha$ and $q_\alpha (\text{mod } P_\alpha) \in \mathfrak{P}_\alpha$. Multiplying each such equation by a square in $\bigcap_{i \neq \alpha} P_i - P_\alpha$, we may assume $h_\alpha \in \bigcap_{i \neq \alpha} P_i - P_\alpha$, and we may assume equality in A, $h_\alpha^2 f = q_\alpha \in \mathfrak{P}$, since $(0) = \cap P_i$ and $\mathfrak{P} \subset A \cap \Pi\, \mathfrak{P}_i$. Adding these equations gives $(\Sigma h_\alpha^2) f = \Sigma q_\alpha \in \mathfrak{P}$ and since $\Sigma h_\alpha^2 \notin P_\beta$ for any minimal P_β, we conclude $f \in \mathfrak{P}_d$. □

The geometric significance of 3.12.1 will come out in Chapter VIII, when we relate derived orders with positivity conditions on certain "non-degenerate" subsets of real varieties. The non-minimal primes correspond to proper subvarieties (hence lower dimensional or "degenerate") of the irreducible components of a variety.

The operator $\mathfrak{P}_p$. Let $(A, \mathfrak{P}) \in$ (PORNN) and let $A \to \Pi A/P_\alpha$ be the canonical injection where the $\{P_\alpha\}$ are the minimal convex primes of A. Consider those elements $f \in A$, such that for some $n \geq 0$, $(f^{2n} + p)f = q$, $p, q \in \mathfrak{P}$. Such f form a set closed under products and contained in the order $A \cap \Pi(\mathfrak{P}/P_\alpha)_d$. It is not clear under what conditions this set will be closed under sums.

In any event, define $\mathfrak{P}_p = \{f \in A \mid f = \Sigma f_i,\ (f_i^{2n_i} + p_i)f_i = q_i$, for some $n_i \geq 0$, $p_i, q_i \in \mathfrak{P}\}$. Then $\mathfrak{P}_p$ is an order and $\mathfrak{P}_s \subset \mathfrak{P}_p$. If A has only finitely many minimal primes P_α, then the zero divisors of A are exactly $\cup P_\alpha$. This can be used to show $\mathfrak{P}_d = A \cap \Pi(\mathfrak{P}/P_\alpha)_d \subset A$. In this case $\mathfrak{P}_p \subset \mathfrak{P}_d$. The operator $\mathfrak{P}_p$ is functorial on $(A,\mathfrak{P}) \in$ (PORNN).

Rewrite $(f^{2n} + p)f = q$ as $f = q/(f^{2n} + p)$. This provides a strong positivity condition on f *off the zero set of* f. In fact, if S(f) is the shadow of f, we have the localization $i = i_{S(f)}: (A,\mathfrak{P}) \to (A_{S(f)},\mathfrak{P}_{S(f)})$ and the condition $(f^{2n} + p)f = q \in \mathfrak{P}$ can be seen to be equivalent to $i(f) \in \mathfrak{P}_{S(f)}$. This is reasonable since $A_{S(f)}$ ought to be a ring of functions off the zeros of f.

It is not difficult to see that any $\mathfrak{P}$-convex prime ideal $Q \subset A$ is $\mathfrak{P}_p$-convex. Suppose $\sum_{i=1}^{k} f_i \in Q$, $(f_i^{2n_i} + p_i)f_i = q_i$, $p_i, q_i \in \mathfrak{P}$. Then $\sum_{j=1}^{k} \left(\prod_{i=1}^{k} (f_i^{2n_i} + p_i)\right) f_j \in Q$, hence each summand $\prod_{i=1}^{k} (f_i^{2n_i} + p_i)f_j \in Q$, since these summands are in $\mathfrak{P}$. It follows that some $f_i \in Q$, hence by induction on k, all $f_i \in Q$.

<u>The operator</u> $\mathfrak{P}_m$. As a last example of an operator, consider those $(A,\mathfrak{P}) \in$ (PORNN) so that the Jacobson radical vanishes,

$$R(A,\mathfrak{P}) = \bigcap_{\substack{Q = \text{maximal} \\ \text{convex ideal}}} Q = (0) .$$

Then

$$\prod_{\text{max. } Q} \rho_Q : (A,\mathfrak{P}) \to \prod_{\text{max. } Q} (A/Q, \mathfrak{P}/Q)$$

is injective and we define

$$\mathfrak{P}_m = A \cap \Big(\prod_{\text{max. } Q} \mathfrak{P}/Q \Big) .$$

That is, $\mathfrak{P}_m$ consists of elements of A positive at each maximal ideal. The analogy with functions is clear, and, in fact, in a later chapter we will see that for finitely generated integral domains over real closed fields $A = R[x_1 \cdots x_n]$, with finitely generated extensions $\mathfrak{P} = \mathfrak{P}_w[g_1 \cdots g_k]$ of the weak order on A, the orders $\mathfrak{P}_p$ and $\mathfrak{P}_m$ *coincide,* and consist of exactly those functions nowhere negative on the semi-algebraic set defined by the relations $f_i(x_1 \cdots x_n) = 0 \in A$ and inequalities $g_j(x_1 \cdots x_n) \geq 0$. In fact, $\mathfrak{P}_p = \mathfrak{P}_m = \{f | (f^{2n} + p)f = q,\ \text{some}\ p,q \in \mathfrak{P},\ n \geq 0\}$. This is essentially a theorem of Stengle [22], and generalizes in several directions the work of Artin on representation of positive functions [4].

It is easy to check that all $\mathfrak{P}$-maximal convex ideals are $\mathfrak{P}_m$-convex and that $\mathfrak{P}_m$ is an idempotent operator $(\mathfrak{P}_m)_m = \mathfrak{P}_m$.

Finally, suppose $\mathfrak{P}'$, $\mathfrak{P}''$ are two orders on A. We state some formulas relating the operators $\mathfrak{P}_s$, $\mathfrak{P}_p$, $\mathfrak{P}_d$ above and intersections.

(a) $(\mathfrak{P}' \cap \mathfrak{P}'')_s \subset \mathfrak{P}'_s \cap \mathfrak{P}''_s$

(b) $(\mathfrak{P}' \cap \mathfrak{P}'')_p \subset \mathfrak{P}'_p \cap \mathfrak{P}''_p$

(c) $(\mathfrak{P}' \cap \mathfrak{P}'')_d = \mathfrak{P}'_d \cap \mathfrak{P}''_d$.

The inclusions (a), (b) are trivial, as is $(\mathfrak{P}' \cap \mathfrak{P}'')_d \subset \mathfrak{P}'_d \cap \mathfrak{P}''_d$. If $f \in \mathfrak{P}'_d \cap \mathfrak{P}''_d$, let $p'f = q'$, $p''f = q''$, $p',q' \in \mathfrak{P}'$, $p'',q'' \in \mathfrak{P}''$, p',p'' not zero divisors. Then $(p'p'')^2 f = p'q'(p'')^2 = p''q''(p')^2 \in \mathfrak{P}' \cap \mathfrak{P}''$. Thus $\mathfrak{P}'_d \cap \mathfrak{P}''_d \subset (\mathfrak{P}' \cap \mathfrak{P}'')_d$.

It is not quite as easy to formulate a result for the operator $\mathfrak{P}_m$, since even if the Jacobson radicals of $\mathfrak{P}'$ and $\mathfrak{P}''$ vanish, it is not clear when the Jacobson radical of $\mathfrak{P}' \cap \mathfrak{P}''$ vanishes. However, in the algebro-geometrical situations to be studied later, our rings $(A,\mathfrak{P})$ will have the property that for each maximal convex ideal Q, A/Q is a fixed real closed field R, therefore with a unique order. In the reduced case, $(A,\mathfrak{P}) \in$ (PORNN), the Jacobson radicals will vanish and $(A,\mathfrak{P}_m)$ identifies with a ring of R-valued functions (ordered by functional values) on the set of maximal convex ideals of $(A,\mathfrak{P})$. For two such orders $\mathfrak{P}',\mathfrak{P}'' \subset A$ the maximal convex ideals of $\mathfrak{P}' \cap \mathfrak{P}''$ will be exactly the union of the maximal convex ideals of $\mathfrak{P}'$ and $\mathfrak{P}''$ by 2.7.2, and clearly $(\mathfrak{P}' \cap \mathfrak{P}'')_m = \mathfrak{P}'_m \cap \mathfrak{P}''_m$.

IV · Some categorical notions

4.1. Fibre Products

<u>Proposition 4.1.1</u>. The categories (POR), (PORCK), (PORNN) admit arbitrary fibre products.

<u>Proof</u>. The construction is identical in all three categories. We first show that direct products exist.

Let $(A_\alpha, \mathfrak{P}_\alpha) \in$ (POR). Consider $(A,\mathfrak{P}) = (\prod_\alpha A_\alpha, \prod_\alpha \mathfrak{P}_\alpha)$. If $f_\alpha\colon (C,\mathfrak{P}_C) \to (A_\alpha, \mathfrak{P}_\alpha)$ is a family of morphisms in (POR), then there is a unique (POR)-morphism $f\colon (C,\mathfrak{P}_C) \to (A,\mathfrak{P})$ such that the following diagram commutes for all α:

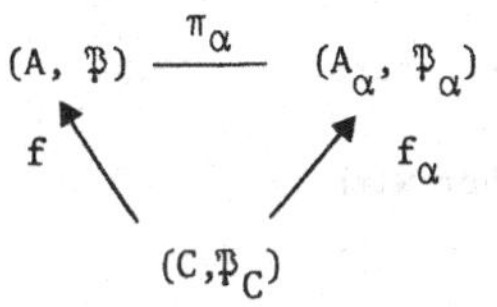

Thus, $(\prod_\alpha A_\alpha, \prod_\alpha \mathfrak{P}_\alpha)$ is the direct product of the family $(A_\alpha, \mathfrak{P}_\alpha)$.

More generally, suppose $g_\alpha\colon (A_\alpha, \mathfrak{P}_\alpha) \to (B, \mathfrak{P}_B)$ is a family of morphisms over the fixed base $(B, \mathfrak{P}_B)$. Then the fibre product of the $(A_\alpha, \mathfrak{P}_\alpha)$ over $(B, \mathfrak{P}_B)$ is the subring $\prod_B A_\alpha \subset \prod A_\alpha$, together with the contraction of the order $\prod \mathfrak{P}_\alpha$. That is, $\prod_{(B,\mathfrak{P}_B)} (A_\alpha, \mathfrak{P}_\alpha)$ is the ring

$$\prod_B A_\alpha = \{(a_\alpha)_\alpha \mid g_\alpha a_\alpha = g_\beta a_\beta \in B, \text{ all } \alpha,\beta\}$$

together with the order

$$\prod_B \mathfrak{P}_\alpha = \prod_B A_\alpha \cap \prod \mathfrak{P}_\alpha . \qquad \square$$

4.2. Fibre Sums

Proposition 4.2.1. The categories (PORNN), (PORCK) and (POR) admit finite fibre sums.

Proof. We first make the construction in (POR). The fibre sums in (PORNN) and (PORCK) require a slight modification of the construction.

Let $(C,\mathfrak{P}_C) \to (A,\mathfrak{P}_A)$ and $(C,\mathfrak{P}_C) \to (B,\mathfrak{P}_B)$ be morphisms in (POR). We begin with the ring $A \underset{C}{\otimes} B$. We would like to impose an order on $A \underset{C}{\otimes} B$ such that the natural ring homomorhisms $A \to A \underset{C}{\otimes} B$ and $B \to A \underset{C}{\otimes} B$ are order preserving. In general, however, this cannot be done.

Instead, we construct the smallest ideal $R \subset A \underset{C}{\otimes} B$ which satisfies the following condition:

4.2.2. $$\begin{cases} \sum_{i=1}^{k} (p_i \otimes q_i)x_i^2 \in R,\ p_i \in \mathfrak{P}_A,\ q_i \in \mathfrak{P}_B,\ x_i \in A \underset{C}{\otimes} B \\ \\ \text{implies } (p_j \otimes q_j)x_j^2 \in R, \qquad 1 \leq j \leq k. \end{cases}$$

This is exactly the generalized extension condition of 1.3 for the pair of maps $A \to A \underset{C}{\otimes} B$, $B \to A \underset{C}{\otimes} B$. The fibre sum $(A,\mathfrak{P}_A) \underset{(C,\mathfrak{P}_C)}{\otimes} (B,\mathfrak{P}_B)$ in (POR) is then the ring $(A \underset{C}{\otimes} B)/R$, together with the order

$$\mathfrak{P} = \{\pi\big(\sum_{i=1}^{k} (p_i \otimes q_i)x_i^2\big) \,|\, p_i \in \mathfrak{P}_A,\ q_i \in \mathfrak{P}_B,\ x_i \in A \underset{C}{\otimes} B\}$$

where $\pi\colon\ A \underset{C}{\otimes} B \to (A \underset{C}{\otimes} B)/R$ is the projection.

We verify that $(A \underset{C}{\otimes} B/R, \mathfrak{P})$ has the desired universal property. Suppose given (POR)-morphisms $f\colon (A,\mathfrak{P}_A) \to (D,\mathfrak{P}_D)$ and $g\colon (B,\mathfrak{P}_B) \to (D,\mathfrak{P}_D)$ such that the diagram below commutes.

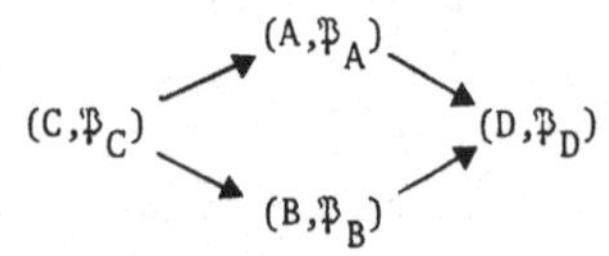

Then there is a unique ring homomorphism $f \otimes g: A \otimes_C B \to D$ such that f is $A \to A \otimes_C B \to D$ and g is $B \to A \otimes_C B \to D$. Let $I = \text{kernel } (f \otimes g)$. Suppose $\sum_{i=1}^{k} (p_i \otimes q_i)x_i^2 \in I$, $p_i \in \mathfrak{P}_A$, $q_i \in \mathfrak{P}_B$, $x_i \in A \otimes_C B$. Since $f \otimes g((p_j \otimes q_j)x_j^2) = f(p_j)g(q_j)(f \otimes g(x_j))^2 \in \mathfrak{P}_D$, $1 \leq j \leq k$, we deduce $(p_j \otimes q_j)x_j^2 \in I$, $1 \leq j \leq k$. Thus, $R \subset I$, and $f \otimes g$ factors $A \otimes_C B \to (A \otimes_C B)/R \to D$. Clearly, $(A \otimes_C B)/R \to D$ is order preserving, hence $((A \otimes_C B)/R, \mathfrak{P}) = (A \otimes \mathfrak{P}_A) \otimes_{(C,\mathfrak{P}_C)} (B,\mathfrak{P}_B)$.

In the category (PORNN), the ideal $R \subset A \otimes_C B$ must satisfy $R = \sqrt{R}$, as well as 4.2.2. There is a smallest such R, and the rest of the proof is unchanged.

In the category (PORCK), the condition 4.2.2 is replaced by

4.2.3.
$$\begin{cases} \left(\sum_{i=1}^{k} (p_i \otimes q_i)x_i^2\right)y \in R, \ p_i \in \mathfrak{P}_A, \ q_i \in \mathfrak{P}_B, \ x_i, y \in A \otimes_C B, \\ \text{implies } (p_j \otimes q_j)x_j^2 y \in R, \quad 1 \leq j \leq k. \end{cases}$$

Again, there is a smallest such R, and the rest of the proof is the same. □

4.3. Direct and Inverse Limits

Let I be a directed, partially ordered set, with relation $\leq$. Let $\{(A_i, \mathfrak{P}_i)\}_{i \in I}$ be a family of partially ordered rings.

Proposition 4.3.1. Suppose given morphisms $g_{kj}: (A_k,\mathfrak{P}_k) \to (A_j,\mathfrak{P}_j)$, $k \leq j$ in I, such that $g_{jj} = \text{Id}$ and $g_{kj}g_{mk} = g_{mj}$, if $m \leq k \leq j$. Then the direct limit $\varinjlim_I (A_i,\mathfrak{P}_i)$ exists in all categories (POR), (PORCK), (PORNN).

Proof. The construction is the same in all categories. Begin with the disjoint union, $\bigvee_{i \in I} A_i$. Identify $a_i \in A_i$ and $a_j \in A_j$ if there is a $k \in I$, $i,j \leq k$, with $g_{ik}(a_i) = g_{jk}(a_j) \in A_k$. The resulting ring is the ring theoretic direct limit $\varinjlim_I A_i = A$.

Impose an order $\mathfrak{P}$ on A by taking as positive elements the images in A of all the $\mathfrak{P}_j \subset A_j$. It is easy to see that $(A,\mathfrak{P}) = \varinjlim_I (A_i,\mathfrak{P}_i)$, in the sense of universal property. □

Proposition 4.3.2. Suppose given morphisms $f_{jk}: (A_j,\mathfrak{P}_j) \to (A_k,\mathfrak{P}_k)$, $k \leq j$ in I, such that $f_{jj} = \mathrm{Id}$ and $f_{km}f_{jk} = f_{jm}$, if $m \leq k \leq j$. Then the inverse limit $\varprojlim (A_i,\mathfrak{P}_i)$ exists in all categories (POR), (PORCK) and (PORNN).

Proof. Again the construction is the same in all three categories. The answer is the usual ring theoretic inverse limit

$$\varprojlim_I A_i \subset \prod_I A_i$$

together with the contracted order

$$\varprojlim_I \mathfrak{P}_i = \prod_I \mathfrak{P}_i \cap \varprojlim_I A_i . \qquad \square$$

4.4. Some Examples

(1) Let $C = \mathbb{Z}[X]$, ordered as a ring of functions on the line. Let $A = B = \mathbb{Z}$, with the unique order. Define $\alpha: \mathbb{Z}[X] \to \mathbb{Z}$ and $\beta: \mathbb{Z}[X] \to \mathbb{Z}$ by $\alpha(X) = 1$, $\beta(X) = -1$. Thus α, β are simply evaluation homomorphisms at the points $+1$, -1, respectively.

In the category of commutative rings with unit, $\mathbb{Z} \otimes_{\mathbb{Z}[X]} \mathbb{Z} \cong \mathbb{Z}/2$, but in (POR), $\mathbb{Z} \otimes_{\mathbb{Z}[X]} \mathbb{Z} = (0)$. Geometrically, this is preferable, since the fibre sum should be a ring of functions on the intersection of the two subspaces $\{+1\}$, $\{-1\}$, which is empty.

(2) Let $(A,\mathfrak{P}) \in$ (POR), $I,J \subset A$ convex ideals. We have projections $(A,\mathfrak{P}) \to (A/I, \mathfrak{P}/I)$ and $(A,\mathfrak{P}) \to (A/J, \mathfrak{P}/J)$. Then

$$(A/I, \mathfrak{P}/I) \underset{(A,\mathfrak{P})}{\otimes} (A/J, \mathfrak{P}/J) = \big(A/H(I+J), \mathfrak{P}/H(I+J)\big) .$$

There are obvious (PORCK) and (PORNN) analogs.

Note that in this example the fibre sum is (0) if there is no maximal convex ideal of A which contains both I and J.

(3) Let $(A,\mathfrak{P}_A) \in$ (POR), $(A,\mathfrak{P}_A) \to (B,\mathfrak{P}_B)$ a (POR)-morphism and $I \subset A$ a convex ideal. Then

$$(A/I, \mathfrak{P}_A/I) \underset{(A,\mathfrak{P}_A)}{\otimes} (B,\mathfrak{P}_B) = \big(B/H(IB), \mathfrak{P}_B/H(IB)\big) .$$

Again, there are obvious (PORCK) and (PORNN) analogs.

(4) If $(A,\mathfrak{P}_A),(B,\mathfrak{P}_B) \in (POR)$, then their direct sum in (POR) is the fibre sum over the unique ring maps $\mathbb{Z} \to A$, $\mathbb{Z} \to B$. We denote the direct sum $(A,\mathfrak{P}_A) \otimes (B,\mathfrak{P}_B)$.

As an example, let $A = Q[X]/(X^2)$, $B = Q[Y]/(Y^2)$, ordered as rings of germs of functions at 0, module those vanishing to second order at 0. In the category of rings, $A \otimes B \cong Q[X,Y]/(X^2,Y^2)$. However, in (POR), $A \otimes B \cong Q[X,Y]/(X^2,Y^2,XY)$, since $H(X^2,Y^2) = (X^2,Y^2,XY)$ by 2.8, Example 7. Moreover, an element $a + bX + cY$ will be strictly positive if and only if $a > 0 \in Q$. So, the direct sum is the ring of germs of functions at $(0,0) \in \mathbb{R}^{(2)}$, modulo those which vanish to second order at $(0,0)$.

(5) We continue with $A = Q[X]/(X^2)$, $B = Q[Y]/(Y^2)$, but with refined orders induced by functions nowhere negative for all sufficiently small *positive* real arguments. As a ring, the direct sum in (POR) will still be $Q[X,Y]/(X^2,Y^2,XY)$. However, now an element $a + bX + cY$ will be positive if and only if $a > 0$ or $a = 0$ and $b,c \geq 0$. This is the order induced by ordering $Q[X,Y]$ as a ring of functions on the first quadrant in $\mathbb{R}^{(2)}$.

(6) Let $(A,\mathfrak{P}) \in (POR)$, $I \subset A$ a convex ideal. There are various inverse systems associated to I. One has the (POR)-inverse system $\big(A/H(I^n), \mathfrak{P}/H(I^n)\big)_{n\geq 1}$, and the (PORCK) inverse system $\big(A/AH(I^n), \mathfrak{P}/AH(I^n)\big)_{n\geq 1}$. If I is a prime convex ideal, we have the symbolic powers $I^{(n)}$ of 3.8.6, and we can form $\big(A/I^{(n)}, \mathfrak{P}/I^{(n)}\big)_{n\geq 1}$. Each of these inverse systems will have a partially ordered inverse limit.

(7) Let $(A,\mathfrak{P}) \in (POR)$, $Q \subset A$ a prime convex ideal. If $a \notin Q$ and $S(a)$ is the shadow of a, then $S(a) \subset A - Q$, since $A - Q$ is concave. Thus there is a natural (POR)-morphism $(A_{S(a)}, \mathfrak{P}_{S(a)}) \to (A_{(Q)}, \mathfrak{P}_{(Q)})$.

We will see in the next chapter that the rings $(A_{S(a)}, \mathfrak{P}_{S(a)})$, $a \notin Q$, form a natural directed system, and that there is an induced isomorphism in (POR), $\varinjlim_{a \notin Q} (A_{S(a)}, \mathfrak{P}_{S(a)}) \xrightarrow{\sim} (A_{(Q)}, \mathfrak{P}_{(Q)})$.

V · The prime convex ideal spectrum

5.1. The Zariski Topology Defined

Let $(A,\mathfrak{P}) \in (POR)$. We denote by $\mathrm{Spec}(A,\mathfrak{P})$ the set of prime convex ideals of A and by $\mathrm{Spec}^m(A,\mathfrak{P})$ the subset of maximal convex ideals.

We define a topology on $\mathrm{Spec}(A,\mathfrak{P})$, called the *Zariski topology*, in the usual way. The closed sets are

$$Z(I) = \{Q \mid Q \in \mathrm{Spec}(A,\mathfrak{P}),\ I \subset Q\}$$

where $I \subseteq A$ is a convex ideal. In fact, if $X \subset A$ is any subset, let

$$Z(X) = \{Q \mid Q \in \mathrm{Spec}(A,\mathfrak{P}),\ X \subset Q\}.$$

Proposition 5.1.1.

(a) $Z(X) = Z(H(X)) = Z(\sqrt{H(X)})$ for any subset $X \subset A$.

(b) $\bigcap_\alpha Z(I_\alpha) = Z(\sum_\alpha I_\alpha)$ for any collection of ideals $I_\alpha \subset A$.

(c) $Z(I_1) \cup Z(I_2) = Z(I_1 \cap I_2)$ for any pair of ideals $I_1, I_2 \subset A$.

(d) $\mathrm{Spec}(A,\mathfrak{P}) = Z((0))$ and $\emptyset = Z(A)$.

(e) If $X,Y \subset A$ are subsets, $Z(X) \subset Z(Y)$ if and only if $Y \subset \sqrt{H(X)}$.

Proof. The proofs are routine, in view of the Corollary 2.3.7, which asserts $\sqrt{H(X)} = \bigcap_{Q \in Z(X)} Q$. □

As a basis for the open sets of $\mathrm{Spec}(A,\mathfrak{P})$ we have the distinguished open sets $D(a)$, $a \in A$, defined as follows:

$$D(a) = \{Q \mid Q \in \mathrm{Spec}(A,\mathfrak{P}),\ a \notin Q\}.$$

That the $\{D(a)\}_{a\in A}$ do form a basis follows from the formulas

$$D(a) \cap D(b) = D(ab)$$

$$\mathrm{Spec}(A,\mathfrak{P}) - Z(X) = \bigcup_{a \in X} D(a)\ .$$

We give $\mathrm{Spec}^m(A,\mathfrak{P}) \subset \mathrm{Spec}(A,\mathfrak{P})$ the subspace topology.

5.2. Some Topological Properties

Proposition 5.2.1. A point $Q \in \mathrm{Spec}(A,\mathfrak{P})$ is closed if and only if Q is a maximal convex ideal. Thus $\mathrm{Spec}(A,\mathfrak{P})$ is not T_1, in general. $\mathrm{Spec}(A,\mathfrak{P})$ is T_0. That is, given two points $P,Q \in \mathrm{Spec}(A,\mathfrak{P})$, at least one is not contained in the closure of the other.

Proof. For any $V \subset \mathrm{Spec}(A,\mathfrak{P})$, the Zariski closure of V is given by

$$\overline{V} = Z(\bigcap_{Q \in V} Q)\ .$$

This proves the first statement, and also the last since $P \in \overline{\{Q\}}$ and $Q \in \overline{\{P\}}$ implies $P \subset Q$ and $Q \subset P$. □

Proposition 5.2.2. The basic open sets $D(a) \subset \mathrm{Spec}(A,\mathfrak{P})$ are compact. In particular, $D(1) = \mathrm{Spec}(A,\mathfrak{P})$ is compact.

Proof. Suppose $D(a) \subseteq \bigcup_i D(a_i)$. Then any convex prime ideal containing all the a_i must also contain a. By Corollary 2.3.7, $a \in \sqrt{H(\{a_i\})}$. By Proposition 2.2.2, there is a finite subset $\{a_1,\ldots,a_k\}$ of $\{a_i\}$ with $a \in \sqrt{H(\{a_1,\ldots,a_k\})}$. Thus $D(a) \subset \bigcup_{1\le j\le k} D(a_j)$, as desired. □

5.3. Irreducible Closed Sets in $\mathrm{Spec}(A,\mathfrak{P})$

Recall that a closed subset of a topological space is *irreducible* if it is not the union of two proper closed subsets.

Proposition 5.3.1. If $V \subset \mathrm{Spec}(A,\mathfrak{P})$ is closed and irreducible, then there is a unique $Q \in V$ with $V = \overline{\{Q\}}$.

Proof. Uniqueness of Q is a consequence of Proposition 5.2.1 of the preceding section. (Specifically, $\mathrm{Spec}(A,\mathfrak{P})$ is T_0.)

We prove existence of Q as follows. Let $V = Z(I)$, where $I = \sqrt{I}$ is a convex ideal. It suffices to prove that I is prime.

Suppose, then, $ab \in I$, $a \notin I$, $b \notin I$. Consider the sets $I \cup \{a\}$, $I \cup \{b\}$. Any prime P containing I contains a or b. Thus

$$Z(I) = Z(I \cup \{a\}) \cup Z(I \cup \{b\}) .$$

On the other hand, since $I = \sqrt{I}$, $a \notin I$ implies there is a prime P' containing I but not a. Thus $Z(I \cup \{a\}) \subsetneq Z(I)$, and similarly, $Z(I \cup \{b\}) \subsetneq Z(I)$, which contradicts the irreducibility of V. □

Corollary 5.3.2. $\mathrm{Spec}(A,\mathfrak{P})$ is irreducible if and only if A has a unique minimal convex prime ideal P. Equivalently, $\mathrm{Spec}(A,\mathfrak{P})$ is irreducible if and only if $\sqrt{(0)}$ is prime. □

Consider the family of convex ideals $I \subset A$ such that $I = \sqrt{I}$. Note that this family of ideals satisfies the ascending chain condition (every increasing chain $I_1 \subset I_2 \subset \ldots$ is eventually constant) if and only if the closed sets of $\mathrm{Spec}(A,\mathfrak{P})$ satisfy the descending chain condition (every decreasing chain $Z_1 \supset Z_2 \supset \ldots$ is eventually constant.) Under this hypothesis it follows that every closed subset of $\mathrm{Spec}(A,\mathfrak{P})$ is a finite union of irreducible closed sets. Translating this in terms of ideals of A we have

Corollary 5.3.3. Suppose the family of convex ideals of A which coincide with their nil radical satisfies the ascending chain condition. Let $J \subset A$ be any proper convex ideal. Then among the convex prime ideals $P \subset A$ containing J, there are finitely many minimal such P, say $P_1,\ldots,P_k$, and

$$\sqrt{J} = \bigcap_{j=1}^{k} P_j .$$ □

5.4. Spec$(A,\mathfrak{P})$ as a Functor

If $f\colon (A,\mathfrak{P}) \to (A',\mathfrak{P}')$ is a morphism in (POR) and $Q' \subset A'$ is a convex prime ideal, then $Q = f^{-1}(Q') \subset A$ is a convex prime ideal. Thus f induces a function $f^*\colon \mathrm{Spec}(A',\mathfrak{P}') \to \mathrm{Spec}(A,\mathfrak{P})$. Clearly, f^* is continuous in the Zariski topologies since $(f^*)^{-1} Z(X) = Z(f(X))$, any $X \subset A$.

As special cases, we have morphisms $\pi\colon (A,\mathfrak{P}) \to (A/I, \mathfrak{P}/I)$, where $I \subset A$ is a convex ideal, and $i_T\colon (A,\mathfrak{P}) \to (A_T,\mathfrak{P}_T)$, where $T \subset A$ is a multiplicative set.

In the case of a projection, $\pi^*\colon \mathrm{Spec}(A/I, \mathfrak{P}/I) \to \mathrm{Spec}(A,\mathfrak{P})$ is a homeomorphism from $\mathrm{Spec}(A/I, \mathfrak{P}/I)$ to the closed subset $Z(I)$ of $\mathrm{Spec}(A,\mathfrak{P})$. This follows from 2.4.

In the case of a localization, $i_T^*\colon \mathrm{Spec}(A_T,\mathfrak{P}_T) \to \mathrm{Spec}(A,\mathfrak{P})$ is a homeomorphism from $\mathrm{Spec}(A_T,\mathfrak{P}_T)$ to the subspace $\bigcap_{t \in T} D(t)$ of $\mathrm{Spec}(A,\mathfrak{P})$. This follows from 3.4. If T is finitely generated as a multiplicative set (that is, if T is the smallest multiplicative set containing, say, $t_1,\ldots,t_k$), then

$$\bigcap_{t \in T} D(t) = \bigcap_{i=1}^{k} D(t_i),$$

and this intersection is an open set in $\mathrm{Spec}(A,\mathfrak{P})$.

We also point out that because the convex prime ideals of A disjoint from T coincide with those disjoint from the shadow $S(T)$, it follows that the natural map $(A_T,\mathfrak{P}_T) \to (A_{S(T)},\mathfrak{P}_{S(T)})$ induces a homeomorphism $\mathrm{Spec}(A_{S(T)},\mathfrak{P}_{S(T)}) \xrightarrow{\sim} \mathrm{Spec}(A_T,\mathfrak{P}_T)$.

5.5. Disconnectedness of Spec$(A,\mathfrak{P})$

In the category (POR), there is not quite as close a relation between disconnectedness of $\mathrm{Spec}(A,\mathfrak{P})$ and decomposition of A as a nontrivial product $A_1 \times A_2$, as there is in the category of commutative rings. We do have the following semi-analogue.

Proposition 5.5.1. $\mathrm{Spec}(A,\mathfrak{P})$ is disconnected if and only if there are elements $e_1, e_2 \in \mathfrak{P} \subset A$, with $1 \le e_1 + e_2$, $e_1 \cdot e_2 = 0$, $e_i \le e_i^2$, $1 \notin H(e_i)$, $i = 1,2$.

Proof. Under the hypothesis, it is clear that $\operatorname{Spec}(A,\mathfrak{P}) = Z(e_1) \cup Z(e_2)$, $Z(e_1) \cap Z(e_2) = \emptyset$, and $Z(e_i) \neq \emptyset$, $i = 1,2$.

Conversely, if $\operatorname{Spec}(A,\mathfrak{P}) = Z(I_1) \cup Z(I_2)$, $Z(I_1) \cap Z(I_2) = \emptyset$, and $Z(I_i) \neq \emptyset$, $i = 1,2$, then $1 \in H(I_1 + I_2)$ and $I_1 \cap I_2$ consists of nilpotent elements. By 2.2.4, we can write $1 \le e_1' + e_2'$, $e_i' \in I_i \cap \mathfrak{P}$. If $(e_1' e_2')^n = 0$, we get

$$1 = 1^{2n-1} \le (e_1' + e_2')^{2n-1} = a_1(e_1')^n + a_2(e_2')^n,$$

with $a_i \in \mathfrak{P}$, $i = 1,2$. Set $e_i = a_i(e_i')^n \in \mathfrak{P}$. Since $e_i' \in I_i$, $1 \notin H(e_i')$, hence $1 \notin H(e_i)$, $i = 1,2$. Also, $e_1 e_2 = 0$, $1 \le e_1 + e_2$ and $e_1 \le e_1^2 + e_1 e_2 = e_1^2$. □

If $(A,\mathfrak{P}) \in$ (PORCK), we can push this result somewhat further. Recall that by 3.8.4 we have an injection $A \to A_{S(1)}$ and an order $\mathfrak{P}_{S(1)} \subset A_{S(1)}$ with $\operatorname{Spec}(A,\mathfrak{P}) = \operatorname{Spec}(A_{S(1)},\mathfrak{P}_{S(1)})$.

Proposition 5.5.2. If $(A,\mathfrak{P}) \in$ (PORCK), then $\operatorname{Spec}(A,\mathfrak{P})$ is disconnected if and only if $(A_{S(1)},\mathfrak{P}_{S(1)}) \cong (A_1,\mathfrak{P}_1) \times (A_2,\mathfrak{P}_2)$ for suitable subrings $A_1, A_2 \subset A_{S(1)}$, $A_i \neq (0)$.

Proof. In 5.5.1 we proved that if $\operatorname{Spec}(A,\mathfrak{P})$ is disconnected, there are elements $e_1, e_2 \in \mathfrak{P} \subset A$, $1 \le e_1 + e_2 = u$, $e_i \le e_i^2$, $e_1 e_2 = 0$, $1 \notin H(e_i)$. Now, in $A_{S(1)}$, u is invertible and $1 = e_1/u + e_2/u = e_1' + e_2'$.

Consider the natural projection

$$\pi\colon\ (A_{S(1)},\mathfrak{P}_{S(1)}) \to (A_{S(1)}/AH(e_1'),\ \mathfrak{P}_{S(1)}/AH(e_1')) \times (A_{S(1)}/AH(e_2'),\ \mathfrak{P}_{S(1)}/AH(e_2')).$$

The kernel of π is $AH(e_1') \cap AH(e_2')$. But if $x' \in AH(e_1') \cap AH(e_2')$, then, since $e_1' e_2' = 0$, we have $e_1' x' = e_2' x' = 0$. This follows from 2.5.3, which gives $(0 : e_i') = (0 : AH(e_i'))$. Thus $x' = (e_1' + e_2')x' = 0$, and π is injective.

If $b',c' \in A_{S(1)}$, then $b' = b'e_1' + b'e_2'$, $c' = c'e_1' + c'e_2'$, and hence

$$\pi(b'e_2' + c'e_1') = (b',c') \in A_{S(1)}/AH(e_1') \times A_{S(1)}/AH(e_2') .$$

Thus, π is surjective.

Next, if $b',c' \in \mathfrak{P}_{S(1)}$, then $b'e_2' + c'e_1' \in \mathfrak{P}_{S(1)}$, since $e_i' \in \mathfrak{P}_{S(1)}$. Thus, π^{-1} is order preserving and π is an isomorphism in (PORCK).

Finally, $1 \notin H(e_i) \subset A$ implies $1 \notin AH(e_i) \subset A$, since $\sqrt{H(e_i)} = \sqrt{AH(e_i)}$. Then, also, $1 \notin AH(e_i') \subset A_{S(1)}$, since $AH(e_i') = AH(e_i)A_{S(1)}$ and $A \cap AH(e_i)A_{S(1)} = AH(e_i)$. Thus the factors $A_{S(1)}/AH(e_i')$ are nonzero. □

5.6. The Structure Sheaf, I - A First Approximation on Basic Open Sets

We will eventually impose on $\mathrm{Spec}(A,\mathfrak{P})$ a sheaf of partially ordered rings. The classical idea is this: on a basic open set $D(a) \subset \mathrm{Spec}(A,\mathfrak{P})$, $a \in A$, we know "a is never zero". That is, if $Q \in D(a)$, then $a \notin Q$, hence $\bar{a} \neq \bar{0} \in A/Q$. So we localize and invert the element a over $D(a)$.

But in our partially ordered ring A we know more than just $a \notin Q$ if $Q \in D(a)$. Specifically, let $S(a)$ be the shadow of the multiplicative set $\{a^i\}_{i \geq 0}$. If $b \in S(a)$, $Q \in D(a)$, then $b \notin Q$. The proof is easy if a is positive, since then $S(a) = \{b \mid 0 \leq a^n \leq b, \text{ some } n\}$. It is not much harder in the general case, but since $D(a) = D(a^2)$ and $(A_{S(a^2)},\mathfrak{P}_{S(a^2)}) \xrightarrow{\sim} (A_{S(a)},\mathfrak{P}_{S(a)})$, we may as well assume a is positive. (see 3.2 for details).

Thus it is natural to begin by trying to invert all elements of $S(a)$ over the basic open set $D(a)$. That is, to the basic open set $D(a) \subset \mathrm{Spec}(A,\mathfrak{P})$, we assign the localized (POR) $(A_{S(a)},\mathfrak{P}_{S(a)})$. Since $\mathrm{Spec}(A,\mathfrak{P}) = D(1)$, this assigns $(A_{S(1)},\mathfrak{P}_{S(1)})$ to $\mathrm{Spec}(A,\mathfrak{P})$.

<u>Lemma 5.6.1</u>. If $D(a_1) = D(a_2)$, $a_1,a_2 \in A$, then there is a canonical isomorphism in (POR), $(A_{S(a_2)},\mathfrak{P}_{S(a_2)}) \xrightarrow{\sim} (A_{S(a_1)},\mathfrak{P}_{S(a_1)})$.

<u>Proof</u>. $D(a_1) = D(a_2)$ implies that the primes containing a_1 and the primes containing a_2 coincide. Thus $\sqrt{H(a_1)} = \sqrt{H(a_2)}$. Since $a_1,a_2 \in \mathfrak{P}$, we have by 2.2.3

$$0 \leq a_1^{r_1} \leq pa_2$$

$$0 \leq a_2^{r_2} \leq qa_1$$

for suitable integers $r_1, r_2 > 0$ and $p, q \in \mathfrak{P}$. The proposition follows since now all elements of $S(a_2)$ are invertible in $A_{S(a_1)}$ and vice versa. □

In fact, this argument really shows that if $D(a_1) \subseteq D(a_2)$, then $a_1 \in \sqrt{H(a_2)}$, hence $0 \leq a_1^{r_1} \leq pa_2$, and there is a canonical morphism in (POR), $\varphi = \varphi_{D(a_2),D(a_1)}: (A_{S(a_2)}, \mathfrak{P}_{S(a_2)}) \to (A_{S(a_1)}, \mathfrak{P}_{S(a_1)})$. φ is explicitly defined by $\varphi(a/b) = ap^r/bp^r \in A_{S(a_1)}$, where $a \in A$, $b \geq a_2^r$ and $pa_2 \geq a_1^{r_1}$, so that $p^r b \geq (pa_2)^r \geq a_1^{rr_1}$. These morphisms $\varphi_{D(a_2),D(a_1)}$ are our candidates for the presheaf restriction maps corresponding to the inclusions of open sets $D(a_1) \subseteq D(a_2)$.

<u>Lemma 5.6.2</u>. If $D(a_0) \subseteq D(a_1) \subseteq D(a_2)$, then the diagram below commutes

$$\begin{array}{ccc}
(A_{S(a_2)}, \mathfrak{P}_{S(a_2)}) & \xrightarrow{\varphi_{2,1}} & (A_{S(a_1)}, \mathfrak{P}_{S(a_1)}) \\
& {\scriptstyle \varphi_{2,0}} \searrow \quad \swarrow {\scriptstyle \varphi_{1,0}} & \\
& (A_{S(a_0)}, \mathfrak{P}_{S(a_0)}) &
\end{array}$$

where $\varphi_{j,i} = \varphi_{D(a_j),D(a_i)}$.

<u>Proof</u>. This is easy using the universal property of the localizations $(A_{S(a_i)}, \mathfrak{P}_{S(a_i)})$ (rather than the explicit formulae for $\varphi_{j,i}$). □

If $Q \subset A$ is a convex prime ideal, $a \notin Q$, then $Q \in D(a)$ and the concave multiplicative set $A - Q$ contains a, hence also $S(a)$. There is thus a natural map

$$(A_{S(a)}, \mathfrak{P}_{S(a)}) \longrightarrow (A_{(Q)}, \mathfrak{P}_{(Q)}) .$$

Lemma 5.6.3. The induced map

$$\varinjlim_{a \notin Q} (A_{S(a)}, \mathfrak{P}_{S(a)}) \xrightarrow{\sim} (A_{(Q)}, \mathfrak{P}_{(Q)})$$

is an isomorphism in (POR). □

We leave the proof to the reader, with only the reminder that one must not forget to check that the inverse ring isomorphism is *order preserving*.

Thus, from Lemma 5.6.3 we expect the stalk of our structure sheaf at the point $Q \in \mathrm{Spec}(A,\mathfrak{P})$ to be the usual local ring $A_{(Q)}$ with the natural order $\mathfrak{P}_{(Q)}$.

5.7. The Structure Sheaf, II - The Sheaf Axioms for Basic Open Sets

We formulate the "first sheaf axiom" for basic open sets in the following.

Lemma 5.7.1(a). Let $(A,\mathfrak{P}) \in (\mathrm{POR})$, $D(a) = \bigcup_\alpha D(a_\alpha)$, a, $a_\alpha \in \mathfrak{P}$. Suppose $x/b \in A_{S(a)}$ and $\varphi_{D(a),D(a_\alpha)}(x/b) = 0 \in A_{S(a_\alpha)}$ all α. Then $x/b = 0 \in A_{S(a)}$.

Lemma 5.7.1(b). From 5.7.1(a), the map $(A_{S(a)}, \mathfrak{P}_{S(a)}) \to \prod_\alpha (A_{S(a_\alpha)}, \mathfrak{P}_{S(a_\alpha)})$ is injective. We claim moreover that $\mathfrak{P}_{S(a)} = (\prod_\alpha \mathfrak{P}_{S(a_\alpha)}) \cap A_{S(a)}$. That is, the natural order $\mathfrak{P}_{S(a)} \subset A_{S(a)}$ is the contraction of the natural product order $\prod_\alpha \mathfrak{P}_{S(a_\alpha)}$ on $\prod_\alpha A_{S(a_\alpha)}$.

Proof 5.7.1.(a). First, $D(a)$ is compact, hence there is a finite subset of the a_α, say, $a_1,\dots,a_k$ with $D(a) = \bigcup_{1\le i\le k} D(a_i)$.

Secondly, $D(a_i) \subset D(a)$ means that $a_i \in \sqrt{H(a)}$, hence $0 \le a_i^{r_i} \le p_i a$ for some $p_i \in \mathfrak{P}$ and integers $r_i > 0$. We can replace all the p_i by the element $p = 1 + p_1 + \cdots + p_k$ in these inequalities.

Since $x/b \in A_{S(a)}$, we assume $b \ge a^s$, some integer $s > 0$. Then $\varphi_{D(a),D(a_i)}(x/b) = xp^s/bp^s$. ($bp^s \ge a^s p^s \ge a_i^{r_i s}$, hence $bp^s \in S(a_i)$.) Moreover, since $p \ge 1$, $a^s p^s \ge a^s$, hence $bp^s \ge a^s$ and we can write

$x/b = xp^s/bp^s \in A_{S(a)}$. To assert that $\varphi_{D(a),D(a_i)}(xp^s/bp^s) = 0 \in A_{S(a_i)}$ means that there exists a $c_i \geq a_i^{t_i}$, with $xp^s c_i = 0 \in A$.

We now use the fact that $D(a) = \bigcup_{i=1}^{k} D(a_i)$ implies $a \in \sqrt{H(\{a_1^{t_1},\ldots,a_k^{t_k}\})}$, hence $0 \leq a^r \leq \sum_{i=1}^{k} q_i a_i^{t_i} \leq \sum_{i=1}^{k} q_i c_i = c$ for suitable $q_i \in \mathfrak{P}$, and some integer $r > 0$. But now, $xp^s c = 0$, and since $c \in S(a)$, we have proved $x/b = xp^s/bp^s = 0 \in A_{S(a)}$ as desired.

Proof 5.7.1(b). The proof is a slight modification of the argument above. We continue with the same notation.

If $xp^s/bp^s \geq 0$ in all $A_{S(a_i)}$, $1 \leq i \leq k$, then there exist elements $c_i' \geq a_i^{t_i'}$ with $xbp^{2s}(c_i')^2 \geq 0$ in A. If $0 \leq a^{r'} \leq \sum_{i=1}^{k} q_i' q_i^{2t_i'} \leq \sum_{i=1}^{k} q_i'(c_i')^2 = c'$ $0 \leq q_i'$, we see that $xbp^{2s}(c') \geq 0$ in A, hence also $xbp^{2s}(c')^2 \geq 0$ in A. But $c' \in S(a)$, so we conclude $xp^s/bp^s \geq 0$ in $A_{S(a)}$. □

The "second sheaf axiom" for basic open sets would say this: Suppose $D(a) = \bigcup_{\alpha} D(a_\alpha)$ and suppose $x_\alpha/b_\alpha \in A_{S(a_\alpha)}$ have the property that $\varphi_{D(a_\alpha),D(a_\alpha a_\beta)}(x_\alpha/b_\alpha) = \varphi_{D(a_\beta),D(a_\alpha a_\beta)}(x_\beta/b_\beta) \in A_{S(a_\alpha a_\beta)}$, for all α,β. (Note $D(a_\alpha a_\beta) = D(a_\alpha) \cap D(a_\beta)$.) Then there is an element $x/b \in A_{S(a)}$ such that $\varphi_{D(a),D(a_\alpha)}(x/b) = x_\alpha/b_\alpha \in A_{S(a_\alpha)}$ for all α.

This assertion does not seem to be true in general. However, the failure of the second sheaf axiom does not really affect the way one defines the structure sheaf. What it does affect is the evaluation of the global sections. We will return to this point in the next section.

What we can show is that in the category (PORCK), the second sheaf axiom for basic open sets is true. Specifically

Lemma 5.7.2. Let $(A,\mathfrak{P}) \in$ (PORCK), $D(a) \in \mathrm{Spec}(A,\mathfrak{P})$ a basic open set, $a \in \mathfrak{P}$, and let $D(a) = \bigcup_{\alpha} D(a_\alpha)$ be a cover, $a_\alpha \in \mathfrak{P}$. Suppose the elements $x_\alpha/b_\alpha \in A_{S(a_\alpha)}$ have the property that $\varphi_{D(a_\alpha),D(a_\alpha a_\beta)}(x_\alpha/b_\alpha) = \varphi_{D(a_\beta),D(a_\alpha a_\beta)}(x_\beta/b_\beta) \in A_{S(a_\alpha a_\beta)}$ for all α,β. Then there exists

$x/b \in A_{S(a)}$ with $\varphi_{D(a),D(a_\alpha)}(x/b) = x_\alpha/b_\alpha \in A_{S(a_\alpha)}$ for all α.

<u>Proof</u>. The proof is essentially the usual proof for commutative rings. First, $D(a)$ is compact, so $D(a) = \bigcup_{i=1}^{k} D(a_i)$, for some finite subset of the a_α. If $x/b \in A_{S(a)}$ restricts to x_i/b_i, $1 \le i \le k$, then x_α/b_α and x/b have the same images in all $A_{S(a_\alpha a_i)}$. But $D(a_\alpha) = \bigcup_{i=1}^{k} D(a_\alpha a_i)$, and we already know from the "first sheaf axiom", Lemma 5.7.1, that an element in $A_{S(a_\alpha)}$ is uniquely determined by its restrictions to the $A_{S(a_\alpha a_i)}$. Thus x/b must restrict to x_α/b_α, all α. This reduces our problem to the case of a finite cover.

From the hypothesis, there are elements $c_{ij} \in A$, $0 \le (a_i a_j)^r \le c_{ij}$ such that $x_i \cdot b_j \cdot c_{ij} - x_j \cdot b_i \cdot c_{ij} = 0$, all i,j. Since $(A,\mathfrak{P}) \in$ (PORCK), $x_i b_j (a_i a_j)^r - x_j b_i (a_i a_j)^r = 0$. Replacing x_i/b_i by $x_i a_i^r / b_i a_i^r$, we may assume $x_i b_j = x_j b_i$. We may also assume that $0 \le a_i^s \le b_i$, $1 \le i \le k$. Since the $D(a_i)$ cover $D(a)$, $a \in \sqrt{H\{a_1 \cdots a_k\}}$, and we can find positives $p_i \in \mathfrak{P}$ such that

$$0 \le a^t \le \sum_{i=1}^{k} p_i a_i^s \le \sum_{i=1}^{k} p_i b_i = b \ .$$

In $A_{S(a)}$ we have $1 = b/b = \sum_{i=1}^{k} (\frac{p_i}{b}) b_i$. Consider the element $x/b = \sum_{i=1}^{k} (\frac{p_i}{b}) \cdot x_i \in A_{S(a)}$. Then, in $A_{S(a)}$, $(x/b) \cdot b_j = \sum_{i=1}^{k} (\frac{p_i}{b}) x_i b_j = \sum_{i=1}^{k} (\frac{p_i}{b}) b_i x_j = x_j$. This proves that $\varphi_{D(a),D(a_j)}(x/b) = x_j/b_j \in A_{S(a_j)}$, $1 \le j \le k$, as desired. □

5.8. <u>The Structure Sheaf, III - Definition</u>

Let $X = \mathrm{Spec}(A,\mathfrak{P})$ with the Zariski topology. We will define a sheaf $\mathcal{O}_X$, of partially ordered rings on X.

If $U \subset X$ is open, consider the product $\prod_{Q \in U} (A_{(Q)}, \mathfrak{P}_{(Q)})$, which projects naturally to $U \subset X$. A section $s\colon U \to \prod_{Q \in U} A_{(Q)}$, say $s(Q) = x_Q/b_Q \in A_{(Q)}$, $Q \in U$, is *continuous* if there exists a cover $U = \bigcup_\alpha D(a_\alpha)$ and elements $x_\alpha/b_\alpha \in A_{S(a_\alpha)}$, such that for all $Q \in D(a_\alpha)$, $\varphi(x_\alpha/b_\alpha) =$

$x_Q/b_Q \in A_{(Q)}$, where $\varphi\colon (A_{S(a_\alpha)}, \mathfrak{P}_{S(a_\alpha)}) \to \varinjlim_{a \notin Q} (A_{S(a)}, \mathfrak{P}_{S(a)}) = (A_{(Q)}, \mathfrak{P}_{(Q)})$ is the natural map.

<u>Definition 5.8.1</u>. Define $(\Gamma(U,\mathcal{O}_X), \mathfrak{P}(U,\mathcal{O}_X)) \in (\text{POR})$ by

$$\Gamma(U,\mathcal{O}_X) = \{s\colon U \to \prod_{Q \in U} A_{(Q)} \mid s = \text{continuous section}\}$$

$$\mathfrak{P}(U,\mathcal{O}_X) = \Gamma(U,\mathcal{O}_X) \cap \prod_{Q \in U} \mathfrak{P}_{(Q)}.$$

<u>Lemma 5.8.2</u>. Let $V \subset U \subset X$ be open sets. There is a canonical map $\varphi_{U,V}\colon (\Gamma(U,\mathcal{O}_X), \mathfrak{P}(U,\mathcal{O}_X)) \to (\Gamma(V,\mathcal{O}_X), \mathfrak{P}(V,\mathcal{O}_X))$ induced by the obvious projection $\prod_{Q \in U} (A_{(Q)}, \mathfrak{P}_{(Q)}) \to \prod_{Q \in V} (A_{(Q)}, \mathfrak{P}_{(Q)})$.

<u>Proof</u>. Suppose $s\colon U \to \prod_{Q \in U} A_{(Q)}$, $s(Q) = x_Q/b_Q \in A_{(Q)}$, is continuous, say with respect to a cover $U = \bigcup_\alpha D(a_\alpha)$, and elements $x_\alpha/b_\alpha \in A_{S(a_\alpha)}$. Cover V, $V = \bigcup_\beta D(a'_\beta)$. Since $V \subset U$, we have $V = \bigcup_{\alpha,\beta} D(a_\alpha a'_\beta)$. Then the section $s|_V\colon V \to \prod_{Q \in V} A_{(Q)}$ is defined locally by the elements $\varphi_{D(a_\alpha), D(a_\alpha a'_\beta)}(x_\alpha/b_\alpha) \in A_{S(a_\alpha a'_\beta)}$. Thus $s|_V$ is continuous and we define $\varphi_{U,V}(s) = s|_V \in \Gamma(V,\mathcal{O}_X)$. $\varphi_{U,V}$ is obviously order preserving. □

<u>Proposition 5.8.3</u>. The $(\Gamma(U,\mathcal{O}_X), \mathfrak{P}(U,\mathcal{O}_X))$ and $\varphi_{U,V}\colon (\Gamma(U,\mathcal{O}_X), \mathfrak{P}(U,\mathcal{O}_X)) \to (\Gamma(V,\mathcal{O}_X), \mathfrak{P}(V,\mathcal{O}_X))$ form a sheaf on X.

<u>Proof</u>. The proof is routine. The first sheaf axiom is a consequence of Lemma 5.7.1. The second sheaf axiom is a consequence of the definitions. □

<u>Remark 1</u>. The significance of the failure of the second sheaf axiom for basic open sets is that we cannot easily describe *global sections* $\Gamma(X,\mathcal{O}_X)$, $X = \text{Spec}(A,\mathfrak{P})$, or more generally $\Gamma(D(a),\mathcal{O}_X)$, $D(a) \subset \text{Spec}(A,\mathfrak{P})$ a basic open set.

Lemma 5.7.1 implies that $(A_{S(a)}, \mathfrak{P}_{S(a)}) \to (\Gamma(D(a),\mathcal{O}_X), \mathfrak{P}(D(a),\mathcal{O}_X))$ is

injective and that $\mathfrak{P}_{S(a)} = A_{S(a)} \cap \mathfrak{P}(D(a),\mathcal{O}_X)$ is the contracted order. However, $\Gamma(D(a),\mathcal{O}_X)$ may be bigger than $A_{S(a)}$. In any case, one shows in the usual way that there is a natural sheaf isomorphism

$$\mathcal{O}_{\mathrm{Spec}(A,\mathfrak{P})}\Big|_{D(a)} = \mathcal{O}_{\mathrm{Spec}(A_{S(a)},\mathfrak{P}_{S(a)})}.$$

Remark 2. It is perhaps good that $\Gamma(X,\mathcal{O}_X)$ is larger than $A_{S(1)}$. The natural map $(A,\mathfrak{P}) \to (A_{S(1)},\mathfrak{P}_{S(1)})$ is *not* always injective in (POR). In such case, $\Gamma(X,\mathcal{O}_X)$ may contain information about $(A,\mathfrak{P})$ lost in the localization $(A_{S(1)},\mathfrak{P}_{S(1)})$.

In the category (PORCK), the second sheaf axiom for basic open sets, Lemma 5.7.2, does allow computation of the global sections of the structure sheaf. We have

Proposition 5.8.4. Let $(A,\mathfrak{P}) \in$ (PORCK), $X = \mathrm{Spec}(A,\mathfrak{P})$. Then the sections of the structure sheaf $\mathcal{O}_X$ over the basic open set $D(a) \subset X$ is the ring $\Gamma(D(a),\mathcal{O}_X) = (A_{S(a)},\mathfrak{P}_{S(a)})$. In particular, the global sections is the ring $\Gamma(X,\mathcal{O}_X) \cong (A_{S(1)},\mathfrak{P}_{S(1)})$. □

The Zariski topology on the prime convex ideal spectrum of a partially ordered ring has the following property not shared by unordered rings. A finite union of basic open sets is still a basic open set. The proof is easy since $D(f) = D(f^2)$ and $\cup D(f_i^2) = D(\Sigma f_i^2)$, which is symmetrical with $\cap D(f_i^2) = D(\Pi f_i^2)$. If, further, our ring is Noetherian (more generally, if the radical convex ideals satisfy the ascending chain condition), then every Zariski open set is a basic open set.

VI · Polynomials

6.1. Polynomials as Functions

Let A be a ring, $A[X_1 \dots X_n]$ the polynomial ring in n-variables over A. Let $A^{(n)}$ denote affine n-space over A. Each polynomial $f \in A[X_1 \dots X_n]$ defines a function $f\colon A^{(n)} \to A$ by evaluation.

If $(A,\mathfrak{P})$ is a partially ordered ring, then so is the ring of functions A^X, from X to A, for any set X. Namely, define $\mathfrak{P}_X \subset A^X$ by $f \in \mathfrak{P}_X$ if for all $x \in X$, $f(x) \in \mathfrak{P}$. Note $A \subset A^X$ (the constants) and $\mathfrak{P}_X$ extends $\mathfrak{P}$. If $(A,\mathfrak{P})$ belongs to (PORCK) or (PORNN), then so does $(A^X, \mathfrak{P}_X)$.

A very natural question is: when is the ring homomorphism $A[X_1 \dots X_n] \to A^{A^{(n)}}$ injective? The following at least gives an easy sufficient condition.

Proposition 6.1.1. If the ring A is $\mathbb{Z}$-torsion free, then $A[X_1 \dots X_n] \to A^{\mathbb{Z}^{(n)}}$ is injective.

Proof. We first establish the result for $n = 1$. Suppose $f(X) = a_0 + a_1X + \cdots + a_dX^d$ is a polynomial such that $f(k) = 0$ for all $k \in \mathbb{Z} \subseteq A$. in particular, $f(0) = a_0 = 0$. Next, let $k = 1,2,\dots,d$. We obtain d equations in the coefficients $a_1,\dots,a_d$

$$\begin{array}{ccccccc} a_1 & + & a_2 & + \cdots + & a_d & = & 0 \\ 2a_1 & + & 4a_2 & + \cdots + & 2^d a_d & = & 0 \\ \vdots & & \vdots & & \vdots & & \vdots \\ da_1 & + & d^2a_2 & + \cdots + & d^d a_d & = & 0\,. \end{array}$$

Since

$$\det\begin{pmatrix} x_1 & x_1^2 & \cdots & x_1^d \\ x_2 & x_2^2 & \cdots & x_2^d \\ \vdots & \vdots & & \vdots \\ x_n & x_n^2 & \cdots & x_n^d \end{pmatrix} = (x_1 \cdot \cdots \cdot x_n) \prod_{1 \le i < j \le n} (x_j - x_i) ,$$

we deduce that for some non-zero integer m, $ma_i = 0$, $1 \le i \le d$. Namely,

$$m = \det\begin{pmatrix} 1, & 1, & \ldots, & 1 \\ 2, & 2^2, & \ldots, & 2^d \\ \vdots & \vdots & & \vdots \\ d, & d^2, & \ldots, & d^d \end{pmatrix}$$

Thus $a_i = 0$, $1 \le i \le d$, and hence $f(X) = 0$.

If $n > 1$, we use induction.

$$\begin{aligned} A[X_1 \cdots X_n] &= A[X_1 \ldots X_{n-1}][X_n] \\ &\subseteq A[X_1 \ldots X_{n-1}]^{\mathbb{Z}} \\ &\subseteq A^{\mathbb{Z}^{(n-1)}\,\mathbb{Z}} \\ &= A^{(\mathbb{Z}^{(n)})} . \end{aligned}$$ □

<u>Corollary 6.1.2</u>. If A is $\mathbb{Z}$-torsion free and $B \subseteq A$ is any subring, then the homomorphism $A[X_1 \ldots X_n] \to A^{B^{(n)}}$ is injective. In particular, $A[X_1 \ldots X_n] \to A^{A^{(n)}}$ is injective. □

<u>Corollary 6.1.3</u>. If $(A,\mathfrak{P}) \in$ (PORCK), then A is $\mathbb{Z}$-torsion free, hence $A[X_1 \ldots X_n] \to A^{A^{(n)}}$ is injective. □

From Corollary 6.1.3 we see that if $(A,\mathfrak{P}) \in$ (PORCK), then there is a natural (PORCK) order on $A[X_1 \ldots X_n]$, namely the contraction of the order $\mathfrak{P}_{A^{(n)}}$ on $A^{A^{(n)}}$ to $A[X_1 \ldots X_n]$. We denote this contracted order by the same symbol $(A[X_1 \ldots X_n], \mathfrak{P}_{A^{(n)}})$.

Next suppose $(A,\mathfrak{P}_A) \subseteq (B,\mathfrak{P}_B)$, both in (PORCK). Suppose as an algebra B is finitely generated over A, say $B = A[y_1 \dots y_n]$, $y_i \in B$. Let $A[T_1 \dots T_n]$ be the polynomial ring in n-indeterminates, so that $B \cong A[T_1 \dots T_n]/I$ where $I = \{f \in A[T_1 \dots T_n] \mid f(y_1 \dots y_n) = 0 \in B\}$.

Proposition 6.1.4. If $(A,\mathfrak{P}_A)$, $(B,\mathfrak{P}_B) \in$ (PORCK), then the projection $(A[T_1 \dots T_n],\mathfrak{P}) \to (B,\mathfrak{P}_B)$ is order preserving, where $\mathfrak{P}$ is the weakest order on $A[T_1 \dots T_n]$ extending $\mathfrak{P}_A$. In particular, $I \subset A[T_1 \dots T_n]$ is $\mathfrak{P}$-absolutely convex.

Proof. A typical element of $\mathfrak{P}$ is $\Sigma p_i g_i^2(\vec{T})$ where $p_i \in \mathfrak{P}_A$, $g_i \in A[T_1 \dots T_n]$. The image in B is $\Sigma p_i g_i^2(\vec{y}) \in \mathfrak{P}_B$. □

Remark. In words, the proposition shows that if $(A,\mathfrak{P}_A) \subset (B,\mathfrak{P}_B)$ is a finitely generated algebraic extension, then $(B,\mathfrak{P}_B)$ is obtained from a pure polynomial extension of A by first dividing by a convex ideal and then, possibly, refining the order. In the case A a totally ordered field, B a totally ordered integral domain over A, this procedure can be reversed. That is, first the order $\mathfrak{P}$ on $A[T_1 \dots T_n]$ can be refined to a total order, inducing $(B,\mathfrak{P}_B)$. This assertion will be established in Chapter VIII.

We conclude this section with an example. Consider the ring $\mathbb{Z}[T]$, ordered as a ring of real valued functions on the real line. Then $(2T,T^2)$ is a convex ideal. Let $A = \mathbb{Z}[T]/(2T,T^2)$. Elements of A have unique expressions $n + \varepsilon T$, $\varepsilon = 0$ or 1, $n \in \mathbb{Z}$. The polynomial $f(X) = TX + TX^2 \in A[X]$ vanishes identically on $A^{(1)}$, so we cannot regard A[X] as a ring of A-valued functions on $A^{(1)}$. On the other hand, A[X] does admit an order, extending the order on A.

6.2. Adjoining Roots

Let $(A,\mathfrak{P}) \in$ (POR), $a \in A$, $A \to A[T]/(T^2-a)$ the obvious inclusion. We ask if there is an order on $A[T]/(T^2-a)$, extending $\mathfrak{P}$. The answer is similar to well-known results for (totally) ordered fields, and extends the results of 1.4. Recall from Chapter I the notation $D\mathfrak{P} = \{a \in A \mid pa \in \mathfrak{P}$, some $p \in \mathfrak{P}^+\}$. We established that $-a \notin D\mathfrak{P}$ implies $\mathfrak{P}[a]$ is an order on A. We now refine this result.

Proposition 6.2.1.

(a) If $-a \notin D\mathfrak{P}$, then $B = A[T]/(T^2-a)$ can be ordered extending $\mathfrak{P}$, and if B can be ordered extending $\mathfrak{P}$, then $\mathfrak{P}[a]$ is an order on A.

(b) If either A has no nilpotent elements or $a \in A$ is not a zero divisor, then the existence of an order on B extending $\mathfrak{P}$ is equivalent to $\mathfrak{P}[a]$ being an order on A.

(c) If $(A,\mathfrak{P}) \in$ (PORNN), $\mathfrak{P}[a]$ is an order on A and a is not a zero divisor, then $B = A[T]/(T^2-a)$ has no nilpotent elements.

(d) If A is an integral domain, then either $A[T]/(T^2-a)$ or $A[T]/(T^2+a)$ admits an order extending $\mathfrak{P}$.

Proof.

(a) The second statement is obvious since if B can be ordered extending $\mathfrak{P}$, then $a = T^2$ is positive in B, hence the contraction of the order to A contains $\mathfrak{P}[a]$. For the first statement, note that elements of the ring B have unique representatives $b + cT$, $b,c \in A$. If B can't be ordered extending $\mathfrak{P}$, then we have a nontrivial relation

$$(*) \qquad 0 = \sum_{i=1}^{k} p_i(b_i + c_iT)^2$$

$$= \sum_{i=1}^{k} (p_i(b_i^2 + c_i^2a) + 2p_ib_ic_iT)$$

with $p_i \in \mathfrak{P} \subset A$. Thus, in A,

$$(**) \qquad 0 = \Big(\sum_{i=1}^{k} p_ib_i^2\Big) + \Big(\sum_{i=1}^{k} p_ic_i^2\Big)a .$$

We will prove $\sum_{i=1}^{k} p_ic_i^2 \neq 0$ in A.

Assuming the contrary, then $p_jc_j^2 = 0$ and $p_jb_j^2 = 0$, $1 \leq j \leq k$. Also $0 = \sum_{i=1}^{k} 2p_ib_ic_i$, by (*). On the other hand, $0 \leq p_j(b_j + c_j)^2 = p_j(b_j^2 + c_j^2) + 2p_jb_jc_j$ so the terms $2p_jb_jc_j$ are all positive. Since their sum vanishes, $2p_jb_jc_j = 0$,

$1 \le j \le k$. This contradicts the assumption that (*) is a nontrivial relation.

(b) If $\mathfrak{P}[a]$ is an order and if there is a relation (*) in $A[T]/(T^2-a)$, then from (**), $p_j b_j^2 = 0$, hence $p_j b_j = 0$, if A has no nilpotent elements. Also, $p_j c_j^2 a = 0$, hence $p_j(b_j + c_j T)^2 = p_j(b_j^2 + c_j^2 a) + 2p_j b_j c_j T = 0$ and relation (*) is trivial.

If we assume that $\mathfrak{P}[a]$ is an order on A and a is not a zero divisor, then a relation (*) in $B = A[T]/(T^2-a)$ implies $p_j b_j^2 = 0$ and $p_j c_j^2 = 0$, $1 \le j \le k$. The argument now proceeds just as in the proof of (a).

(c) If $(b + cT)^2 = b^2 + c^2 a + 2bcT = 0 \in B$, then the assumptions imply $b^2 = c^2 a = 0$, hence $b = ca = 0$ and $c = 0$.

(d) If A is an integral domain, then $D\mathfrak{P} \subset A$ is an order on A. Hence if $a \neq 0$, either $-a \notin D\mathfrak{P}$ or $a \notin D\mathfrak{P}$. Alternatively, since $\mathfrak{P}$ admits total refinements, either $\mathfrak{P}[a]$ or $\mathfrak{P}[-a]$ is an order on A. These two assertions have already been discussed in 1.6. Thus (a) or (b) implies (d). □

Remark. It does not seem clear whether there is a reasonable (PORCK) analogue of parts (a) and (b) of the proposition. One would hope that if $(A,\mathfrak{P}) \in$ (PORCK), $-a \notin D_{CK}(\mathfrak{P}^+)$, then one can adjoin a square root of a. That is, there ought to exist a (PORCK) order on $A[T]/(T^2 - a)$, extending $\mathfrak{P}$. However, imitating the proof of (a) in the (PORCK) case does not lead to a simple relation like (**), but rather to a matrix equation

$$(***) \qquad \begin{pmatrix} r & sa \\ s & r \end{pmatrix}\begin{pmatrix} x \\ y \end{pmatrix} = \begin{pmatrix} 0 \\ 0 \end{pmatrix}$$

where $r = \sum_{i=1}^{k} p_i(b_i^2 + c_i^2 a)$ and $s = \sum_{i=1}^{k} 2p_i b_i c_i$, $p_i \in \mathfrak{P}$, $p_i, c_i, x, y \in A$. The usefulness of (***) is not apparent.

Proposition 6.2.2. Let $(A,\mathfrak{P}) \in$ (POR), A an integral domain. Let $f(T) \in A[T]$ an irreducible polynomial of odd degree. Then there exist orders on $A[T]/(f(T))$ extending $\mathfrak{P}$.

Proof. This is a familiar result in the Artin-Schreier theory of ordered fields. First, we assume $\mathfrak{P}$ is a total order on A, and we let K be the field of fractions of A. Then K is also totally ordered, and we proceed to show K[T]/(f(T)) can be ordered, extending the order on K. If not, there would be a relation

$$-1 = \Sigma g_i^2(T) + h(T)f(T) \in K[T]$$

where degree (g_i) < degree (f). The g_i cannot all be constant, because then h = 0, and we would contradict the fact that K is ordered. Now, degree $(\Sigma g_i^2(T)) \leq 2n-2$, and is even, hence h(T) has odd degree, no greater than n-2. By induction we have a contradiction since some irreducible factor of h(T) also has odd degree. □

6.3. A Universal Bound on the Roots of Polynomials

Let A be a ring. For simplicity in what follows, we assume $\mathbb{Q} \subset A$, where $\mathbb{Q}$ is the field of rational numbers.

Proposition 6.3.1. Let $f(T) = T^{2n} + a_1T^{2n-1} + \cdots + a_{2n} \in A[T]$ be a monic polynomial of even degree. Then there are universal polynomials $\beta^+(u_1 \ldots u_{2n})$, $\beta^-(u_1 \ldots u_{2n}) \in \mathbb{Q}[u_1 \ldots u_{2n}]$ and polynomials $h_i^+(T)$, $h_i^-(T) \in A[T]$ such that

$$T - \beta^+(a_1 \ldots a_{2n}) + \sum_{i=1}^{k} h_i^+(T)^2 = f(T)$$

and

$$\beta^-(a_1 \ldots a_{2n}) - T + \sum_{i=1}^{k} h_i^-(T)^2 = f(T) .$$

Corollary 6.3.2. If $(A,\mathfrak{P}) \in$ (POR), $\mathbb{Q} \subseteq A$ and $f(T) = T^{2n} + a_1T^{2n-1} + \cdots + a_{2n}$ is a monic polynomial of even degree, and if $f(\alpha) \leq 0$, $\alpha \in A$, then

$$\beta^-(a_1 \ldots a_{2n}) \leq \alpha \leq \beta^+(a_1 \ldots a_{2n}) .$$

Proof. Substitute α for T in the polynomial equalities of the proposition. □

Corollary 6.3.3. Same assumptions as above. Let $g(T) = T^{2n-1} + b_1T^{2n-2} + \cdots + b_{2n-1}$ be a monic polynomial of odd degree. If $g(\alpha) = 0$, $\alpha \in A$, then

$$\beta^{-1}(b_1 \ldots b_{2n-1}, 0) \le \alpha \le \beta^{+}(b_1 \ldots b_{2n-1}, 0).$$

Moreover, if $\alpha \in \mathfrak{P}$ and $g(\alpha) \le 0$, then

$$0 \le \alpha \le \beta^{+}(b_1, \ldots, b_{2n-1}, 0).$$

Proof. Apply Corollary 6.3.2 to $f(T) = Tg(T)$. □

Proof of Proposition 6.3.1. The proof is by induction on n. If $n = 1$,

$$\begin{aligned} f(T) &= T^2 + a_1T + a_2 \\ &= \left(T + \tfrac{a_1-1}{2}\right)^2 + T + \left(a_2 - \left(\tfrac{a_1-1}{2}\right)^2\right) \\ &= \left(T + \tfrac{a_1+1}{2}\right)^2 - T + \left(a_2 - \tfrac{a_1+1}{2}\right)^2\right). \end{aligned}$$

Thus

$$h_1^{\pm}(T) = T + \frac{a_1 \mp 1}{2}$$

and

$$\beta^{\pm}(a_1, a_2) = \pm \left(\left(\tfrac{a_1 \mp 1}{2}\right)^2 - a_2\right).$$

Assume the proposition for polynomials of degree $2i < 2n$. Let $X = T + (a_1/2n)$, and write

$$f(T) = g(X) = X^{2n} + b_2X^{2n-2} + b_3X^{2n-3} + \cdots + b_{2n}$$

where the b_j are polynomials in the a_i with rational coefficients. Let

$$\tilde{h}_1^{\pm}(X) = X^n + \left(\tfrac{b_2-1}{2}\right)X^{n-2}$$

so that

$$\tilde{h}_1^{\pm}(X)^2 = X^{2n} + b_2X^{2n-2} - X^{2n-2} + \left(\frac{b_2-1}{2}\right)^2 X^{2n-4}$$

and

$$g(X) - \tilde{h}_1^{\pm}(X)^2 = X^{2n-2} + b_3X^{2n-3} + \left(b_4 - \left(\frac{b_2-1}{2}\right)^2\right)X^{2n-4} + b_5X^{2n-5} + \cdots + b_{2n} .$$

Now apply induction to the right-hand side to find $\tilde{\beta}^{\pm}$, $\tilde{h}_j^{\pm}$ such that

$$\pm\left(X - \tilde{\beta}^{\pm}\left(b_3, b_4 - \left(\frac{b_2-1}{2}\right)^2, b_5 \ldots b_{2n}\right)\right) + \sum_{j=2}^{k} \tilde{h}_j^{\pm}(X)^2 = g(X) - \tilde{h}_1^{\pm}(X)^2 .$$

Then, since $X = T + (a_1/2n)$, we set

$$\beta^{\pm}(a_1 \ldots a_{2n}) = \tilde{\beta}^{\pm}\left(b_3, b_4 - \left(\frac{b_2-1}{2}\right)^2, b_5 \ldots b_{2n}\right) - \frac{a}{2n}$$

$$h_j^{\pm}(T) = \tilde{h}_j^{\pm}\left(T + \frac{a_1}{2n}\right)$$

and check the desired formulas. □

Remark. The assumption $\mathbb{Q} \subset A$ is not too important. In any partially ordered ring A we have $\mathbb{Z} \subset A$, and one can include $\mathbb{Q}$ by a simple localization. If $ma > 0$ implies $a > 0$, $m \in \mathbb{Z}^+$, $a \in A$, even this is unnecessary. For, one can go through the proof above, "clearing denominators". In any event, one obtains bounds in A for some integral multiple $m\alpha$ of a root α of $f(T)$, where m depends only on the degree of $f(T)$.

6.4. A "Going-Up Theorem" for Semi-Integral Extensions

We apply the result of 6.3 to prove an analogue in (POR) of a well-known theorem in commutative algebra. Specifically, let $A \subset B$, B integral over A. Then each prime ideal of A is the intersection with A of a prime ideal of B.

If we now assume an order $\mathfrak{P}_B \subset B$, extending $\mathfrak{P}_A \subset A$, it is natural to ask if the convex prime ideals of A are intersections of A with convex prime ideals of B. In this generality, the answer is surely no, because one could have a very weak order on A and a strong order on B. Thus it is philosophically sound to make some additional assumption on the orders $\mathfrak{P}_A$ and $\mathfrak{P}_B$.

In categories of partially ordered rings, there is perhaps a more natural notion than integral extension, at least for certain purposes. If $(A,\mathfrak{P}_A) \subset (B,\mathfrak{P}_B)$, we say that B is *semi-integral* over A if for all $b \in B$ there is a monic, even degree polynomial $f(T) = T^{2n} + a_1T^{2n-1} + \cdots + a_{2n}$, $a_i \in A$, with $f(b) \leq 0$ in B. Clearly, integral implies semi-integral.

We first characterize semi-integral elements in several ways. For simplicity, assume $\mathbb{Q} \subset A$.

Proposition 6.4.1. Let $(A,\mathfrak{P}_A) \subset (C,\mathfrak{P}_C)$. Then the following subsets of C coincide.

(i) $\{b|a' < b < a''$ some $a', a'' \in A\}$

(ii) $\{b|-a < b < a$ some $a \in \mathfrak{P}_A\}$

(iii) $\{b|0 \leq b^2 \leq a$ some $a \in \mathfrak{P}_A\}$

(iv) $\{b|f(b^2) \leq 0$ some monic polynomial $f(T) \in A[T]\}$

(v) $\{b|f(b) \leq 0$ some monic, even degree $f(T) \in A[T]\}$

Moreover, if $B \subset C$ is the subset so described, then B is a convex subring of C.

Proof.

(i) ⊂ (ii) We have $0 \leq (a'' - \tfrac{1}{2})^2$, so $a'' < (a'')^2 + 1$. Thus, if $a' < b < a''$, then $-a < b < a$ with $a = 1 + (a'')^2 + (a')^2$.

(ii) ⊂ (iii) If $-a < b < a$, then $a^2 - b^2 = (a-b)(b+a) \geq 0$, hence $b^2 \leq a^2$.

(iii) ⊂ (iv) If $b^2 \leq a$, let $f(T) = T - a$

(iv) ⊂ (v) Obvious.

(v) ⊂ (i) This is exactly Corollary 6.3.2.

Finally, B is a subring because (i) is obviously closed under sums and (iii) is closed under products and negatives. Also, (iii) (or (ii) or (i)) is clearly convex. □

Remark. The extremely simple characterizations (i), (ii), (iii) of semi-integral elements make the original definition (v) seem unnatural. However, geometrically, semi-integral extensions are quite analogous to the integral extensions of algebraic geometry, as we will see in later chapters. Also, in some situations, it is exactly the characterizations (iv) or (v) which are easiest to verify.

We now turn to our going-up theorem. The proposition below is surely capable of some improvement and clarification. We will present a different approach for integral domains in the next chapter.

Proposition 6.4.2. Assume $\mathbb{Q} \subset A$.

(a) Let $(A,\mathfrak{P}_A) \subset (B,\mathfrak{P}_B)$, B semi-integral over A, $\mathfrak{P}_A = A \cap \mathfrak{P}_B$. Then for any convex prime $P \subset A$, there exists a convex prime $Q \subset B$ with $A \cap Q = P$.

(b) Suppose $P_0 \subset P_1 \subset A$ are convex primes, and $Q_0 \subset B$ is a convex prime of B above P_0. Assume further that the orders $(\mathfrak{P}_B/Q_0) \cap (A/P_0)$ and $\mathfrak{P}_A/P_0$ coincide on A/P_0. (Only $\mathfrak{P}_A/P_0 = (\mathfrak{P}_B \cap A/P_0) \subset (\mathfrak{P}_B/Q_0)$ is clear.) Then there exists a prime $Q_1 \subset B$ above P_1 with $Q_0 \subset Q_1$.

(c) If $Q \subset B$ is a maximal convex prime, $P = Q \cap A$, and $(\mathfrak{P}_B/Q) \cap A/P = \mathfrak{P}_A/P$, then $P \subset A$ is a maximal convex prime.

(d) If $A \subset B$ is an *integral* ring extension, $P \subset A$ a maximal convex prime and $Q \subset B$ a convex prime above P, then Q is maximal convex.

Proof.

(a) Let $S = A - P$, a multiplicative set. Then $(A_S, \mathfrak{P}_{A_S}) \subset (B_S, \mathfrak{P}_{B_S})$ is a semi-integral extension and $\mathfrak{P}_{B_S} \cap A_S = \mathfrak{P}_{A_S}$. Thus we may assume $P \subset A$ is a maximal convex ideal.

We now claim $\sqrt{H(PB)} \subset B$ is a *proper* convex ideal. In fact, $\sqrt{H(PB)} \cap A = P$. To see this, let (for purposes of this proof) P^2 denote the set $\{y^2 | y \in P \subset A\}$. Then $\sqrt{H(PB)} = \sqrt{H(P^2)} = \{z | 0 \leq z^{2n} \leq \Sigma p_i y_i^2,\ p_i \in \mathfrak{P}_B,\ y_i \in P\}$. Since $(B,\mathfrak{P}_B)$ is semi-integral over $(A,\mathfrak{P}_A)$, let $0 \leq p_i \leq a_i$, $a_i \in A$. Then $0 \leq z^{2n} \leq \Sigma a_i y_i^2$ in $\mathfrak{P}_A = \mathfrak{P}_B \cap A$. If $z \in A$, then convexity of P implies $z \in P$.

Finally, with our assumption that $P \subset A$ is a maximal convex prime ideal, it is only necessary to choose $Q \subset B$ any prime convex ideal containing $\sqrt{H(PB)}$.

(b) Apply (a) to the semi-integral extension $(A/P_o, \mathfrak{P}_A/P_o) \subset (B/Q_o, \mathfrak{P}_B/Q_o)$.

(c) Apply (b) to deduce a contradiction if $P \underset{\neq}{\subset} P_1 \subset A$, P_1 convex prime.

(d) $A/P \subset B/Q$ is an integral extension and $(A/P, \mathfrak{P}_A/P)$ is a semi-field. If $S = S(1) \subset A/P$ is the shadow of 1, then $(A/P)_S$ is a field and $(B/Q)_S$ is integral over $(A/P)_S$. Thus $(B/Q)_S$ is a field, hence $(B/Q, \mathfrak{P}_B/Q)$ is a semi-field (see Proposition 3.6.1). □

We conclude with a counter-example to a "going-down" property for semi-integral extensions. Let R be a real closed field $A = R[X,Y]$, $B = R[X,Y,Z]$. Let $S = \{(x,y,z) \in R^3 | 0 \leq x,y,z,\ z \leq 1,\ xz \leq y\}$. We partially order B as a ring of functions on S, and contract this order to A. Then $(A,\mathfrak{P}_A) \subset (B,\mathfrak{P}_B)$ is certainly a semi-integral extension since $0 \leq Z \leq 1$ (rel $\mathfrak{P}_B$). Consider

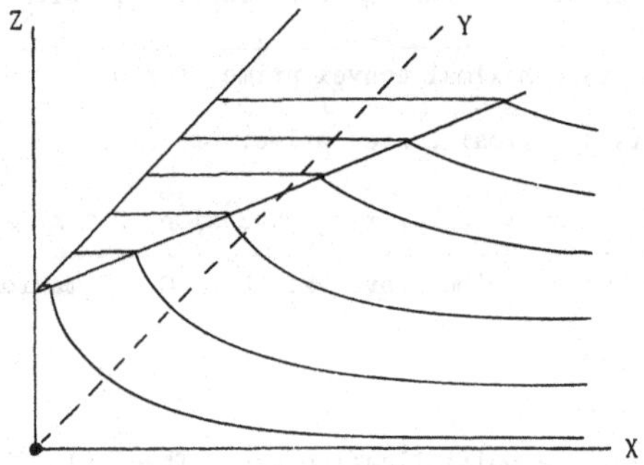

prime ideals $(Y) = P_o \subset P_1 = (X,Y) \subset A$ and $Q_1 = (X,Y) \subset B$. Then $Q_1 \cap A = P_1$, but there is no $\mathfrak{P}_B$ convex prime $Q_o \subset Q_1$ with $Q_o \cap A = P_o$. Namely, if such a Q_o did exist, $P_o = (Y) \subset Q_o \subset Q_1 = (X,Y)$ gives $Q_o = (Y)$, by a trivial argument. But $0 \leq XZ \leq Y$ (rel $\mathfrak{P}_B$) implies (Y) is not $\mathfrak{P}_B$-convex.

The primes $Q_1 \subset B$ and $P_o,P_1 \subset A$ are convex for $\mathfrak{P}_B$, $\mathfrak{P}_A$ respectively. Geometrically, P_o corresponds to the (positive) X-axis, P_1 corresponds to the origin and Q_1 corresponds to the (positive) Z-axis. A potential

$Q_0 \subset Q_1$ lying over P_0 would correspond to a 2-dimensional surface in S containing some of the Z-axis and intersecting the XY-plane in some of the X-axis. The only possible such surface is the XZ-plane which has only a 1-dimensional intersection with S.

VII · Ordered fields

7.1. Basic Results

In the first few sections of this chapter, we recall the basic results of Artin-Schreier on ordered fields. The last few sections develop the notion of *signed place*, which seems very natural in real algebraic geometry. A signed place of a field is a place with values in a totally ordered field and for which we distinguish $+\infty$ and $-\infty$ in an arithmetically coherent fashion. In the next chapter, we apply these ideas to establish the basic results of affine semi-algebraic geometry. Our organization follows expositions of Lang [13], [66], up to a point. In particular, Lang emphasized the usefulness of real places in real algebraic geometry.

To begin, then, recall that a field E is *real* if -1 is not a sum of squares in E. From Chapter I, this is equivalent to the existence of partial orders on E. If $A \subset E$ is any ring which has E as field of fractions, then partial orders on E coincide bijectively with derived partial orders on A $(\mathfrak{P} = \mathfrak{P}_d)$. Any partial order $\mathfrak{P}$ on E admits total order refinements. In fact, if $a, -a \notin \mathfrak{P}$, then both $\mathfrak{P}[a]$ and $\mathfrak{P}[-a]$ are refinements of $\mathfrak{P}$.

A real field R is *real closed* if no proper algebraic extension of R is real. A simple Zorn's lemma argument shows that any real E embeds in a real closed R, algebraic over E. A more precise understanding of this situation stems from the following, which we established in Chapter VI.

Lemma 7.1.1. Let E be a totally ordered field. Let $f(x) \in E[X]$ be an irreducible polynomial, either of odd degree or a quadratic $X^2 + aX + b$ with $a^2 - 4b > 0$. Let α be a root of f. Then $E(\alpha)$ can be partially (hence totally) ordered, extending the order on E. □

Proposition 7.1.2. The following conditions are equivalent on a field R:

(i) R real closed

(ii) R real, every element or its negative is a square in R, and every polynomial of odd degree has a root in R.

(iii) R not algebraically closed, but $R[\sqrt{-1}]$ algebraically closed. □

We omit the proof, which can be found in many elementary texts [64], [65], [66], [67]. The main points are that Lemma 7.1.1 provides the necessity of the conditions in (ii), and Galois theory and the solvability of 2-groups gives the implication (ii) ⇒ (iii).

The following corollaries of 7.1.1 and 7.1.2 are routine, the first two by Zorn's lemma, the third directly from 7.1.2 (ii).

Corollary 7.1.3. Let E be a totally ordered field. Then E can be algebraically extended to a real closed field $\bar{E}$, whose order contracts to the original order on E. That is, each positive element of E is a square in $\bar{E}$. □

Corollary 7.1.4. Let K be an algebraically closed field of characteristic 0. Then there are real closed fields $R \subset K$ with $K = R[\sqrt{-1}]$. □

Corollary 7.1.5. Let E be a real field $E \subset R$, R real closed. Then $\bar{E}$, the algebraic closure of E in R, is a real closure of E. □

Remark 7.1.6. Artin and Schreier showed that if $K = \mathbb{C}$, the complex numbers, then there are uncountably many *pairwise non-isomorphic* real closed subfields R_α of $\mathbb{R}$, with $\mathbb{C}$ isomorphic (algebraically) to $R_\alpha[\sqrt{-1}]$. Thus the R of 7.1.4 is highly non-unique.

Remark 7.1.7. Among other results of Artin-Schreier theory is the fantastic characterization of real closed fields as the *only* fields of finite codimension under their algebraic closure. We will not use this result.

7.2. Function Theoretic Properties of Polynomials

If E is a totally ordered field, any polynomial $f(T) \in E[T]$ can be interpreted as a function $f: E \to E$. We will use the usual symbols (a,b), $[a,b]$, $[a,b)$, $(a,b]$, for open, closed, and half open intervals in E. We also allow $a = -\infty$, $b = +\infty$, with the obvious meaning. Similarly, it is clear what me mean by f (strictly) increasing or (strictly) decreasing on an interval and by f positive or non-negative on an interval. If $a \in E$, we define $|a| \in E$ by $|a| = \pm a$, $|a| \geq 0$. The following is proved in general just as it is for real numbers.

Lemma 7.2.1(a). Let $f(T) = a_0 T^n + a_1 T^{n-1} + \cdots + a_n \in E[T]$, E a totally ordered field. If $x, b \in E$, $|x| \leq |b|$, then

$$|f(x)| \leq \sum_{i=0}^{n} |a_i||b|^{n-i} .$$

(b) If $\alpha \in E$ is a root of f, then

$$|\alpha| \leq \sum_{i=0}^{n} |a_i|/|a_0| . \qquad \square$$

Corollary 7.2.2. Polynomials are "continuous". □

Corollary 7.2.3. If $f(T) \in E[T]$ is a monic polynomial of odd degree, then f is negative on some interval $(-\infty, a)$ and f is positive on some interval (b, ∞). □

If $p \in E$ is positive, the polynomial $T^2 - p$ is certainly negative at $T = 0$ and positive if $T > 1 + p$. From this fact, 7.1.2 and 7.2.3 we obtain the "if" direction in the following characterization of real closed fields.

Proposition 7.2.4. A totally ordered field R is real closed if and only if for every polynomial $f(T) \in R[T]$ and $a,b \in R$ with $f(a) < 0 < f(b)$, there is a root $c \in R$ of f between a and b.

Proof: The algebraic closure of a real closed R is $R[\sqrt{-1}]$. We thus can factor

$$f(T) = \Pi(T-r_j)^{m_j} \Pi((T-\alpha_j)^2 + \beta_j^2)^{n_j}$$

where the $r_j \in R$ are the real roots and the $\alpha_j \pm \beta_j \sqrt{-1}$ are the conjugate pairs of non-real roots. It is then clear that if $f(a) < 0 < f(b)$, a and b must lie on opposite sides of a real root r_j of odd multiplicity m_j. □

Remark: The existence of this simple factorization also easily implies that if $f(T) \in R[T]$ is nowhere negative on R, R real closed, then $f(T) = g_1^2(T) + g_2^2(T)$ for suitable $g_1, g_2 \in R[T]$.

Next we consider the geometric behavior of a polynomial function $f(x)$ near a point $x = a$, over any totally ordered field E. We can write

$$f(x) - f(a) = \sum_{i=1}^{d} \frac{f^{(i)}(a)}{i!} (x-a)^i .$$

Using the estimates of Lemma 7.2.1, one can verify the following.

Proposition 7.2.5. Let $f(T) \in E[T]$, E totally ordered, and let $m \geq 1$ be smallest such that $f^{(m)}(a) \neq 0$, $a \in E$. Then there is $\varepsilon > 0$ in E such that:

(i) If m even, $f^{(m)}(a) > 0$, then $f(x) - f(a)$ is non-negative on $(a - \varepsilon, a + \varepsilon)$, decreasing on $(a - \varepsilon, a)$ and increasing on $(a, a + \varepsilon)$.

(ii) If m odd, $f^{(m)}(a) > 0$, then $f(x) - f(a)$ is increasing on $(a - \varepsilon, a + \varepsilon)$, negative on $(a - \varepsilon, a)$ and positive on $(a, a + \varepsilon)$.

Moreover, on these ε-intervals around a, the derivative f' of f is positive when f is increasing, negative when f is decreasing. □

We obtain statements for the two cases when $f^{(m)}(a) < 0$ by replacing $f(T)$ by $-f(T)$. We draw the four possibilities for local behavior below.

Combining this local behavior of polynomials with the characterization 7.2.4 of real closed fields, we can deduce certain global results about the behavior of polynomials over real closed fields.

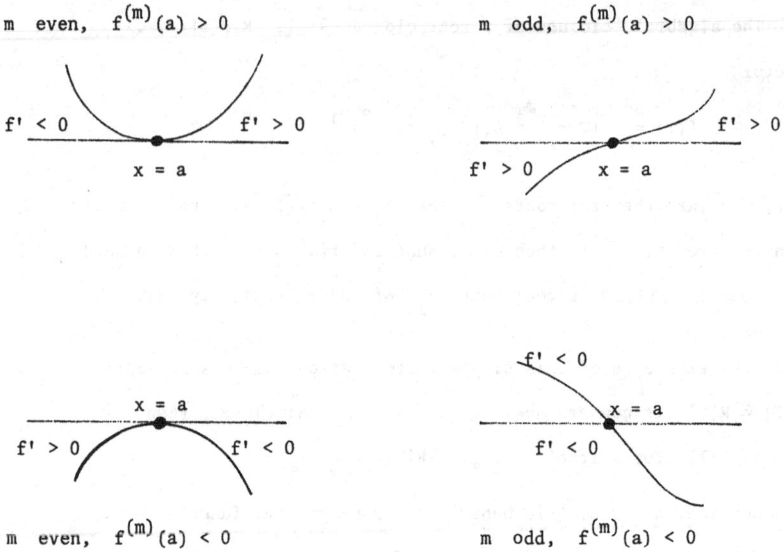

Figure 7.2.6

Proposition 7.2.7. If R is real closed, $f(T) \in R[T]$, $f(a) = f(b) = 0$, $a < b$ in R, then the derivative f' of f has a root c, $a < c < b$. More generally, for any $f(T) \in R[T]$, there exists $c \in (a,b)$ with

$$f'(c) = \frac{f(b) - f(a)}{b - a}.$$

Proof: The second statement (Mean Value Theorem) follows from the first (Rolle's Theorem) in the usual way. To prove Rolle's theorem, one first assumes $f(T)$ has no roots between a and b, by replacing b by another root closer to a if necessary. By 7.2.3, f does not change sign on (a,b). Then a short case-by-case consideration of the local possibilities for f near a and b, drawn in Fig. 7.2.6, implies that for suitable $\varepsilon > 0$, $f'(a+\varepsilon)$ and $f'(b-\varepsilon)$ have opposite sign. Now 7.2.4 completes the proof. □

Corollary 7.2.8. Let R be real closed, $f(T) \in R[T]$. Then f is monotonic on any interval (a,b) not containing roots of f'. On any closed interval $[a,b]$, f assumes maximum and minimum values, either at the endpoints a,b or at roots of f' in (a,b). □

7.3. Sturm's Theorem

Let E be a totally ordered field, $f(T) \in E[T]$ an irreducible polynomial. A priori, the number, or even the existence, of roots of f in a real closure of E might depend on the choice of real closure. Sturm's theorem provides an algorithm *in* E for counting the real roots of $f(T)$ in any real closed field extending E. We first show how this implies the uniqueness of the real closure of E.

Proposition 7.3.1. Let E be a totally ordered field, R_1,R_2 two real closures of E. Then there exists a unique homomorphism $R_1 \to R_2$ over E.

Proof: Uniqueness follows from existence, since, first, symmetry gives $R_2 \to R_1$ over E. But R_1 admits no non-trivial endomorphisms over E, since any $R_1 \to R_1$ maps squares to squares, hence preserves order, but also induces a permutation of the roots in R_1 of any polynomial over E, which are, of course, finite in number.

Sturm's theorem implies that if $\alpha \in R_1$, with minimal polynomial $f(T) \in E[T]$, then f has roots in R_2. Thus we obtain algebraic maps $E(\alpha) \to R_2$ over E. Let $\alpha_0 \cdots \alpha_k \in R_1$ be all the roots of f, and let $\alpha_0 < \alpha_1 < \cdots < \alpha_k$ with $\alpha_{i+1} - \alpha_i = \gamma_{i+1}^2$. The theorem of the primitive element implies $E(\alpha_0 \cdots \alpha_k, \gamma_1 \cdots \gamma_k) = E(\beta)$, and by the above, we can find $E(\beta) \to R_2$. This gives by restriction $E(\alpha_0 \cdots \alpha_k) \to R_2$, keeping the roots of f in order. There is clearly a unique such map over E, by symmetry, since f has no more roots in R_2 than in R_1. On the other hand, if $0 < y \in E(\alpha_0 \cdots \alpha_k)$ and $y = \gamma^2 \in R_1$, then we get an extension $E(\beta,\gamma) \to R_2$, hence $E(\alpha_0 \cdots \alpha_k) \to R_2$ is order preserving. Zorn's lemma now gives $R_1 \to R_2$. □

We turn now to Sturm's algorithm. Beginning with an irreducible $f(T) \in E[T]$, or more generally, any f with $(f,f') = 1$, let $f_0 = f$, $f_1 = f'$ and define f_i, $i \geq 2$ by $f_{i-2} = g_{i-1}f_{i-1} - f_i \in E[T]$, $\deg(f_i) < \deg(f_{i-1})$. Since $(f,f') = 1$, we get a sequence $f_0,f_1,\ldots,f_k$ of polynomials of decreasing degree with f_k a non-zero constant. Let $a,b \in E$ be two points which are not roots of any f_i. For each $x \in [a,b]$, not a root of any f_i, let $\delta(x)$ be the number of sign changes in the sequence $\{f_0(x),f_1(x),\ldots,f_k(x)\}$. From the defining equations $f_{i-1} = g_i \cdot f_i - f_{i+1}$ and the fact that f_k is a non-zero constant, we see that if α is a root of f_i, $i \geq 1$, then $f_{i-1}(\alpha)f_{i+1}(\alpha) < 0$.

In some real closure R of E, let $\alpha_1 < \cdots < \alpha_n$ be all the roots of all the f_i in the interval (a,b). Then $\delta(x)$ is constant on (α_j,α_{j+1}), and on $[a,\alpha_1)$, $(\alpha_n,b]$. We ask the relation between $\delta(x')$ and $\delta(x'')$, if $x' < x''$ and x',x'' are separated by a single α_j. If α_j is not a root of $f = f_0$, the sequences defining $\delta(x')$ and $\delta(x'')$ may differ at $\{\cdots f_{i-1}(x'),f_i(x'),f_{i+1}(x') \cdots\}$ and $\{\cdots f_{i-1}(x''),f_i(x''),f_{i+1}(x'') \cdots\}$. But even if f_i changes sign at its root α_j, so that $f_i(x')$ and $f_i(x'')$ have opposite sign, one checks that the *number* of sign changes, reading across the two sequences, is invariant, $\delta(x') = \delta(x'')$. On the other hand, if α_j is a root of $f = f_0$, then the irreducibility of f implies f changes sign across α_j but f' does not. Comparing $\{f_0(x'),f_1(x'),\cdots\}$ with $\{f_0(x''),f_1(x''),\cdots\}$, we see, in fact, $\delta(x') - \delta(x'') = 1$. (The local picture 7.2.6 shows the two sign sequences are $\{+,-,\cdots\}$, $\{-,-,\cdots\}$ or $\{-,+,\cdots\}$, $\{+,+,\cdots\}$.) We conclude then, with the notation as above:

<u>Proposition 7.3.2</u>. The number of roots of f in (a,b) is $\delta(a) - \delta(b)$, hence is independent of the real closure R of E chosen. □

Since we can first compute all the f_i, and then choose (a,b) containing all roots of all f_i by 7.2.1(b), we can count all the real roots of f by this algorithm.

We state a consequence of 7.3.2.

<u>Proposition 7.3.3</u>. For each $n \geq 1$, there is a finite number of

systems of finitely many inequalities $S_i = \{g_{ij}(t_1 \ldots t_n) > 0\}$, with $g_{ij} \in Q[t_1 \ldots t_n]$, such that given any totally ordered field E and irreducible $f(T) = T^n + a_1 T^{n-1} + \cdots + a_n \in E[T]$ (or, more generally, $f(T)$ with no multiple roots), then f has real roots over E if and only if at least one of the systems S_i holds in E with $t_k = a_k$, that is, for some i, all $g_{ij}(a_1 \ldots a_n) > 0$. □

Remark 7.3.4. The Sturm algorithm itself would give a rather inefficient set of systems S_i. Of course, 7.3.3 could be improved to a statement concerning the precise number of real roots in a given interval (a,b), again using nothing but the Sturm algorithm.

Despite the totally elementary character of the proof of Sturm's theorem, (which is basically no harder than the statement itself), it seems reasonable to believe that added insight into 7.3.3 could be derived from Galois theory. The coefficients a_i of $f(T) \in E[T]$ are the elementary symmetric functions of all the roots in $R[\sqrt{-1}]$, and somehow the *signs* in E of various rational combinations of the a_i should determine not only the real roots and their location, but also the location of the real and imaginary parts of the complex roots and other information about real subfields of the splitting field of f.

The determination of whether a quadratic has real roots is trivial: $T^2 + aT + b$ has real roots if $d = a^2 - 4b > 0$. For the cubic $T^3 + pT + q$, with $d = -(4p^3 + 27q^2)$, we have $d < 0$ or $d > 0$ accordingly as the cubic has one or three real roots. It turns out that a quartic $T^4 + pT^2 + qT + r$ has two real roots if the discriminat $d < 0$ and four real roots if $d > 0$, $p < 0$ and $8pr - 2p^3 - 9q^2 > 0$.

7.4. Dedekind Cuts; Archimedean and Non-Archimedean Extensions

Let E be a totally ordered field. By a *cut* we mean a subset $D \subset E$ such that if $b \in D$, $a < b$, then $a \in D$. If $D = \emptyset$, we write $D = -\infty$, and if $D = E$, we write $D = +\infty$. All other cuts are called *finite*. A cut D is *rational* if D has a maximal element or $E - D$ has a minimal element

$b \in E$. We write $D \approx b$ if D is rational at b. Note there are exactly two cuts rational at b, one contains b the other does not. We write $D \approx D'$ if the two cuts D, D' differ by at most a single element of E.

Let $E \subset F$ be totally ordered fields, the order on F extending E. Each $\alpha \in F$ defines a cut $D_E(\alpha)$ of E by

$$D_E(\alpha) = \{a \in E | a < \alpha \text{ in } F\}.$$

We call a cut D of E *algebraic* if $D = \pm\infty$ or if $D \approx D_E(\alpha)$ for some $\alpha \in \bar{E}$, the real closure of E. All other cuts are called *transcendental*.

Proposition 7.4.1. The following are equivalent for an extension of ordered fields $E \subset F$:

(i) If $\alpha \in F$ and $D_E(\alpha)$ is rational, then $\alpha \in E$.

(ii) $D_E(\alpha)$ is finite for all $\alpha \in F$.

(iii) For all $\varepsilon > 0$ in F there exists $e > 0$ in E with $\varepsilon > e$.

Proof: (i) ⇒ (ii) Suppose some $D_E(\beta) = \pm\infty$. Then $\beta \notin E$, but $D_E(1/\beta) \approx 0$.

(ii) ⇒ (iii) Suppose $\varepsilon > 0$ in F was smaller than all positive elements of E. Then $D_E(1/\varepsilon) = +\infty$.

(iii) ⇒ (i) Suppose $D_E(\alpha) \approx a \in E$, $\alpha \notin E$. If $\alpha < a$, then $1/a - \alpha$ is positive, but is smaller than all positives of E. If $a < \alpha$, the same conclusion holds for $1/\alpha - a$. □

Extensions satisfying the conditions of 7.4.1 will be called *Archimedean*.

Lemma 7.4.2.

(a) If $E \subset F$ and F is algebraic over E, then F is Archimedean over E.

(b) If $E \subset F$ is Archimedean and $F \subset K$ is Archimedean, then $E \subset K$ is Archimedean.

Proof: The first results follows from the bounds on algebraic quantities given in 7.2.1. The second result is clear from any of the conditions in 7.4.1. □

Remark 7.4.3. $\mathbb{Q} \subset \mathbb{R}$, the real numbers over the rationals, contains all Archimedean extensions of $\mathbb{Q}$ and $\mathbb{R}$ admits no proper Archimedean extension. $\mathbb{R}$ is *unique* with this latter property. If E is a totally ordered field properly containing $\mathbb{R}$, then one cannot make sense out of arithmetic of Dedekind cuts of E. For example, the smallest cut D of E containing $\mathbb{Q}$ cannot be distinguished from $2D$, whatever $2D$ means. Also, Cauchy sequences in E are not generally interesting, for example, if E has uncountable transcendence degree over $\mathbb{R}$.

Remark 7.4.4. Although the real closure of E, say $\overline{E}$, is Archimedean over E, it *does not* follow that E is order dense in $\overline{E}$. This fact makes certain topological reasoning precarious. For example, given any ordering of $\mathbb{R}(t)$, there are irreducible polynomials in $\mathbb{R}(t)[X]$ which are strictly positive as functions on $\mathbb{R}(t)$, yet which have roots and change sign over the real closure of $\mathbb{R}(t)$. (See 8.2 for specific examples.)

We now point out how an extension $E \subset F$ of ordered fields defines a valuation ring in F. This construction and its uses goes back to Krull [12]. We define

$$A_E = \{\alpha \in F \mid D_E(\alpha) \text{ is finite}\}.$$

If $\beta \notin A_E$, then $D_E(\beta) = \pm\infty$, hence $D_E(1/\beta) \approx 0$, so A_E is a valuation ring.

Proposition 7.4.5. A_E is a totally ordered valuation ring in F. The maximal ideal is $Q_E = \{\alpha \in F \mid D_E(\alpha) \approx 0\}$, and Q_E is a convex ideal. The residue field $\Delta_E = A_E/Q_E$ is totally ordered, contains E and is Archimedean over E. □

This result is checked routinely, so we omit the details.

Note that given $E \subset F$, we have $A_E = F$ if and only if F is Archimedean over E. If $E \subset K \subset F$ and K is Archimedean over E, then $A_E = A_K$. Zorn's Lemma implies that any $K \subset F$, Archimedean over E, is contained in a subfield maximal with this property. We say E is *Archimedean closed* in F if E is maximal Archimedean over itself in F. In particular, E Archimedean closed in F implies E algebraically closed in F.

Proposition 7.4.6. If E is Archimedean closed in F, then the residue field $A_E/Q_E = \Delta_E$ is algebraic over E. Conversely, if E is algebraically closed in F and Δ_E is algebraic over E, then E is Archimedean closed in F.

Proof: Suppose $t \in A_E$ and the image $\bar{t}$ in Δ_E is transcendental over E. Then $E \subset E[t]$ and $Q_E \cap E[t] = (0)$, hence $E(t) \subset A_E$. Thus $E(t)$ is Archimedean over E, contradiction.

Conversely, if $t \in F - E$ and $E(t)$ is Archimedean over E, then $Q_E \cap E(t) = (0)$, hence Δ_E contains the transcendental extension $E(\bar{t})$. □

7.5. Orders on Simple Field Extensions

We will classify all total orders on a simple field extension $E(x)$, extending a fixed total order on E. Let R be the real closure of E.

If x is algebraic over E, with minimal polynomial $f(t) \in E[t]$, then E and $E(x)$ have the same real closure R. The orderings on $E(x)$ correspond to embeddings of $E(x)$ in R, that is, to the real roots of $f(t)$. This is not completely obvious, because $f(t)$ may have two roots (or more) in R not separated by any element of E (see 7.4.4). Thus, there may exist several embeddings of $E(x)$ in R, which define the same cut of E, $D_E(x)$.

However, let x_1, x_2 be two roots of $f(t)$ in R. Of course, if the two fields $E(x_1)$ and $E(x_2)$ *coincide* in R, we don't distinguish the orders on $E(x)$ induced by the two maps taking x to x_1 and x_2, respectively. However, if $E(x_1)$ and $E(x_2)$ are distinct, any order isomorphism between them over E would extend to a non-identical endomorphism of R over E, by the proof of uniqueness of real closure. But this is impossible, by the same proof.

We now turn to orders on the transcendental extension $E(t)$. Any ordering on $E(t)$ has a real closure, which contains $R(t)$, in a natural way. Thus orderings on $E(t)$ are seen to correspond bijectively to orderings on $R(t)$. Recall total orderings on $R(t)$ correspond to orderings of $R[t]$.

Proposition 7.5.1. A total order $\mathfrak{P}$ on $R[t]$, extending R, R real closed, is uniquely determined by the cut $D = D_R(t)$ of R. The possibilities are:

Case (i) D infinite. Then:

If $D = \infty$, $\mathfrak{P} = \mathfrak{P}_\infty = \{f \in R[t] \mid f((b,\infty)) > 0,$ some $b \in R\}$

If $D = -\infty$, $\mathfrak{P} = \mathfrak{P}_{-\infty} = \{f \in R[t] \mid f((-\infty,b)) > 0,$ some $b \in R\}$.

Case (ii) D rational, say $D \approx a \in R$. If $a \in D$ (that is, $a < t$ in $R(t)$), then

$\mathfrak{P} = \mathfrak{P}_{a,+} = \{f \in R[t] \mid f((a,a+\varepsilon)) > 0,$ some $\varepsilon > 0$ in $R\}$.

If $a \notin D$ (that is, $t < a$ in $R(t)$), then

$\mathfrak{P} = \mathfrak{P}_{a,-} = \{f \in R[t] \mid f((a-\varepsilon, a)) > 0,$ some $\varepsilon > 0$ in $R\}$.

Case (iii) D transcendental. Then

$\mathfrak{P} = \mathfrak{P}_D = \{f \in R[t] \mid$ there exists $a,b \in R$, $a < D < b$, such that $f((a,b)) > 0\}$.

Moreover, $R(t)$ is Archimedean over R if and only if D is transcendental

Proof: First one checks the orders $\mathfrak{P}_\infty, \mathfrak{P}_{-\infty}, \mathfrak{P}_{a,+}, \mathfrak{P}_{a,-}, \mathfrak{P}_D$ are, in fact, total orders on $R(t)$.

Secondly, the case D infinite is dealt with by replacing t by $1/t$, with $D(1/t) \approx 0$, hence rational.

Thirdly, suppose $D \approx a$, as in case (ii), with $a < t$. Given $f(t) \in R[t]$, which we may assume monic, factor $f(t)$ over R. Thus

$$f(t) = \prod_{i,j} (t - r_i)((t-\alpha_j)^2 + \beta_j^2)$$

where r_i are the real roots and $\alpha_j \pm \sqrt{-1}\,\beta_j$ are the non-real roots. (Recall $R(\sqrt{-1})$ is the algebraic closure of R.) Obviously, $(t - r_i) \in \mathfrak{P}$ if and only if $r_i \leq a$ in R, if and only if $(t - r_i) \in \mathfrak{P}_{a,+}$. Thus $\mathfrak{P} = \mathfrak{P}_{a,+}$

as asserted. If $t < a$ in $\mathfrak{P}$, then $(t - r_i) \in \mathfrak{P}$ if and only if $r_i < a$, if and only if $(t - r_i) \in \mathfrak{P}_{a,-}$.

We sketch a second proof that $\mathfrak{P} = \mathfrak{P}_{a,+}$ if $a < t$, $D = D_R(t) \approx a$. Begin by writing $f(t) = f(a) + f^{(1)}(a)(t-a) + \cdots + (f^{(n)}(a)/n!)(t-a)^n = (f^{(m)}(a)/m!)(t-a)^m(1 + \text{terms in } (t-a))$, for some $m \geq 0$, with $f^{(m)}(a) \neq 0$. In the abstract order $\mathfrak{P}$ on $R[t]$, $t - a$ is positive, but infinitesimally small relative to R. Thus the $\mathfrak{P}$-sign of $f(t)$ is simply the sign in R of $f^{(m)}(a)$. By the discussion of local behavior of functions of 7.2, we deduce $\mathfrak{P} = \mathfrak{P}_{a,+}$, as asserted. The case $t < a$ is analogous.

We point out that the replacement of t by $1/t$, converting behavior of functions "at $\pm \infty$" into behavior "to the right or left of 0" is easily formulated in terms of coefficients. Namely, if $D(t) = +\infty$, and $f(t) = a_0t^n + \cdots + a_n$, then $f \in \mathfrak{P}_\infty$ if and only if $a_0 > 0$ in R. If $D(t) = -\infty$, then $f \in \mathfrak{P}_{-\infty}$ if n is even, $a_0 > 0$ or n odd, $a_0 < 0$. One simply divides through by t^n.

Next, suppose $D = D_R(t)$ is transcendental. For each $f \in R[t]$, choose $a < D < b$ such that the function f has constant sign on (a,b). (Simply choose a and b closer to D than all the roots of f.) If $f(t)$ had the opposite sign as its values on (a,b) *in the abstract order* $\mathfrak{P}$ on $R(t)$, then the polynomial f would have a real root between a and b in some real field containing R. By Sturm's algorithm, f would therefore have a root in R between a and b, contradicting our choices. Thus $\mathfrak{P} = \mathfrak{P}_D$.

A simpler proof that $\mathfrak{P} = \mathfrak{P}_D$ in the case D transcendental can be based on the factorization $f(t) = \prod_{i,j} (t-r_i)((t-\alpha_j)^2 + \beta_j^2)$. Obviously, $(t-r_i) \in \mathfrak{P}$ if and only if $r_i \in D$, if and only if $(t-r_i) \in \mathfrak{P}_D$.

Finally, let $A_R \subset R(t)$ be the valuation ring associated to $R \subset R(t)$. If $D = D_R(t)$ is infinite or rational, then $A_R \neq R(t)$ and $R(t)$ is non-Archimedean over R. Conversely, assume $A_R \neq R(t)$. Either t or $1/t \in A_R$, so we may assume $t \in A_R$. If $Q_R \subset A_R$ is the maximal ideal, then $Q_R \cap R[t]$ is a non-trivial, *convex* prime ideal of $R[t]$. Necessarily, then, $Q_R \cap R[t] = ((t-a))$, some $a \in R$, hence $D = D_R(t) \approx a$. □

Corollary 7.5.2. Let $R(t)$ be partially ordered, R real closed. Suppose $g_1,\dots,g_k \in R[t]$ are finitely many elements positive in the order. Then there is an interval of values (a,b) in R with $g_i((a,b)) > 0$, $1 \leq i \leq k$.

Proof: Extend the partial order to a total order and consider the classification of total orders directly. □

Corollary 7.5.3. Let E be totally ordered with real closure R. Suppose E is order dense in R and $f(t)$ is non-negative as a function on E, $f(t) \in E[t]$. Then $f(t)$ is a sum of squares in $E(t)$.

Proof: If there is no equation $h^2 f = \Sigma g_i^2$ in $E[t]$, then $E[t]$ can be ordered with $f(t) < 0$. By 7.5.2, $f(t)$ is strictly negative on some interval $(a,b) \subset R$, hence f is negative at some point of E, which contradicts our assumption. □

Remark. It is known that if $f(t)$ is a sum of squares in $E(t)$, then $f(t)$ is actually a sum of squares in $E[t]$. See [4], [5], [6] for detailed discussions of this point.

Corollary 7.5.4.

(a) Let E be totally ordered with real closure R, and suppose $E(t)$ is ordered, non-Archimedean over E, with valuation ring $A_E \subset E(t)$. The residue field $\Delta = A_E/Q_E$ is then a simple algebraic extension of E, $\Delta = E(\alpha)$.

(b) If $t \in A_E$, let $(f(t)) = Q_E \cap E[t]$, $f(t)$ irreducible, $f(t) > 0$ in $E(t)$. Then $\Delta = E[t]/(f(t))$, and the ordering on Δ induced by that of $E[t]$ corresponds to a choice of a root of $f(t)$, $\alpha \in R$.

(c) If $g(t) \in E[t]$ and $g(t) = f(t)^m(g_0(t) + g_1(t)f(t) + \cdots + g_k(t)f^k(t))$, with degree $g_i(t) <$ degree $f(t)$, $g_0(t) \neq 0$, then $g(t) > 0$ in $E(t)$ if and only if $g(\alpha) > 0$ in R.

Proof: Assertion (b) implies (a) since either t or $1/t \in A_E$. Since $E[t]$ is a principal ideal domain, the localized ring $E[t]_{(f(t))}$ *is* the valuation ring A_E, and $Q_E = (f(t))E[t]_{(f(t))}$. This proves (b). Finally, (c) is more or less obvious. □

7.6. Total Orders and Signed Places

Just as valuation rings in fields correspond to places, we will show that totally ordered valuation rings with convex maximal ideal correspond to a special type of place, in which we distinguish $+\infty$ and $-\infty$. A signed place on a field F immediately yields a total order of F, in fact, a totally ordered valuation ring in F, with convex maximal ideal, and conversely, such a valuation ring yields an equivalence class of signed places.

Let Δ be a totally ordered field. We adjoin symbols ∞ and $-\infty$ to Δ and extend the operations of addition, subtraction, multiplication, and division between ∞, $-\infty$ and elements $a \in \Delta$ as follows. (These rules then justify the notation $-\infty$.)

Addition

$$\infty + \infty = \infty$$
$$(-\infty) + (-\infty) = -\infty$$
$$a + \infty = \infty$$
$$a + (-\infty) = -\infty$$

Subtraction

$$\infty - (-\infty) = \infty$$
$$(-\infty) - \infty = -\infty$$
$$a - \infty = -\infty$$
$$a - (-\infty) = \infty$$

Multiplication

$$\infty \cdot \infty = (-\infty) \cdot (-\infty) = \infty$$
$$\infty \cdot (-\infty) = -\infty$$
$$a \cdot \infty = \begin{cases} \infty & \text{if } a > 0 \\ -\infty & \text{if } a < 0 \end{cases}$$
$$a \cdot (-\infty) = \begin{cases} -\infty & \text{if } a > 0 \\ \infty & \text{if } a < 0 \end{cases}$$

Division $\qquad a/\infty = a/-\infty = 0$

$$\infty/a = (1/a)\cdot\infty \qquad \text{if } a \neq 0$$

$$-\infty/a = (1/a)\cdot(-\infty) \qquad \text{if } a \neq 0$$

The symbols $0\cdot(\pm\infty)$, $\pm\infty/0$, $\infty+(-\infty)$, $\infty-\infty$, $(-\infty)-(-\infty)$ and all of the four possibilities $\pm\infty/\pm\infty$ are undefined. Of course, we want the usual commutative and associative rules to hold, in all defined expressions. Note that we do not distinguish the signs of a/∞ and $a/-\infty$, $a \neq 0$, but call both 0. These elements can be distinguished by "inverting".

If K is a field by a *signed place*, with the values in Δ, we mean a function $p: K \to \Delta, \pm\infty$, such that $p(x+y) = p(x)+p(y)$, $p(xy) = p(x)p(y)$, and $p(1) = 1$, whenever the terms are all defined. The last condition simply guarantees $p \not\equiv 0$ and $p \not\equiv \infty$.

Proposition 7.6.1. Let $p: K \to \Delta, \pm\infty$ be a signed place on K. Then the ring $A_p = \{x \in K | p(x) \in \Delta\}$ is a valuation ring. Moreover, K is totally ordered if the strictly positive elements are defined to be $\mathfrak{P}^+ = \{x \in K | p(x) = \infty$, $p(1/x) = \infty$, or $p(x) > 0$ in $\Delta\}$. If the order on K is restricted to A_p, the maximal ideal $Q_p \subset A_p$ is convex.

Proof: First, if we simply suppress the distinction between $+\infty$ and $-\infty$, $p: K \to \Delta, \infty$ becomes a place on K. Thus A_p is a valuation ring.

Secondly it is clear that for all non-zero $x \in K$, $x^2 \in \mathfrak{P}^+$, and exactly one of $x, -x \in \mathfrak{P}^+$. It remains to show that $\mathfrak{P}^+$ is closed under sums and products xy and $x+y$, for $x,y \in \mathfrak{P}^+$.

The possibilities for $p(x)$, $p(y)$ are covered by (i) $p(x), p(y) > 0$, (ii) $p(x) = \infty$, $p(y) > 0$, (iii) $p(\frac{1}{x}) = \infty$, $p(y) > 0$, (iv) $p(x) = \infty$, $p(y) = \infty$, (v) $p(\frac{1}{x}) = \infty$, $p(\frac{1}{y}) = \infty$, (vi) $p(x) = \infty$, $p(\frac{1}{y}) = \infty$. In all cases except (vi) it is trivial to check directly that $xy \in \mathfrak{P}^+$. In case (vi), one cannot evaluate either $p(xy)$ or $p(1/xy)$ directly. By symmetry, we may assume $p(xy) \neq 0$. If $p(xy) < 0$ or $p(xy) = -\infty$, then $p(x) = p(xy\frac{1}{y}) = p(xy)p(\frac{1}{y}) = -\infty$. This contradiction shows $p(xy) > 0$ or $p(xy) = \infty$. The six cases for sums are all trivial except (v). But $x+y = xy(\frac{1}{x} + \frac{1}{y})$, so this case is covered by product cases, and case (iv) for sums.

Finally, $Q_p \subset A_p$ is convex as the kernel of an order preserving homomorphism $A_p \to \Delta$. □

A signed place p: $K \to \Delta, \pm\infty$ determines a total order on K. Conversely, given a valuation ring A in a totally ordered field K, with convex maximal ideal $Q \subset A$, one recovers a signed place p: $K \to \Delta, \pm\infty$ where $\Delta = A/Q$. Specifically, if $1/x \in Q$, then we set $p(x) = \infty$ if x is positive in K and $p(x) = -\infty$ if x is negative. Verification of $p(x+y) = p(x) + p(y)$ and $p(xy) = p(x)p(y)$ consists of a routine case-by-case discussion. Note that since A is a valuation ring, we may always write $x+y = x(1+(y/x))$ with $y/x \in A$. This shows that if $p(x) = \infty$, $p(y) = \infty$, then $0 \leq 1 \leq 1+(y/x)$ in A, hence $1+(y/x) \notin Q$ and $1/(x+y) = (1/x)(1+(y/x))^{-1} \in Q$. Thus $p(x+y) = \infty$. This is perhaps the trickiest verification.

We say signed places p_i: $K \to \Delta_i, \pm\infty$, $i = 1,2$, are *equivalent* if there is an order preserving isomorphism σ: $p_1(K), \pm\infty \xrightarrow{\sim} p_2(K), \pm\infty$ such that $p_2 = \sigma p_1$. Here $p_i(K)$ means the subfield $p_i(A_{p_i}) \subseteq \Delta_i$, where $A_{p_i} \subset K$ is the valuation ring of p_i. We then see that there is a bijective correspondence between equivalence classes of signed places of K and totally ordered valuation rings in K with convex maximal ideal. A signed place p of K is *trivial* if its valuation ring A_p is all of K.

Remark 7.6.2. In commutative algebra, equivalence classes of places on a field K and valuation rings in K correspond to a third concept, that of a (Krull) *valuation* of K. A valuation is a function v: $K^* \to \Gamma$, where Γ is a totally ordered (additive) abelian group and $K^* = K - (0)$, with $v(xy) = v(x)+v(y)$ and $v(x+y) \geq \min(v(x),v(y))$. If $A \subset K$ is a valuation ring with $Q \subset A$ the maximal ideal, one sets $\Gamma = K^*/A^*$, $A^* = A - Q$. The multiplication in K^* becomes the "addition" in Γ and $1 \in K^*$ becomes "0" $\in \Gamma$. We order Γ by $x \leq y$ if $y/x \in A$. The natural projection v: $K^* \to \Gamma$ is then a valuation of K, $A - (0) = \{x \in K | v(x) \geq 0\}$ and $Q - (0) = \{x \in K | v(x) > 0\}$. For any valuation, $v(x) = v(-x)$.

Now, if K is totally ordered and $A \subset K$ is a valuation ring with *convex* maximal ideal, then the sign homomorphism $\sigma\colon K^* \to \{\pm 1\}$, defined by $\sigma(x) = +1$ if x is positive and $\sigma(x) = -1$ if x is negative, is related to $v\colon K^* \to \Gamma$ in the following manner. If $v(x_i) \geq 0$, $\sigma(x_i) = +1$, and $v(\Sigma x_i) > 0$, then $v(x_i) > 0$. Translating, this condition simply says if $x_i \in A$, x_i positive and $\Sigma x_i \in Q$, then $x_i \in Q$. In fact, we can state if $\sigma(x_i) = +1$ and $v(\Sigma x_i) > 0$, then $v(x_i) > 0$ for any $x_i \in K$. Namely, if the x_i of minimum value, say x_0, did not belong to Q, then $v(1/x_0) \geq 0$, hence $v(\Sigma(x_i/x_0)) > 0$. But then $1 \leq \Sigma(x_i/x_0)$ in A, which contradicts convexity of Q.

Conversely, given a field K, and a sign homomorphism $\sigma\colon K^* \to \{\pm 1\}$ with $\sigma(-1) = -1$, $\sigma(xy) = \sigma(x)\sigma(y)$ and $\sigma(x+y) = 1$ if $\sigma(x) = \sigma(y) = 1$, one recovers a total order on K by $\mathfrak{P}^+ = \{x \in K^* | \sigma(x) = +1\}$. If $v\colon K^* \to \Gamma$ is a valuation such that $\sigma(x_i) = +1$ and $v(\Sigma x_i) > 0$ implies $v(x_i) > 0$, then the valuation ring A of v has convex maximal ideal Q. We call such a pair (v,σ) a *split valuation* on K, since σ serves to distinguish the elements $+x$, $-x$ not distinguished by v.

Although the residue fields $A/Q = \Delta$ and the domain K of such valuations are certainly restricted (they are totally ordered), no general restrictions are implied on the value group Γ of a split valuation.

We have seen in 7.4 that if K is totally ordered and $E \subset K$ is a subfield, then E defines an (ordered) valuation ring $A_E \subset K$, hence a signed place $p_E\colon K \to \Delta, \pm\infty$ where $\Delta = A_E/Q_E$. Moreover, $p_E = p_{E'}$ if $E \subset E' \subset K$ and E' is Archimedean over E. We show now that any signed place p on K is equivalent to p_E for some subfield $E \subset K$.

Let, then, $p\colon K \to \Delta, \pm\infty$ be a signed place. We may assume p is surjective, that is, $p(A_p) = \Delta$. Consider the family of subfields $E \subset K$ such that the restriction $p|_E$ is trivial. For example, $\mathbb{Q} \subset K$ is such a subfield. Zorn's lemma implies that there exist maximal such subfields $E \subset K$, and we fix such an E.

<u>Proposition 7.6.3.</u>

(a) Δ is algebraic over its subfield $p(E) \subset \Delta$.

(b) The place p is equivalent to p_E.

(c) E is Archimedean closed in K.

Proof:

(a) Let $t \in A_p - E$, so $p(t) \in \Delta$. Then $p: E[t] \to \Delta$ is a non-trivial homomorphism by the choice of E, hence $p(t)$ is algebraic over $p(E)$.

(b) We prove $A_p = A_E$. The signed place $p: K \to \Delta, \pm\infty$ orders K and the induced order on $E \subset K$ of course coincides with the induced order on $p(E) \subset \Delta$. If $x \in A_p$, then $p(x) \in \Delta$ is algebraic over $p(E)$ by (a), hence $p(-e) < p(x) < p(e)$ for some $e \in E$. Thus, $x \in A_E$. Conversely, if $x \in A_E$, $-e < x < e$ for some $e \in E$, hence $p(-e) < p(x) < p(e)$ in Δ and $x \in A_p$.

(c) This is a consequence of (b) since if $E' \subset K$ were Archimedean over E, then $p_E = p$ would be trivial on E', hence $E = E'$, by the maximality of E. □

Proposition 7.6.4. Suppose Δ is totally ordered, $p: K \to \Delta, \pm\infty$ is a signed place on K and $q: L \to K, \pm\infty$ is a signed place on L, relative to the order on K induced by p. Then $p \circ q: L \to \Delta, \pm\infty$ is a signed place on L, which induces the same order on L as $q: L \to K, \pm\infty$. □

The proof is routine. On the other hand, suppose $q': L \to K, \pm\infty$ is a signed place on K relative to some *new* order on K. If q' is non-trivial and surjective, then $r = p \circ q': L \to \Delta, \pm\infty$ will *never* be a signed place on L. (If $q'(a) \in A_p \subset K$ "changes sign" and $q'(b) = \infty$, then $r(ab) \neq r(a)r(b)$.) Nonetheless, if we suppress the distinction between $+\infty$ and $-\infty$, $p \circ q': L \to \Delta, \infty$ is a place on L and we will see in the next section that L can be ordered so as to convert any real valued place into a signed place.

7.7. Existence of Signed Places

We have seen in previous sections how signed places of a field K correspond closely with total orders on K together with a subfield $E \subset K$. In this section we show how, beginning with less data, signed places can be constructed. The first main result, which is due to Krull [12] (and in a different form to R. Baer, according to Krull) is that if $p\colon K \to \Delta, \infty$ is an ordinary place of K and Δ is a real field, then given any total ordering of Δ, p can be refined to a signed place, $p'\colon K \to \Delta, \pm\infty$. The second main result is that if $(K,\mathfrak{P})$ is a *partially* ordered field, $A \subset K$ a subring, $P \subset A$ a $\mathfrak{P} \cap A$-convex prime ideal, then there exist signed places $p\colon K \to \Delta, \pm\infty$ with center P on A (that is, $A \subset A_p$, $Q_p \cap A = P$ where (A_p, Q_p) is the valuation ring and maximal ideal of p). Moreover, the induced order on K can be arranged to refine $\mathfrak{P}$ and the residue field $\Delta = A_p/Q_p$ will be Archimedean over the fraction field of A/P.

Lemma 7.7.1. Let $p\colon K \to \Delta, \infty$ be a place of a field K with valuation ring (A,Q). Suppose Δ is a real field. Then K is real and all prime ideals of A are $\mathfrak{P}_w \cap A$-convex, where $\mathfrak{P}_w$ is the weak order on K (so $\mathfrak{P}_w \cap A = (\mathfrak{P}_w)_d$, the *derived* order of the weak order on A).

Proof: Suppose $\Sigma x_i^2 = 0$, $x_i \in K$. Let x_o have minimum value for p, that is, $x_i/x_o \in A$. Then $\Sigma(x_i/x_o)^2 = 1 + \sum_{i\neq 0} (x_i/x_o)^2 = 0$ in A, hence $0 = 1 + \sum_{i\neq 0} p(x_i/x_o)^2$ in Δ. Since Δ is real, we must have had all $x_i = 0$ originally. Thus K is real.

Next suppose $\Sigma x_i^2 \in Q$, $x_i \in K$. If we show all $x_i \in Q$, it follows that Q is $\mathfrak{P}_w \cap A$-convex. If some $x_i \notin Q$, then the term of minimum value, say x_o, is not in Q. Thus $1/x_o \in A$, hence $(1/x_o)^2(\Sigma x_i^2) = \Sigma(x_i/x_o)^2 = 1 + \sum_{i\neq 0} (x_i/x_o)^2 \in Q$. But, again, $1 + \Sigma p(x_i/x_o)^2 = 0 \in \Delta$ is a contradiction.

Finally, let $P \subset Q$ be a prime ideal of A. A standard result of valuation theory is that the localized ring $A_{(P)} \subset K$ is a valuation ring with maximal ideal P, that is, $PA_{(P)} = P$. Let $\Delta_P = A_{(P)}/P$. The pair (A/P, Q/P) is a valuation ring and maximal ideal in Δ_P, with residue field Δ. Thus, by the first part of the proof, Δ_P is a real field. Since Δ_P is the residue field of the place of K with valuation ring $(A_{(P)}, P)$, we deduce from the second part of the proof that P is $\mathfrak{P}_w \cap A_{(P)}$-convex in $A_{(P)}$, hence is also $\mathfrak{P}_w \cap A$-convex in A. (In other terms, we are using the specialization chain $K \to \Delta_P, \infty \to \Delta, \infty$ corresponding to the pair of valuation rings $A \subset A_{(P)}$ in K.

But see the Remark following the proof of the next proposition.) □

Proposition 7.7.2. (Krull) Let $p\colon K \to \Delta, \infty$ be a place of a field K, with valuation ring (A,Q). Let $\mathfrak{P} \subset K$ be a partial order on K, with Q a $\mathfrak{P} \cap A$-convex ideal, and $\overline{\mathfrak{P}} = \mathfrak{P}/Q \subset \Delta = A/Q$ the induced order. Let $\overline{\mathfrak{P}}' \supset \overline{\mathfrak{P}}$ be any total order refinement of $\overline{\mathfrak{P}}$ on Δ. Then there exist total order refinements $\mathfrak{P}' \supset \mathfrak{P}$ on K such that Q is $\mathfrak{P}' \cap A$-convex and $\mathfrak{P}'/Q = \overline{\mathfrak{P}}'$. Moreover, all prime ideals $P \subset A$ are $\mathfrak{P}' \cap A$-convex.

Proof: Let Q^{-1} denote the subset of elements $1/x \in K$ with $x \in Q$. We will define a function $\sigma\colon K^* \to \{\pm 1\}$, then refine $p\colon K \to \Delta, \infty$ to a signed place $p'\colon K \to \Delta, \pm\infty$, relative to $\overline{\mathfrak{P}}' \subset \Delta$, by $p'(x) = p(x)$, $x \in A$ and $p'(y) = \sigma(y) \cdot \infty$, $y \in Q^{-1}$. The order $\mathfrak{P}' \subset K$ will be the order induced by p'.

Let A^* denote $A - Q$, the units in A. We say $x,y \in K^*$ are *associates* if $x = yz$, with $z \in A^*$. A subset $S \subset K^*$ is $\mathfrak{P}$-*independent* if no finite product $x_1 \cdot \ldots \cdot x_k$, $x_i \in S$, $x_i \neq x_j$, is an associate of an element of $\mathfrak{P}$. By Zorn's lemma, there exist maximal $\mathfrak{P}$-independent subsets $S \subset K^*$ although possibly $S = \emptyset$, and certainly $S \cap A^* = \emptyset$.

In any event, let $\sigma\colon S \to \{\pm 1\}$ be an arbitrary function. If $y \in K^* - S$, then there is a relation $yx_1 \cdots x_k = pz$, with $x_i \in S$, $p \in \mathfrak{P}$, $z \in A^*$. (Possibly no x_i occur.) Define $\sigma(z) \in \{\pm 1\}$ to be the sign of z in $(\Delta, \overline{\mathfrak{P}}')$ and define $\sigma(y) \in \{\pm 1\}$ by $\sigma(y)\sigma(x_1)\cdots\sigma(x_k) = \sigma(z)$. The point is, these choices are *forced* on us if $\mathfrak{P}'$ is to refine $\mathfrak{P}$ and if we are to have $\mathfrak{P}'/Q = \overline{\mathfrak{P}}'$ as orders on Δ.

We first check that $\sigma\colon K^* \to \{\pm 1\}$ is well defined. If also $yx_1' \cdots x_n' = p'z'$ with $x_j' \in S$, $p' \in \mathfrak{P}$, $z' \in A^*$, then $x_1 \cdots x_k x_1' \cdots x_n'$ is an associate of pp'/y^2 $\mathfrak{P}$. This can occur only if $\{x_i\} = \{x_j'\}$ by our choice of S. Thus $pz = p'z'$, $z = (p'/p)z'$, $(p'/p) \in A^*$, hence $\sigma(z) = \sigma(z')$, and it follows that $\sigma(y)$ is independent of the choices made.

At this point we have a function $p'\colon K \to \Delta, \pm\infty$ and we only need verify that p' is a signed place, relative to the total order $\overline{\mathfrak{P}}'$ on Δ. Verification of $p'(xy) = p'(x)p'(y)$ is quite simple and verification of

$p'(x+y) = p'(x) + p'(y)$ is also easy if use is made of the identity $x+y = x(1+\frac{y}{x})$.

Finally, let $P \subset A$ be a prime ideal, $\Delta_P = A_{(P)}/P$. Restriction of the signed place $p'\colon K \to \Delta, \pm\infty$ to $A_{(P)} \subset K$ induces a signed place $\Delta_P \to \Delta, \pm\infty$. The positive elements in Δ_P are clearly the images of the positive elements in $A_{(P)} \subset K$. But this just says $P \subset A_{(P)}$ is $\mathfrak{P}' \cap A_{(P)}$ - convex. □

Remark: The last paragraph in the proof can be established more directly. In fact, if (A,Q) is a valuation ring, $\mathfrak{P} \subset A$ is any partial order with $\mathfrak{P} = \mathfrak{P}_d$ and Q $\mathfrak{P}$-convex, then *any* ideal $I \subset A$ is $\mathfrak{P}$-absolutely convex. Namely, suppose $0 \leq x \leq y$, $yz \in I$. Since $\mathfrak{P} = \mathfrak{P}_d$, $0 \leq 1 \leq y/x$, and since Q is $\mathfrak{P}$-convex, $y/x \notin Q$. Thus $x/y \in A$, hence $x = y(x/y)$ and $xz = y(x/y)z \in I$.

Example: Let E be a totally ordered field, $K = E(x)$, $f(x) \in E[x]$ an irreducible polynomial. We obtain the f-adic place $p\colon K \to \Delta, \infty$, where $\Delta = E[x]/(f(x))$. If $f(x)$ has real roots over E, we obtain orderings of Δ. If $\mathfrak{P} = \mathfrak{P}_w$ is the weak order on K, a maximal $\mathfrak{P}$-independent set will consist of a single element $\{f(x)\}$. Thus there are exactly two orderings on K over each ordering of Δ, determined by whether $f(x) \in K$ is made positive or negative.

We turn now to our second main existence theorem, the construction of signed places on a field K with preassigned center on a subring $A \subset K$.

Lemma 7.7.3. Let $(K, \mathfrak{P})$ be a partially ordered field, $A \subset K$ a subring $P \subset A$ a prime $\mathfrak{P} \cap A$ - convex ideal. Then there exist total order refinements $\mathfrak{P}' \supset \mathfrak{P}$ on K such that P is $\mathfrak{P}' \cap A$ - convex.

Proof: We use the basic result of A. Klapper, Proposition 2.7.1, that if a prime ideal is convex for an intersection of two orders, it is convex for one of them separately. If $x \in K$, $x, -x \notin \mathfrak{P}$, then both simple refinements $\mathfrak{P}[x]$ and $\mathfrak{P}[-x]$ are (derived) orders on K. Moreover, $\mathfrak{P}[x] \cap \mathfrak{P}[-x] = \mathfrak{P}$ since if $y = p + qx = p' - q'x$, $p, p', q, q' \in \mathfrak{P}$, then $(q+q')y = pq' + p'q$, hence $y \in \mathfrak{P}$. Thus our prime ideal $P \subset A$ is either $\mathfrak{P}[x] \cap A$ - convex or $\mathfrak{P}[-x] \cap A$-convex. Zorn's lemma now implies 7.7.3. □

Proposition 7.7.4. Let $(K,\mathfrak{P})$ be a totally ordered field, $A \subset K$ a subring, $P \subset A$ a $\mathfrak{P} \cap A$ - convex prime ideal $\Delta = A_{(P)}/PA_{(P)}$, with induced order $\mathfrak{P}_\Delta = \mathfrak{P} \cap A_{(P)}/PA_{(P)}$. Then there exists a valuation ring $A' \subset K$ with $\mathfrak{P}$ - convex maximal ideal Q', such that $A \subset A'$, $P = Q' \cap A$, and residue field $\Delta' = A'/Q'$, $\mathfrak{P}' = \mathfrak{P} \cap A'/Q'$, Archimedean ordered over $(\Delta, \mathfrak{P}_\Delta)$.

Proof: We may as well assume A is a local ring with maximal ideal P, by passing to $(A_{(P)}, PA_{(P)})$ if necessary. Given $x \in K$, we claim either $\sqrt{HPA[x]}$ or $\sqrt{HPA[1/x]}$ is a proper ideal. First, we lose no generality by assuming $x > 0$. Then suppose $1 \leq a_0 + a_1x + \cdots + a_nx^n$ and $1 \leq b_0 + b_1/x + \cdots + b_m/x^m$ with $a_i, b_j \in P$, $m \leq n$ and n least possible. Then $x^m(1 - b_0) \leq b_1x^{m-1} + \cdots + b_m$. We must have $b_0 < 1$ in the total order $\mathfrak{P}$ since P is $\mathfrak{P} \cap A$ - convex. Thus $1 - b_0$ is a positive unit in A and $x^m \leq b_1'x^{m-1} + \cdots + b_m'$, $b_j' = b_j/(1-b_0)$. Finally, since n is least possible, we must have $a_n > 0$, or the term a_nx^n could simply be omitted from $1 \leq a_0 + a_1x + \cdots + a_nx^n$. But now $1 \leq a_0 + a_1x + \cdots + a_{n-1}x^{n-1} + a_n(b_1'x^{n-1} + \cdots + b_m'x^{n-m})$, which is a contradiction of minimality of n.

Now we simply choose any maximal convex ideal $P_1 \subset A_1 = A[x]$, say, with $P \subset P_1$. Since A is a local ring, $P_1 \cap A = P$. Zorn's lemma now implies the existence of a valuation ring (A',Q') in K with the properties $Q' \cap A = P$ and Q' $\mathfrak{P}$ - convex.

It remains to show that we may construct $\Delta' = A'/Q'$ Archimedean over $\Delta = A/P$. Return to the inductive step, where we passed from (A,P) to (A_1,P_1) where $A_1 = A[x]$ or $A[1/x]$. Now, say, $A_1/P_1 = \Delta[\bar{x}]$, where $\bar{x} = x$ modulo P_1. Then $\Delta[\bar{x}]$ is a totally ordered *semi-field* since $P_1 \subset A_1$ is maximal convex. If $\bar{x}$ is algebraic over Δ, certainly the field $\Delta[\bar{x}]$ is Archimedean over Δ. If $\bar{x}$ is transcendental, consider the cut of $\Delta, D = D_\Delta(\bar{x})$, defined by $\bar{x}$. If D is a transcendental cut of Δ, then from 7.5 we conclude $\Delta(\bar{x})$ is Archimedean over Δ. If D is an algebraic cut of Δ, then again from 7.5, $\Delta[\bar{x}]$ would *not* be a semi-field unless $D = \infty$. But if $D = \infty$, then we return and work with $A_1 = A[1/x]$ rather than $A[x]$. This is justified because if $\sqrt{HPA[1/x]}$ were improper, we would have $1 \leq b_0 + b_1/x + \cdots + b_m/x^m$ or $x^m \leq b_1'x^{m-1} + \cdots + b_m'$, $b_j' = b_j/(1 - b_0)$. But then from 6.3, x would be bounded by an element of A, so we could not have $D_\Delta(\bar{x}) = \infty$. This completes the proof of 7.7.4. □

Combining 7.7.3 and 7.7.4, we see that if $(K,\mathfrak{P})$ is a *partially* ordered field, and $P \subset A$ is a $\mathfrak{P} \cap A$-convex prime ideal of a subring of K, then we can find a signed place $p: K \to \Delta', \pm\infty$ such that the induced total order on K refines $\mathfrak{P}$, p is finite on A with center P and Δ' is Archimedean over the fraction field Δ of A/P. One might be interested in knowing if first the order $\mathfrak{P} \cap A/P$ on A/P can be refined to a total order in an arbitrary way, then a total order on K constructed lifting this choice. For example, if (A,P) is already a valuation ring and maximal ideal in K, Proposition 7.7.2 says this is the case. We will see in the next chapter that this is also the case if $A = R[X_1 \cdots X_n]$ is a polynomial ring over a totally ordered ground field R, and $K = R(X_1 \cdots X_n)$ is the field of rational functions. However, in general, one *cannot* first assign the order on A/P. A typical difficulty is that one could have an equation $p^2 f = q^2 + r^2 g$ in A, with $p,q,r \in P$. Thus, if g is made positive in A, so must be f. However, this relation between f and g disappears in A/P. The geometry of this problem is interesting, and will be clarified by results in the next chapter.

Our proof of 7.7.4 is directly modeled after the standard proof of Chevalley of the place extension theorem in algebra. That proof has the additional conclusion that the residue field Δ' can be constructed *algebraic* over $\Delta = A_{(P)}/PA_{(P)}$. Now, in our context, this cannot always be done. The difference is, a simple extension $\Delta[\bar{x}]$ of a field Δ can never be a field unless $\bar{x}$ is algebraic over Δ, whereas $\Delta[\bar{x}]$ can be a semi-field if $\bar{x}$ defines a transcendental cut of the ordered field Δ.

Specific examples arise as follows. First, suppose $A = \mathbb{Q}$, the rationals, $P = (0)$, $K = \mathbb{Q}(\pi)$, $\pi = 3.14159\ldots \in \mathbb{R}$, with $\mathfrak{P}$ the total order on K induced by the embedding in $\mathbb{R}$. Then K is Archimedean over $\mathbb{Q}$, hence any signed place of K will be trivial. Secondly, suppose $A = \mathbb{Q}[x,y]$ is a polynomial ring in two indeterminates over $\mathbb{Q}$, $P = (x,y)$. The extension field K could contain enough elements so that even the weak order on K forces $p(x/y) = \pi \in \mathbb{R}$, where $p: K \to \mathbb{R}, \pm \infty$ is any signed place with center P on A. Thirdly, even if K is smaller, say $K = \mathbb{Q}(x,y)$, our method of proof of the signed place extension theorem involves first choosing a total order on K so that $(x,y) \subset \mathbb{Q}[x,y]$ is convex. If this choice is made "carelessly", it is possible that again $p(x/y) = \pi \in \mathbb{R}$, for the resulting signed place on K.

In a positive direction, we can at least say this. Suppose $A = R[x_1 \cdots x_n]$ is a finitely generated integral domain over an ordered ground field R, $K = R(x_1 \cdots x_n)$ is the fraction field of A, $P \subset A$ is a convex prime relative to a partial order $\mathfrak{P} \subset K$, with $\text{tr.deg.}_R(A/P) = \text{tr.deg.}_R(A) - 1$. (So P is a minimal non-zero prime.) Then from classical valuation theory, the residue field Δ' of any place p on K with center $P \subset A$ must have $\text{tr.deg.}_R(\Delta') = \text{tr.deg.}_R(A/P)$, since otherwise p would necessarily be the trivial place. Thus Δ' will be algebraic over Δ, the fraction field of A/P. Moreover, [68 , Vol. II, Chapter VI, §14, Theorem 31], the residue field Δ' will actually be a finite algebraic extension of Δ, that is, a function field over R, of transcendence degree over R one less than that of K.

We next give some applications of the signed place extension theorem to the notion of semi-integral elements introduced in 6.4.

Proposition 7.7.5. Let $(K,\mathfrak{P})$ be a partially ordered field, $A \subset K$ a subring. Then the following subrings of K coincide

(i) The ring of elements semi-integral over A.

(ii) The ring of elements bounded by elements of A in all total order refinements of $\mathfrak{P}$ on K.

(iii) The ring of elements finite in all signed places of K finite on A, and refining the order $\mathfrak{P}$.

(iv) The intersection of all valuation rings of K containing A with $\mathfrak{P}$-convex maximal ideal.

Proof: (i) $\subset$ (ii) since semi-integral elements over A are already bounded by elements of A in the order $\mathfrak{P}$.

(ii) $\subset$ (iii) since a signed place induces an order.

(iii) $\subset$ (iv) since if an element was not in some such valuation ring containing A, it would be infinite in the associated place and in any signed place refinement.

(iv) $\subset$ (i) since if an element $x \in K$ is not semi-integral over A, then the ideal $\sqrt{H(1/x^2)A[1/x^2]} \subset A[1/x^2]$ is proper. Otherwise, there would be an equation $1 \leq a_1/x^2 + \cdots + a_n/x^{2n}$, or $x^{2n} \leq a_1 x^{2n-2} + \cdots + a_n$, $a_i \in A$. Thus, there exist by 7.7.3 and 7.7.4 valuation rings (A',Q') in K, Q' convex, with $A[1/x^2] \subset A'$, $1/x^2 \in Q'$. Thus $x^2 \notin A'$ hence $x \notin A'$. □

The following corollary was proved more generally in 6.4.

<u>Corollary 7.7.6</u>. Let $(B,\mathfrak{P})$ be a partially ordered integral domain, $\mathfrak{P} = \mathfrak{P}_d$ a derived order, $A \subset B$ a subring, such that $(B,\mathfrak{P})$ is semi-integral over $(A, \mathfrak{P} \cap A)$. Then for each $\mathfrak{P} \cap A$ - convex prime ideal $P \subset A$, there exist $\mathfrak{P}$ - convex prime ideals $Q \subset B$ with $Q \cap A = P$.

<u>Proof</u>: Let K denote the field of fractions of B, $\mathfrak{P} \subset K$ the order defined by $\mathfrak{P} \subset B$. (Since $\mathfrak{P} = \mathfrak{P}_d$ on B, the notation is justified.) Using 7.7.3, 7.7.4, construct a signed place p of K, with center $P \subset A$. By 7.7.5 $B \subset A_p$, where A_p is the valuation ring of p. Let $Q = B \cap Q_p$, where $Q_p \subset A_p$ is the maximal ideal. □

<u>Proposition 7.7.7</u>. Let $(K,\mathfrak{P})$ be a partially ordered field, $A \subset K$ a *local* ring, with $\mathfrak{P}$ - convex maximal ideal $P \subset A$. Then the semi-integral closure of A in K is the intersection of all valuation rings in K, (A',Q') with Q' $\mathfrak{P}$ - convex, $A \subset A'$, and $P = Q' \cap A$. Moreover, the elements x which satisfy a semi-integral equation $x^{2n} + b_1 x^{2n-1} + \cdots + b_{2n} \leq 0$ with $b_i \in P$ are exactly the elements in the intersection of the above maximal ideals $Q' \subset A'$.

<u>Proof</u>: Suppose $x \in K$ is not semi-integral over A. Then $\sqrt{H(P,\ 1/x^2)A[1/x^2]} \subset A[1/x^2]$ is proper. Otherwise $1 \leq b_0 + a_1/x^2 + \cdots + a_n/x^{2n}$, with $b_0 \in P$, $a_i \in A$, hence $x^{2n} \leq a_1' x^{2n-2} + \cdots + a_n'$, where $a_j' = a_j/(1-b_0) \in A$, since A is a local ring. (Strictly speaking, we should first refine the

order $\mathfrak{P}$ to insure that $1 - b_o$ is positive.) Thus, we can construct a valuation ring (A',Q') with $A[1/x^2] \subset A'$, $(P, 1/x^2) \subset Q'$. This shows x does not belong to all valuation rings over A with maximal ideal over P.

The last statement follows since if $x^{2n} + b_1x^{2n-1} + \cdots + b_{2n} \leq 0$, $b_i \in P$ and $p\colon K \to \Delta, \pm\infty$ is a signed place finite on A and zero on P, then $p(x) \in \Delta$ since x is semi-integral over A, hence $p(x^{2n}) \leq 0$ and $p(x) = 0$. Conversely, if x satisfies no such equation, then one argues as above that $\sqrt{HPA[1/x^2]} \subset A[1/x^2]$ is proper. Thus a valuation ring (A',Q') can be constructed with $Q' \cap A = P$ and $1/x^2 \in A'$, hence $x \notin Q'$. □

Our final results in this section will concern *chains* of prime ideals in an integral domain $P_0 \subset P_1 \subset \cdots \subset P_n \subset A$. In particular, if K is a field containing A, we are interested in specialization chains of signed places of K,

$$K \to \Delta_o, \pm\infty \to \Delta_1, \pm\infty \to \cdots \to \Delta_n, \pm\infty$$

with p_i finite on A and P_i the center of p_i on A. Here p_i denotes the composition $K \to \Delta_i, \pm\infty$. If (A_i,Q_i) is the valuation ring and maximal ideal of p_i, then $A_n \subset A_{n-1} \subset \cdots \subset A_o$ and $Q_o \subset Q_1 \subset \cdots \subset Q_n$. In fact, $(A_j,Q_j) = ((A_n)_{(Q_j)}, Q_j(A_n)_{(Q_j)})$.

The following general result provides some added insight into the close relationship between valuation rings and totally ordered integral domains.

Proposition 7.7.8. Let $(A,\mathfrak{P})$ be a totally ordered ring. Then, if $I,J \subset A$ are two convex ideals, either $I \subset J$ or $J \subset I$. All convex ideals are absolutely convex.

Proof: Suppose $I \not\subset J$, say $x \in I - J$. We may assume $x > 0$. Since J is convex, no element $y \in J$ can be greater than x. Thus $y < x$ for all elements of J, in particular for all positive elements. Since I is convex, $y \in I$, hence $J \subset I$.

Also, if I is convex, $0 < x < y$ and $yz \in I$, then we may assume z positive, hence $0 < xz < yz$ hence $xz \in I$ and I is absolutely convex. □

Next let $(K,\mathfrak{P})$ be a partially ordered field, $A \subset K$ a subring and

P_i, $0 \le i \le n$, $\mathfrak{P} \cap A$-convex prime ideals of A. When is there a total order refinement of $\mathfrak{P}$, keeping all P_i convex? Necessarily, the P_i must form a chain, say $P_0 \subset P_1 \subset \cdots \subset P_n$. Also, all $P_j/P_i \subset A/P_i$, $i < j$, are convex for a total order refinement of $\mathfrak{P}/P_i$. Thus P_j/P_i must be $(\mathfrak{P}/P_i)_d$-convex. But now if $i < j < k$, then $P_k/P_i \supset P_j/P_i$ are $(\mathfrak{P}/P_i)_d$-convex ideals of A/P_i and $P_k/P_j = (P_k/P_i)/(P_j/P_i) \subset (A/P_i)/(P_j/P_i) = A/P_j$ is convex for a total order refinement of $(\mathfrak{P}/P_i)_d/(P_j/P_i) \subset A/P_j$. If we abuse notation slightly and write this last order as $(\mathfrak{P}/P_i)_d/P_j$, then we deduce $P_k/P_j \subset A/P_j$ is necessarily $((\mathfrak{P}/P_i)_d/P_j)_d$-convex. Continuing, we find the following necessary conditions for the existence of a total order refinement of $(K,\mathfrak{P})$ which keeps all $P_i \subset A$ convex:

(*) If $i \le j$, then $P_j/P_i \subset A/P_i$ is $(\cdots((\mathfrak{P}/P_0)_d/P_1)_d \cdots /P_i)_d$-convex.

<u>Proposition 7.7.9</u>. In the notation above, the conditions (*) are necessary and sufficient for the existence of a total order refinement $\mathfrak{P}' \supset \mathfrak{P}$ on K such that all $P_i \subset A$ are $\mathfrak{P}' \cap A$-convex.

<u>Proof</u>: Let $x \in K$, with $x, -x \notin \mathfrak{P}$. Let $\mathfrak{P}' = \mathfrak{P}[x]$, $\mathfrak{P}'' = \mathfrak{P}[-x]$ as orders on K. As in the proof of 7.7.3, $\mathfrak{P}' \cap \mathfrak{P}'' = \mathfrak{P}$. Now we will prove generally that if $\mathfrak{P}, \mathfrak{P}', \mathfrak{P}''$ are orders on A with $\mathfrak{P}' \cap \mathfrak{P}'' = \mathfrak{P}$ and such that conditions (*) hold for $\mathfrak{P}$, then conditions (*) hold for $\mathfrak{P}'$ or for $\mathfrak{P}''$. The proposition then follows from Zorn's lemma.

We distinguish two cases, case (I) if $P_0 \subset A$ is not $\mathfrak{P}''$-convex, and case (II) if P_0 is both $\mathfrak{P}'$ and $\mathfrak{P}''$-convex. In case (I) we will prove conditions (*) hold for $\mathfrak{P}'$. In case (II), we will pass to the ring $A/P_0 = \overline{A}$ with prime ideals $P_i/P_0 = \overline{P}_i$, and orders $\overline{\mathfrak{P}}' = (\mathfrak{P}'/P_0)_d$, and $\overline{\mathfrak{P}}'' = (\mathfrak{P}''/P_0)_d$. Then we will establish conditions (*) for the order $\overline{\mathfrak{P}}' \cap \overline{\mathfrak{P}}''$ on $\overline{A}$ relative to the chain of prime ideals $\overline{P}_i$, $1 \le i \le n$. The assertion that (*) holds for either $\mathfrak{P}'$ or $\mathfrak{P}''$ on A then follows in case (II) by induction on the number of prime ideals P_i. (We do *not* assert $\overline{\mathfrak{P}}' \cap \overline{\mathfrak{P}}'' = (\mathfrak{P}/P_0)_d$ as orders on $\overline{A} = A/P_0$. If this last assertion were true, case (II) would be easy.)

Consider first case (I). Fix an equation $0 \le u \le v$ (rel $\mathfrak{P}''$), $v \in P_0$, $u \notin P_0$. First, we check that P_i/P_0 are all $(\mathfrak{P}'/P_0)_d$-convex. Note this makes sense because 2.7.1 implies if P_0 is not $\mathfrak{P}''$ convex, then P_0 is $\mathfrak{P}'$ convex. Suppose P_i/P_0 was not $(\mathfrak{P}'/P_0)_d$ - convex. There would then be an equation

$$0 \le ax + z_1 \le ay + z_2 \quad (\text{rel } \mathfrak{P}')$$

with $z_k \in P_0$, $a \in \mathfrak{P}'$, $a \notin P_0$, $y \in P_i$, $x \notin P_i$. Combined with $0 \le u \le v$ (rel $\mathfrak{P}''$), we obtain as in the proof of 2.7.1

$$0 \le u^2(ax+z_1)^2 \le u^2(ay+z_2)^2 + v^2(ax+z_1)^2 \quad (\text{rel } \mathfrak{P})$$

But $v \in P_0$, $z_k \in P_0$, so this equation simplifies to

$$0 \le u^2a^2x^2 \le u^2a^2y^2 + w \quad (\text{rel } \mathfrak{P})$$

with $w \in P_0$. But $u, a \notin P_0$, and we deduce therefore $0 \le x^2 \le y^2$ (rel $(\mathfrak{P}/P_0)_d$) in the ring A/P_0. But P_i/P_0 is *assumed* $(\mathfrak{P}/P_0)_d$-convex and $y \in P_i$ hence $x \in P_i$ which is our contradiction.

Next, we verify the "second level" conditions of (*), namely that if $i \le j$, then P_j/P_i is $((\mathfrak{P}'/P_0)_d/P_i)_d$-convex. This makes sense because of the first part of the proof above. Assuming our desired conclusion false, we would obtain

$$0 \le ax + z_1 \le ay + z_2 \quad (\text{rel} (\mathfrak{P}'/P_0)_d),$$

an inequality in A/P_0, with $a \notin P_i$, $a \in \mathfrak{P}'/P_0$, $z_k \in P_i/P_0$, $y \in P_j/P_0$, $x \notin P_j/P_0$. From this we obtain an inequality in A,

$$0 \le b(ax+z_1) + z_1' \le b(ay+z_2) + z_2' \quad (\text{rel } \mathfrak{P}')$$

where $b \notin P_0$, $b \in \mathfrak{P}'$, $z_k' \in P_0$. (By slight abuse of notation, we are using the same symbol for an element of A and its image in A/P_0.) Now we combine this with $0 \le u \le v$ (rel $\mathfrak{P}''$) to obtain as in the proof of 2.7.1

$$0 \le u^2b^2(ax+z_1)^2 \le u^2b^2(ay+z_2)^2 + w \quad (\text{rel } \mathfrak{P})$$

with $w \in P_o$. Since $u,b \notin P_o$, we have

$$0 \leq a^2x^2 \leq a^2y^2 + w' \qquad (\mathrm{rel}(\mathfrak{P}/P_o)_d)$$

where $w' \in P_i/P_o \subset A/P_o$, since $z_k \in P_i/P_o$. From this follows

$$0 \leq x^2 \leq y^2 \qquad (\mathrm{rel}((\mathfrak{P}/P_o)_d/P_i)_d)$$

in the ring A/P_i. But $y \in P_j/P_i$ and P_j/P_i is assumed $((\mathfrak{P}/P_o)_d/P_i)_d$-convex (more precisely, (*) easily implies this), hence $x \in P_j/P_i$, contradicting our original choice of x.

The general technique involved here should now be clear. Assuming P_j/P_i is not $((\cdots((\mathfrak{P}'/P_o)_d/P_1)_d\cdots/P_m)_d/P_i)_d$ convex, $m < i$, we unravel a suitable inequality in A/P_i and get a (complicated) inequality (rel $\mathfrak{P}'$) in A. This is combined with the inequality $0 \leq u \leq v$ (rel $\mathfrak{P}''$), $v \in P_o$, $u \notin P_o$ to obtain a complicated inequality (rel $\mathfrak{P}$) in A. Finally, this is unraveled again to get an inequality $\mathrm{rel}((\cdots((\mathfrak{P}/P_o)_d/P_1)_d\cdots/P_m)_d/P_i)_d$ which contradicts the assumed convexity of P_j/P_i given by (*). We will leave the actual details of this induction to the reader.

We next consider case II, when P_o is both $\mathfrak{P}'$ and $\mathfrak{P}''$-convex. We need to establish conditions (*) for the order $(\mathfrak{P}'/P_o)_d \cap (\mathfrak{P}''/P_o)_d$ on A/P_o, relative to the chain of prime ideals $P_1/P_o \subset \cdots \subset P_n/P_o$. We first argue that each P_i/P_o is $(\mathfrak{P}'/P_o)_d \cap (\mathfrak{P}''/P_o)_d$- convex, or, what is the same by 2.7.1, that each P_i/P_o is either $(\mathfrak{P}'/P_o)_d$-convex or $(\mathfrak{P}''/P_o)_d$-convex. Assuming otherwise, we would have inequalities

$$0 \leq ax + z_1 \leq ay + z_2 \ (\mathrm{rel}\ \mathfrak{P}')$$

$$0 \leq bu + w_1 \leq bv + w_2 \ (\mathrm{rel}\ \mathfrak{P}'')$$

where $w_k, z_k \in P_o$, $y,v \in P_i$, $x,u \notin P_i$, $a \in \mathfrak{P}'$, $a \notin P_o$, $b \in \mathfrak{P}''$, $b \notin P_o$. From this

$$(ax+z_1)^2(bu+w_1)^2 \leq (bu+w_1)^2(ay+z_2)^2 + (ax+z_1)^2(bv+w_2)^2 \quad (\mathrm{rel}\ \mathfrak{P})$$

or, simplifying,

$$0 \le a^2b^2x^2u^2 \le a^2b^2(u^2y^2 + x^2v^2) + t \quad (\text{rel } \mathfrak{P})$$

for some $t \in P_o$. Then $a,b \notin P_o$ implies

$$0 \le x^2u^2 \le u^2y^2 + x^2v^2 \quad (\text{rel } (\mathfrak{P}/P_o)_d)$$

in the ring A/P_o, which implies since $y,v \in P_i$ and P_i/P_o is $(\mathfrak{P}/P_o)_d$ convex, that either $x \in P_i$ or $u \in P_i$, contradiction.

The second level verification in case (II) consists of showing that for $i \le j$, P_j/P_i is $(((\mathfrak{P}'/P_o)_d \cap (\mathfrak{P}''/P_o)_d)/P_i)_d$-convex. Assuming otherwise, we get in A/P_o

$$0 \le ax + z_1 \le ay + z_2 \quad (\text{rel}(\mathfrak{P}'/P_o)_d \cap (\mathfrak{P}''/P_o)_d)$$

where $z_k \in P_i$, $y \in P_j$, $x \notin P_j$, $a \notin P_i$, $a \in (\mathfrak{P}'/P_o)_d \cap (\mathfrak{P}''/P_o)_d$. This inequality becomes two inequalities in A

$$0 \le b'(ax+z_1) + z_1' \le b'(ay+z_2) + z_2' \quad (\text{rel } \mathfrak{P}')$$

$$0 \le b''(ax+z_1) + z_1'' \le b''(ay+z_2) + z_2'' \quad (\text{rel } \mathfrak{P}'')$$

where $z_k', z_k'' \in P_o$, $b', b'' \notin P_o$, $b' \in \mathfrak{P}'$, $b'' \in \mathfrak{P}''$. Then we get $(\text{rel } \mathfrak{P})$ in A

$$0 \le (b')^2(b'')^2(ax+z_1)^4 \le 2(b')^2(b'')^2(ax+z_1)^2(ay+z_2)^2 + t$$

with $t \in P_o$. This gives in A/P_o

$$0 \le a^4x^4 \le 2a^4x^2y^2 + u \quad (\text{rel}(\mathfrak{P}/P_o)_d)$$

where $u \in P_i$ since $z_k \in P_i$. Finally, in A/P_i

$$0 \le x^4 \le 2x^2y^2 \quad (\text{rel}((\mathfrak{P}/P_o)_d/P_i)_d)$$

since $a \notin P_i$. But $y \in P_j$ and convexity conditions (*) for P_j/P_i imply $x \in P_j$ which is our contradiction.

Again, we leave the rest of the details of case (II) to the reader. The general step is to show that P_j/P_i is $((\cdots(((\mathfrak{P}'/P_o)_d \cap (\mathfrak{P}''/P_o)_d)/P_1)_d \cdots P_m)_d/P_i)_d$

convex for $m < i$. Assuming otherwise ultimately contradicts the assumed $(\cdots(((\mathfrak{P}/P_o)_d/P_1)_d/P_1)_d \cdots /P_m)_d/P_i)_d$ - convexity of P_j/P_i. So ends our proof of 7.7.9. □

Remark: The geometric significance of 7.7.9 for function fields over a fixed real closed ground field R is roughly as follows. Convex prime ideals $P \subset R[X_1 \ldots X_n]$ will have d-dimensional zero set in affine space R^n, where $d = \text{tr.deg.}_R(R[X_1 \ldots X_n]/P)$. However, P may also have "degenerate" zeros, of local dimension less than d. Passing to derived orders has the effect of eliminating these degenerate zeros from the maximal convex ideal spectrum of $R[X_1,,,X_n]/P$. Similarly, total orders on $R[X_1 \ldots X_n]/P$ can be interpreted as orders obtained by evaluating functions on (infinitesimally) small open sets in the non-degenerate part of the zeros of P. Thus the conditions in 7.7.9 that assure a chain of convex prime ideals P_i can be kept convex in a total order amount to adherence conditions on the non-degenerate zeros of the P_i. All this will be clarified by the results in the next chapter.

Proposition 7.7.10. Suppose $(K,\mathfrak{P})$ is a partially ordered field, $A \subset K$ a subring $P_o \subset P_1 \subset \cdots \subset P_n$ a chain of $\mathfrak{P} \cap A$-convex prime ideals satisfying the conditions (*) of 7.7.9. Then there is a specialization chain of signed places of K,

$$K \to \Delta_o, \pm\infty \to \Delta_1, \pm\infty \to \cdots \to \Delta_n, \pm\infty,$$

with valuation rings of $K \to \Delta_i, \pm\infty$ denoted by (A_i, Q_i) such that $A \subset A_i$ and $Q_i \cap A = P_i$. Moreover, each residue field $\Delta_i = A_i/Q_i$ is Archimedean over the field of fractions of A/P_i.

Proof: First, choose a total order on K, keeping all P_i convex. Now apply 7.7.4 to get a signed place $p_o\colon K \to \Delta_o, \pm\infty$, with center $P_o \subset A$ and Δ_o Archimedean over A/P_o. We now have the convex prime $P_1/P_o \subset A/P_o \subset \Delta_o$ and we apply 7.7.4 again to get a suitable signed place $\Delta_o \to \Delta_1, \pm\infty$, with center $P_1/P_o \subset A/P_o$ and with Δ_1 Archimedean over A/P_1. Continuing by induction gives the desired specialization chain. □

VIII · Affine semi-algebraic sets

8.1. Introduction and Notation

Throughout this chapter we fix a real closed ground field R. A subset $E \subset R^n$ of affine n-space over R is a *semi-algebraic set* if E belongs to the smallest Boolean ring of subsets of R^n which contains the sets

$$U(g) = \{x \in R^n \mid g(x) > 0\}, \qquad g \in R[X_1 \cdots X_n] .$$

We also define sets

$$W(g) = \{x \in R^n \mid g(x) \geq 0\}$$

$$Z(g) = \{x \in R^n \mid g(x) = 0\} .$$

If $g_i \in R[X_1 \cdots X_n]$ is a finite collection of polynomials, define

$$U\{g_i\} = \cap\, U(g_i)$$

$$W\{g_i\} = \cap\, W(g_i)$$

$$Z\{g_i\} = \cap\, Z(g_i) .$$

Any semi-algebraic set $E \subset R^n$ can be expressed as a finite union of intersections of Z's and U's

$$(*) \qquad E = \bigcup_k \left(Z\{f_{ik}\} \cap U\{g_{jk}\}\right) ,$$

for finite collections of polynomials f_{ik}, g_{jk}. Such a representation of E is of course highly non-unique. We are primarily interested in "geometric" properties of the sets E themselves, not in the particular representations (*).

Finite unions of the $U\{g_i\}$ will be called *open semi-algebraic sets*. Their complements in R^n will be called *closed semi-algebraic sets*. A typical closed semi-algebraic set S can be represented

$$S = \cup S_i\ , \quad \text{where}$$
(**)
$$S_i = Z\{f_{ij}\} \cap W\{g_{ik}\}\ .$$

Of course, equalities $f = 0$ could be avoided altogether by writing $f \geq 0$ and $f \leq 0$, but this is psychologically less natural.

Lemma 8.1.1. If $x \in U\{g_i\}$, then there is $g \in R[X_1 \ldots X_n]$ such that $x \in U(g) \subset U\{g_i\}$.

Proof: The estimates of 7.2 can be used to find a ball around the point $x = (x_1 \ldots x_n) \in R^n$, contained in $U\{g_i\}$. That is, $B(x,\varepsilon) \subset U\{g_i\}$, where $B(x,\varepsilon) = \{y \mid \|y-x\| < \varepsilon\}$, $\|y-x\|^2 = \Sigma(y_j-x_j)^2$. A suitable g is then given by $g_{x,\varepsilon}(y) = \varepsilon^2 - \Sigma(y_j-x_j)^2$. □

Of course, Lemma 8.1.1 states that the sets $U(g)$ form a base for a topology on R^n. But we now emphasize that semi-algebraic geometry is precisely not concerned with this strong topology. Instead, only the Boolean ring of semi-algebraic subsets of R^n is relevant. If $E \subset R^n$ is semi-algebraic, then we also get a natural Boolean ring of subsets of E, simply by intersecting E with other semi-algebraic sets.

If E, F are semi-algebraic and $F \subset E$, define the *interior* F^o_E and *closure* $\overline{F}_E$ of F in E as follows.

$$F^o_E = \{x \in E \mid \text{exists } \varepsilon > 0 \text{ such that } y \in E,\ \|y-x\| < \varepsilon \text{ implies } y \in F\}$$

$$\overline{F}_E = \{x \in E \mid \text{for all } \varepsilon > 0, \text{ there exists } y \in F \text{ such that } \|y-x\| < \varepsilon\}.$$

Of course, F^o_E and $\overline{F}_E$ are the ordinary toplogical interior and closure of F in the topology on E with base the sets $U(g) \cap E$. But, for our purposes, the point is that it follows easily from the Tarski-Seidenberg theorem, to be discussed below, that F_E^o and $\overline{F}_E$ are also semi-algebraic. (If $E = R^n$, we will simply write F^o and $\overline{F}$.)

We now make the following confusing (but crucial for understanding the differences between algebra and topology) definition. A semi-algebraic subset $F \subset E$ will be called an *open, semi-algebraic subset* of E if $F = F_E^o$ and F will be called a *closed, semi-algebraic subset* of E if $F = \overline{F}_E$. The commas should not be ignored, at least for now. It is easy to check that the complements of the open, semi-algebraic subsets of E are the closed, semi-algebraic subsets of E. It is also easy to check that finite unions of $U\{g_i\} \cap E$ are open, semi-algebraic and (hence) finite unions of $Z\{f_i\} \cap W\{g_j\} \cap E$ are closed, semialgebraic subsets of E. In other words, open semi-algebraic subsets are open, semi-algebraic subsets. In fact, the converse is also true, which we now state as

Unproved Proposition 8.1.2. Open, semi-algebraic subsets of E are open semi-algebraic subsets of E. □

Thus, the comma can be ignored when talking about open or closed semi-algebraic subsets. *However,* this foundational point seems to be rather difficult to prove. In particular, it does not follow directly from the Tarski-Seidenberg principle, as does the fact that F_E^o and $\overline{F}_E$ are indeed semi-algebraic.

The paragraphs above are a bit of a joke, written in order to emphasize the subtlety of 8.1.2. We will develop in this chapter an extensive theory of semi-algebraic sets *without* using 8.1.2. This will require some care with words, or, at least, with commas. It would be nice to have a simple proof of 8.1.2 right at the beginning. On the other hand, all the results we will prove in order to circumvent 8.1.2 are results we would want anyway.

The Tarski-Seidenberg theorem, [56], [57], [62], is sometimes regarded as a theorem in logic. It asserts the existence of a decision procedure for deciding the truth of any elementary sentence, built up from finitely many inequalities $P_i(x_1 \ldots x_n) > 0$, where the P_i are polynomials over a real closed field, using basic logical connectives "and", "or", "not", and quantifiers "exists x_j", "for all x_j". The decision procedure amounts to checking whether or not certain polynomial inequalities involving the coefficients of the P_i hold. For example, the sentence "there exists $x \in R$ such that $ax^2 + bx + c = 0$" is equivalent to "$b^2 - 4ac \geq 0$". Also, Sturm's theorem in 7.3 is a special case of such a decision procedure.

The proof of the Tarski-Seidenberg theorem is not difficult. We will give Cohen's proof [62] in an appendix. From the algebraists point of view, what is involved is just an argument making use of (1) induction on degrees of polynomials and number of variables and (2) explicit calculations in polynomial rings involving partial derivatives and division algorithms. In other words, elimination theory.

The applications of the theorem are rather striking. We will make a sharp distinction between two types of application. The first type, which is almost a reformulation of the theorem itself, says that any set defined in terms of semi-algebraic sets by an elementary sentence is still a semi-algebraic set. For example, this includes closures and interiors mentioned above, images of semi-algebraic sets under polynomial mappings $R^n \to R^m$, and other frequently used constructions. (However, the theorem gives little insight into the question of whether a set is, say, open.) It is amusing that even the simplest special cases of this type (say the projection to R^n of the zeros in R^{n+1} of a single polynomial, or the closure of the set where a single polynomial is strictly positive) are no easier to analyze than the whole theorem. Thus the Tarski-Seidenberg theorem is a very efficient tool. We emphasize that this type of application actually provides a *proof* that the asserted set is semi-algebraic, simultaneously for all real closed fields, in fact, an elementary proof. The reason is, any single elementary sentence is just a special case of the theorem. The more subtle application is this. Given an elementary sentence, suppose it can be checked in *one* real closed field where it makes sense and is true. For example, it might be checked for the classical real numbers by transcendental methods (use of completeness, possibility of integration, etc.) Then the sentence must be true for *any* real closed field to which it applies, because there *exists* a decision procedure which is independent of the field in which it is carried out. Thus, one might say that this method of application amounts to proving by non-elementary methods in one special case that an elementary proof of the statement in general does exist. It certainly might be very tedious, if not physically impossible, to actually work out this elementary proof.

In this book we absolutely and unequivocally refuse to give proofs of this second type. Every result is proved uniformly for all real closed ground fields. Our philosophical objection to transcendental proofs is that they may logically *prove* a result but they do not *explain* it, except for the special case of real numbers. Also, one of our central themes is that the real numbers are totally irrelevant in algebraic topology, so it would not do to rely on them at some point in our chain of reasoning. Finally, there is already a respectable tradition in this century of finding non-transcendental proofs of purely algebraic results concerning algebraically closed fields. We think real closed fields deserve (at least) equal time and effort.

We do not at all mean that only elementary proofs are acceptable. In fact, we use Dedekind cuts, total orders, and signed places repeatedly. The point is, in the form we use these concepts they apply uniformly to all real closed fields. One advantage to developing such techniques is precisely that one is not tied down to "elementary sentences". There is rich non-elementary theory to be studied for arbitrary real closed fields, and even if a statement turns out to be equivalent to an elementary statement, it may be unnatural to dwell on this fact, and even worse to be forced to depend upon it.

There are obviously aesthetic questions involved in this discussion. We admit that many of our proofs are long and could be replaced by the single phrase "Tarski-Seidenberg and true for real numbers". However, we feel the effort is worthwhile. In fact, we do not use the Tarski-Seidenberg theorem at all, until 8.7.

We now change tacks somewhat and indicate a more invariant approach to semi-algebraic geometry. Suppose $(A, \mathfrak{P}) \in (PORNN)$, and assume A is an R-algebra of finite type. Define a set $X = X(A, \mathfrak{P})$ by

$$X(A, \mathfrak{P}) = \text{Hom}_{(POR)}((A, \mathfrak{P}), (R, \mathfrak{P}_w)).$$

(Of course, $\mathfrak{P}_w$ is the only order on R.) We have the canonical adjoint homomorphism from A to the ring R^X of R-valued functions on X, and the elements $f \in \mathfrak{P}$ define functions nowhere negative on X. If $g_i \in A$, the subsets $U\{g_i\}$, $W\{g_i\}$, $Z\{g_i\} \subset X$ are defined in the obvious way. Interiors and closures of subsets of X are also defined just as above, essentially using the topology X with base the collection of sets $U(g)$, $g \in A$. (The $U(g)$, $g \in A$, are invariants of $(A,\mathfrak{P})$ and replace the ε-balls in the earlier definition.)

If we choose a specific presentation $A = R[X_1 \ldots X_n]/I$, where $I \subset R[X_1 \ldots X_n]$ is some ideal (necessarily radical and $\mathfrak{P}_w$-convex), then X gets identified with a subset of the zeros of I, $Z(I) \subset R^n$. Specifically,

$$X = X(A,\mathfrak{P}) = \{x \in Z(I) \mid g(x) \geq 0 \text{ all } g \in \mathfrak{P}\}.$$

Obviously, X is a closed subset of R^n in the topological sense.

Now, the problem is, the homomorphism $A \to R^X$ need not be injective. In fact, X could be empty. Thus, we make the following definition.

Definition. $(A,\mathfrak{P})$ is an RHJ-algebra (real, Hilbert-Jacobson) if A is an R-algebra of finite type $(A,\mathfrak{P}) \in$ (PORNN), and for each prime $\mathfrak{P}$-convex ideal $P \subset A$, and $g \notin P$, there exists $x \in X$ with $f(x) = 0$ for all $f \in P$ and $g(x) \neq 0$.

Corollary 8.1.3. If $(A,\mathfrak{P})$ is an RHJ-algebra, then

(a) the homomorphism $A \to R^X$ is injective

(b) the set X is identified with the maximal convex ideal spectrum of $(A,\mathfrak{P})$

(c) for any subset $J \subset A$, we have

$$\sqrt{H(J,\mathfrak{P})} = \{f \in A \mid f(x) = 0 \text{ all } x \in Z(J)\},$$

where $Z(J) \subset X$ is the set of zeros of J.

Proof:

(a) The intersection of all $\mathfrak{P}$-convex prime ideals of A is (0), since $(A,\mathfrak{P}) \in$ (PORNN).

(b) Each $x \in X$ corresponds to a surjection $p: A \to R$ with $p(f) \geq 0$, all $f \in \mathfrak{P}$. Thus kernel $(p) \subset A$ is a maximal, $\mathfrak{P}$-convex ideal. Conversely, the definition of RHJ-algebra states that every $\mathfrak{P}$-convex prime ideal P is contained in such a maximal ideal. Thus if P is already maximal, then P corresponds to some $x \in X$.

(c) $\sqrt{H(J,\mathfrak{P})}$ is the intersection of the prime $\mathfrak{P}$-convex ideals which contain J. Again, the definition and (b) says each $\mathfrak{P}$-convex prime P is the intersection of all $\mathfrak{P}$-convex maximal ideals containing P. Thus $\sqrt{H(J,\mathfrak{P})}$ is the set of $f \in A$ which vanish on all zeros of J in X. □

Although the notion of RHJ-algebra is attractive, there is still a problem. Without some control on the *order* $\mathfrak{P} \subset A$, the resulting subsets $X = X(A,\mathfrak{P})$ of affine space (relative to a presentation $A = R[X_1 \ldots X_n]/I$) can be rather chaotic. The control needed in order to guarantee that such an $X(A,\mathfrak{P})$ is semi-algebraic (necessarily *closed*, semi-algebraic) roughly amounts to finiteness conditions on orders naturally associated to $\mathfrak{P}$. Our study of semi-algebraic geometry will essentially amount, then, to identifying a large class of RHJ-algebras $(A,\mathfrak{P})$ with $X(A,\mathfrak{P})$ semi-algebraic and, conversely, given a closed, semi-algebraic set $E \subset R^n$, construct natural RHJ-algebras $(A,\mathfrak{P})$ with $E = X(A,\mathfrak{P})$. This last turns out to be fairly easy if E is closed semi-algebraic, but not so easy if E is only closed, semi-algebraic.

8.2. Some Properties of RHJ-Algebras

In this section we will construct many RHJ-algebras, assuming the following basic real Nullstellensatz, which will be proved in 8.4.

Proposition 8.2.1. The polynomial ring with the weak order $(R[X_1 \ldots X_n],\mathfrak{P}_w)$ is an RHJ-algebra. □

Of course, $X(R[X_1 \ldots X_n],\mathfrak{P}_w) = R^n$. Once we have this one RHJ-algebra, there are natural constructions of others. We indicate several such constructions in the propositions below.

Proposition 8.2.2. If $(A,\mathfrak{P})$ is an RHJ-algebra $X = X(A,\mathfrak{P})$, and $\mathfrak{P}(X) = \{f \in A | f(x) \geq 0 \text{ all } x \in X\}$, then for any order $\mathfrak{P}' \subset A$ with $\mathfrak{P} \subset \mathfrak{P}' \subset \mathfrak{P}(X)$, we have that $(A,\mathfrak{P}')$ is an RHJ-algebra and $X(A,\mathfrak{P}') = X(A,\mathfrak{P})$. In particular, the radical $\mathfrak{P}$-convex ideals and the radical $\mathfrak{P}'$-convex ideals coincide.

Proof: Any $\mathfrak{P}'$-convex ideal is also $\mathfrak{P}$-convex. Since the points $x \in X$ obviously define $\mathfrak{P}(X)$-convex ideals, hence also $\mathfrak{P}'$-convex ideals, the proposition follows easily. □

Corollary 8.2.3. If $(A,\mathfrak{P})$ is an RHJ-algebra, then $(A,\mathfrak{P}_s)$, $(A,\mathfrak{P}_p)$, and $(A,\mathfrak{P}_m)$ are RHJ-algebras. We have $X = X(A,\mathfrak{P}) = X(A,\mathfrak{P}_s) = X(A,\mathfrak{P}_p) = X(A,\mathfrak{P}_m)$, and $\mathfrak{P}_m = \mathfrak{P}(X)$.

Proof: The orders $\mathfrak{P}_s$, $\mathfrak{P}_p$, $\mathfrak{P}_m$ were defined in 3.12. By definition, $\mathfrak{P}_m = \mathfrak{P}(X)$ and it is easy to check that for an RHJ-algebra $(A,\mathfrak{P})$, we have $\mathfrak{P} \subset \mathfrak{P}_s \subset \mathfrak{P}_p \subset \mathfrak{P}(X)$. □

Proposition 8.2.4. If $(A,\mathfrak{P}_i)$ are RHJ-algebras, $1 \leq i \leq k$, then $(A, \cap\mathfrak{P}_i)$ is an RHJ-algebra and $X(A, \cap\mathfrak{P}_i) = \bigcup_i X(A,\mathfrak{P}_i)$.

Proof: The results of 2.7 show that any $\cap\mathfrak{P}_i$-convex prime ideal is $\mathfrak{P}_j$-convex for some j. The result follows easily. □

Proposition 8.2.5. If $(A,\mathfrak{P})$ is an RHJ-algebra and $I \subset A$ is a radical, $\mathfrak{P}$-convex ideal, then $(A/I, \mathfrak{P}/I)$ is an RHJ-algebra and $X(A/I, \mathfrak{P}/I) = Z(I) \subset X(A,\mathfrak{P})$.

Proof: Any prime $\mathfrak{P}/I$-convex ideal of A/I is of the form P/I, where $P \subset A$ is a prime $\mathfrak{P}$-convex ideal containing I. □

Corollary 8.2.6. If A is a reduced, real R-algebra of finite type, then $(A,\mathfrak{P}_w)$ is an RHJ-algebra.

Proof: Write $A = R[X_1 \ldots X_n]/I$, where $I \subset R[X_1 \ldots X_n]$ is a radical,

$\mathfrak{P}_W$-convex ideal. (This is exactly what the hypotheses on A mean.) Now apply 8.2.1 and 8.2.5. □

Proposition 8.2.7. If $(B,\mathfrak{P}_B)$ is an RHJ-algebra, $A \subset B$ a subring such that A is finitely generated over R and $(B,\mathfrak{P}_B)$ is semi-integral over $(A,\mathfrak{P}_A)$, where $\mathfrak{P}_A = A \cap \mathfrak{P}_B$, then $(A,\mathfrak{P}_A)$ is an RHJ-algebra and $X(A,\mathfrak{P}_A)$ = image $(X(B,\mathfrak{P}_B))$ under the natural map $\mathrm{Spec}(B,\mathfrak{P}_B) \to \mathrm{Spec}(A,\mathfrak{P}_A)$ induced by the inclusion $A \subset B$.

Proof: This is immediate from the going up theorem for semi-integral extensions proved in 6.4. □

Proposition 8.2.8. Let A be a reduced, real R-algebra of finite type, let $\mathfrak{P} \subset A$ be an order, and let $P_i \subset A$ be any finite collection of $\mathfrak{P}$-convex primes with $(0) = \cap P_i$. (For example, the P_i could be the minimal primes of A, which are convex for any order on A by 3.11.) Let $(A_i,\mathfrak{P}_i) = (A/P_i, \mathfrak{P}/P_i)$. Then $X(A,\mathfrak{P}) = \bigcup_i X(A_i,\mathfrak{P}_i)$. Moreover, $(A,\mathfrak{P})$ is an RHJ-algebra if and only if each $(A_i,\mathfrak{P}_i)$ is an RHJ-algebra.

Proof: The kernel of any homomorphism $A \to R$ contains some P_i. (In fact, any prime ideal of A contains some P_i.) This implies $X(A,\mathfrak{P}) = \bigcup_i X(A_i,\mathfrak{P}_i)$.

Next, suppose $(A,\mathfrak{P})$ is an RHJ-algebra and $Q_i \subset A_i$ is a $\mathfrak{P}_i$-convex prime $g \notin Q_i$. Then Q_i corresponds to a $\mathfrak{P}$-convex prime $Q \subset A$, with $P_i \subset Q$, and $g \notin Q$. Find $x \in X(A,\mathfrak{P})$ so that x is a zero of Q and $g(x) \neq 0$. Then $x \in X(A_i,\mathfrak{P}_i)$, x is a zero of $Q_i = Q/P_i$, and $g(x) \neq 0$. Thus $(A_i,\mathfrak{P}_i)$ is an RHJ-algebra. The converse is equally routine, using the fact that any prime ideal of A contains some P_i. □

Proposition 8.2.9. Let A be a reduced R-algebra of finite type, let $P_i \subset A$ be a finite collection of primes with $(0) = \cap P_i$. Suppose given orders $\mathfrak{P}_i \subset A_i = A/P_i$ and define an order $\mathfrak{P} \subset A$ by $\mathfrak{P} = A \cap \Pi\, \mathfrak{P}_i$, under the inclusion $A \to \Pi A_i$. Then $X(A,\mathfrak{P}) = \bigcup_i X(A_i,\mathfrak{P}_i)$ and $(A,\mathfrak{P})$ is an RHJ-algebra if and only if each $(A_i,\mathfrak{P}_i)$ is an RHJ-algebra.

Proof: By Proposition 2.7.8, if $Q \subset A$ is any $\mathfrak{P}$-convex prime, then for some i, $P_i \subset Q$ and Q/P_i is $\mathfrak{P}_i$-convex. All parts of the proposition above follow routinely from this fact. □

Proposition 8.2.10. If A is a reduced, real R-algebra of finite type and $\mathfrak{P} = \mathfrak{P}_w[g_j]$ is a finite refinement of the weak order on A, then $(A,\mathfrak{P})$ is an RHJ-algebra.

Proof: Let $P_i \subset A$ be the minimal primes $A_i = A/P_i$. The finite refinement $\mathfrak{P}_w[\bar{g}_j] = \mathfrak{P}_i$ will also be an order on A_i, where $\bar{g}_j = g_j \pmod{P_i}$. Of course, some $\bar{g}_j$ may be 0 in A_i and these can be omitted in studying $\mathfrak{P}_i$. Now, A_i is an integral domain, and assuming all $\bar{g}_j \neq 0$, consider the finite sequence of integral extensions $B_0 = A_i$, $B_{m+1} = B_m[\sqrt{\bar{g}_{m+1}}]$, $m \geq 0$. Each B_{m+1} is a domain, unless $\bar{g}_{m+1}$ is already a square in B_m. The first time this happens, an easy computation shows $\bar{g}_{m+1} \in \mathfrak{P}_w[\bar{g}_1 \ldots \bar{g}_m] \subset A_i$. Thus, we can skip $\bar{g}_{m+1}$ and go on to $B_m[\sqrt{\bar{g}_{m+2}}]$ instead. After finitely many steps, we have a domain $B = \varinjlim B_m$, integral over A_i, of finite type over R, and real, say be Proposition 6.2.1. Moreover, $\mathfrak{P}_i = \mathfrak{P}_w[\bar{g}_j] = A_i \cap \mathfrak{P}_w(B)$, where $\mathfrak{P}_w(B) \subset B$ is the weak order. Applying 8.2.6, $(B,\mathfrak{P}_w(B))$ is an RHJ-algebra. Applying 8.2.7, $(A_i,\mathfrak{P}_i)$ is an RHJ-algebra. Finally, by 8.2.8, $(A,\mathfrak{P})$ is an RHJ-algebra. □

Remark. The proof of the basic Nullstellensatz 8.2.1, to be given in 8.4, will, in fact, yield directly that any order of type $\mathfrak{P}_w[g_j]$ on a finite real integral domain over R gives an RHJ-algebra. This could then be used instead of 8.2.7 in the proof of 8.2.10. On the other hand, both proofs use the device of adjoining square roots.

If $(A,\mathfrak{P})$ is an RHJ-algebra, then so is $(A,\mathfrak{P}_w)$, by 8.2.6, and $X(A,\mathfrak{P}) \subset X(A,\mathfrak{P}_w)$. In fact, $X(A,\mathfrak{P}) = \{x \in X(A,\mathfrak{P}_w) \mid g(x) \geq 0, \text{ all } g \in \mathfrak{P}\}$. Now, if $A = R[X_1 \ldots X_n]/J$, where $J = (f_i) \subset R[X_1 \ldots X_n]$ is a radical, $\mathfrak{P}_w$-convex ideal, then $X(A,\mathfrak{P}_w)$ is identified with the real algebraic set $Z\{f_i\} \subset R^n$. If $\mathfrak{P} = \mathfrak{P}_w[g_j]$ is a finite refinement of the weak order on A, then $X(A,\mathfrak{P}) = Z\{f_i\} \cap W\{g_j\}$. More generally, if $\mathfrak{P} = \cap \mathfrak{P}_k$ where each

$\mathfrak{P}_k = \mathfrak{P}_w[g_{jk}]$ is a finite refinement of the weak order, $1 \leq k \leq m$, then by 8.2.4, $X(A,\mathfrak{P}) = \bigcup_k Z\{f_i\} \cap W\{g_{jk}\}$. Thus orders of this type give rise to closed semi-algebraic sets.

We now want to prove that any closed semi-algebraic set $S \subset R^n$,

$$(**) \qquad S = \bigcup S_i, \qquad S_i = Z\{f_{ij}\} \cap W\{g_{ik}\}$$

can be represented as $S = X(A,\mathfrak{P})$ for a suitable RHJ-algebra $(A,\mathfrak{P})$, with A a quotient of $R[X_1 \dots X_n]$. Let $I(S) \subset R[X_1 \dots X_n]$ denote the ideal of functions which vanish on S. Then $I(S)$ is a radical, $\mathfrak{P}_w$-convex ideal since $A(S) = R[X_1 \dots X_n]/I(S)$ is a ring of R-valued functions on S. Trivially, $I(S) = \cap I(S_i)$, $A(S_i) = R[X_1 \dots X_n]/I(S_i)$ is a ring of functions on S_i, $f_{ij} \in I(S_i)$, and the g_{ik} are non-negative on S_i, hence $\mathfrak{P}_w[g_{ik}]$ is an order on $A(S_i)$. Of course, some g_{ik} may be zero in $A(S_i)$.

Let $P_{i\alpha}$ denote the minimal primes over $I(S_i)$ in $R[X_1 \dots X_n]$, so that $I(S_i) = \cap P_{i\alpha}$. Let $A_{i\alpha} = R[X_1 \dots X_n]/P_{i\alpha}$. Then $\mathfrak{P}_{i\alpha} = \mathfrak{P}_w[g_{ik}]$ is an order on $A_{i\alpha}$, since the $P_{i\alpha}/I(S_i)$ are convex for any order on $R[X_1 \dots X_n]/I(S_i) = A(S_i)$. We now have $I(S) = \bigcap_i I(S_i) = \bigcap_i \bigcap_\alpha P_{i\alpha}$ and thus we have an inclusion $A(S) \to \prod_{i,\alpha} A_{i\alpha}$. Note that all the minimal primes of $A(S)$ occur among the $P_{i\alpha}$. It is possible that some relations $P_{i\alpha} \subset P_{j\beta}$ hold, even $P_{i\alpha} = P_{j\beta}$, but since for fixed i, the $P_{i\alpha}$ are the minimal primes of $I(S_i)$, we know that $P_{i\alpha} \subset P_{j\beta}$ implies $i \neq j$.

Proposition 8.2.11. If $\mathfrak{P} = A(S) \cap \prod_{i,\alpha} \mathfrak{P}_{i\alpha} \subset A(S)$, then $(A(S),\mathfrak{P})$ is an RHJ-algebra and $X(A(S),\mathfrak{P}) = S \subset R^n$.

Proof: Propositions 8.2.9 and 8.2.10 guarantee that $(A(S),\mathfrak{P})$ is an RHJ-algebra, with $X(A(S),\mathfrak{P}) = \bigcup_{i,\alpha} X(A_{i\alpha},\mathfrak{P}_{i\alpha})$. Since clearly $X(A_{i\alpha},\mathfrak{P}_{i\alpha}) \subset S_i$, we have $X(A(S),\mathfrak{P}) \subset S$. Conversely, since $I(S_i) = \bigcap_\alpha P_{i\alpha}$, any zero x of $I(S_i)$ is a zero of $P_{i\alpha}$, some α. If $x \in S_i$, then x is clearly $\mathfrak{P}_{i\alpha}$-convex as an $A_{i\alpha}$-ideal, since $g_{ik}(x) \geq 0$. Thus $S_i \subset \bigcup_\alpha X(A_{i\alpha},\mathfrak{P}_{i\alpha})$, and $S \subset X(A(S),\mathfrak{P})$. □

Of course, the order $\mathfrak{P} \subset A(S)$ constructed above depends on the representation (**) of S, and is a rather weak order. Let $\mathfrak{P}(S) = \{f \in A(S) | f(x) \geq 0 \text{ all } x \in S\}$. Then by 8.2.2, $(A(S),\mathfrak{P}(S))$ is also an RHJ-algebra. We refer to $(A(S),\mathfrak{P}(S))$ as the *affine coordinate ring* of $S \subset R^n$. The fact that it is an RHJ-algebra is geometrically very satisfying.

However, a useful question is, how does one go about deciding *algebraically*, in terms of a representation (**) of S, whether a given $f \in A(S)$ belongs to $\mathfrak{P}(S)$? One answer is provided by a theorem of Stengle, which asserts that for a certain class of RHJ-algebras $(A',\mathfrak{P}')$, with $X' = X(A',\mathfrak{P}')$, we have $f \in \mathfrak{P}(X')$ if and only if there is an equation $(f^{2n} + p)f = q$, for some $n \geq 0$, $p,q \in \mathfrak{P}'$. (The "if" part is trivial.) In particular, this implies $\mathfrak{P}'_m = \mathfrak{P}(X') = \mathfrak{P}'_p$, where $\mathfrak{P}'_m$ and $\mathfrak{P}'_p$ are the operators on orders defined in 3.12. In fact, it also shows that in this case the collection of f satisfying such a formula is *closed under sums*, thus simplifying the construction of $\mathfrak{P}'_p$ in these cases. The most general class of orders satisfying Stengle's theorem is obscure, but it includes at least all $(A,\mathfrak{P})$, where A is a reduced R-algebra of finite type, and $\mathfrak{P} \subset A$ is a finite refinement of the weak order.

In particular, reconsider the closed semi-algebraic set S described by (**) above. We constructed an inclusion $A(S) \to \Pi A_{i\alpha}$, where $A_{i\alpha} = R[X_1 \ldots X_n]/P_{i\alpha}$, $P_{i\alpha}$ prime. We had the orders $\mathfrak{P}_{i\alpha} = \mathfrak{P}_w[g_{ik}] \subset A_{i\alpha}$ and $S = X(A(S),A(S) \cap \Pi \mathfrak{P}_{i\alpha})$. As a consequence of Stengle's theorem, we can now state

<u>Corollary 8.2.12</u>. $\mathfrak{P}(S) = A(S) \cap \Pi(\mathfrak{P}_{i\alpha})_p$, where

$$(\mathfrak{P}_{i\alpha})_p = \{f \in A_{i\alpha} | (f^{2n} + p)f = q, \text{ some } n \geq 0, p,q \in \mathfrak{P}_{i\alpha}\}. \qquad \square$$

This result characterizes algebraically the collection of polynomials $f \in R[X_1 \ldots X_n]$ non-negative on the closed semi-algebraic set $S \subset R^n$. We will prove Stengle's theorem in 8.5.

Note that recovering S from the $(A_{i\alpha},\mathfrak{P}_{i\alpha})$ is analogous to recovering an algebraic set from its irreducible components. In this semi-algebraic setting, where orders $\mathfrak{P}_{i\alpha}$ are carried along with the integral domains

$A_{i\alpha}$, or with the prime ideals $P_{i\alpha} \subset R[X_1 \ldots X_n]$, we can have the same $P_{i\alpha}$ occurring several times with different orders $\mathfrak{P}_w[g_{jk}]$ on $A_{i\alpha}$. This corresponds to several patches on the variety $Z(P_{i\alpha})$ of the form $W\{g_{jk}\} \cap Z(P_{i\alpha})$, all contained in S. We can even have proper inclusions $P_{i\alpha} \subsetneq P_{j\beta}$, or equivalently, $Z(P_{j\beta}) \subsetneq Z(P_{i\alpha})$, with patches on $Z(P_{j\beta})$ also belonging to S.

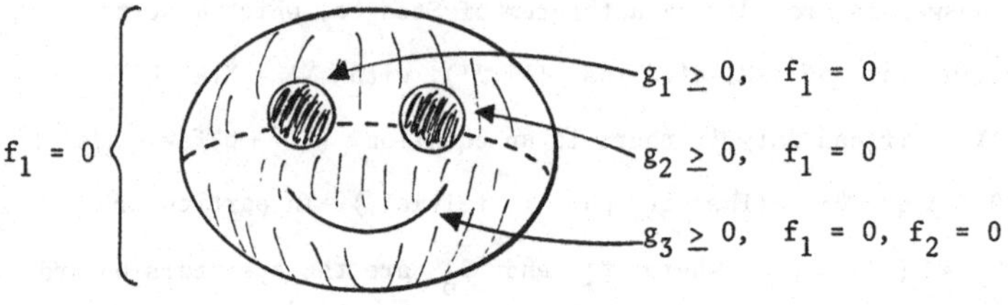

$$S = Z(f_1) \cap W(g_1) \cup Z(f_1) \cap W(g_2) \cup Z(f_1,f_2) \cap W(g_3) .$$

<u>Remark</u>. The orders $\mathfrak{P}_w[g_{ik}]$ on $A_{i\alpha} = R[X_1 \ldots X_n]/P_{i\alpha}$ considered above can be replaced by $\mathfrak{P}_w[g_{ik} | g_{ik} \notin P_{i\alpha}]$. Let $g = \prod_{g_{ik} \notin P_{i\alpha}} g_{ik}$. Then $g \notin P_{i\alpha}$, so we can find $x \in X(A_{i\alpha}, \mathfrak{P}_{i\alpha})$ with $g(x) \neq 0$. It follows that $g_{ik}(x) > 0$, hence the patches $W\{g_{ik}\} \cap Z(\mathfrak{P}_{i\alpha})$ contain points x where the inequalities $g_{ik}(x) \geq 0$ are *strict*, $g_{ik}(x) > 0$, for all $g_{ik} \notin P_{i\alpha}$. Of course, if $g_{ik} \in \mathfrak{P}_{i\alpha}$, the condition $g_{ik} \geq 0$ is superfluous on $Z(P_{i\alpha})$.

We have indicated above how certain operations on orders defining RHJ-algebras yield other RHJ-algebras. Certain other operators, which one might hope do the same, turn out to be more subtle. We will show that there are RHJ-algebras $(A,\mathfrak{P})$ such that some simple refinement of the order $\mathfrak{P}$, $(A,\mathfrak{P}[g])$, is not an RHJ-algebra. Also, $(A[t], i_*\mathfrak{P})$ need not be an RHJ-algebra, where $i: A \to A[t]$ is the inclusion of A in the polynomial ring in one indeterminate over A.

The basic example, which is a good source of counterexamples, is $(A,\mathfrak{P}) = (R[x,y], \mathfrak{P}_w[x] \cap \mathfrak{P}_w[y])$. Thus, $X(A,\mathfrak{P}) = \{(x,y) \in R^2 | x \geq 0 \text{ or } y \geq 0\}$. Suppose $h(x,y)$ is nonnegative on $X(A,\mathfrak{P})$. Since h cannot change sign across either the positive x-axis or the positive y-axis, it follows that

$h(x,y)$ is divisible by an even power of x and y. It then follows that for some $r < 0$, h is nonnegative on a *neighborhood* in R^2 of the two rays $\{(x,0) | x < r\}$, $\{(0,y) | y < r\}$ on the negative x- and y-axes, respectively. A somewhat surprising consequence is that $\mathfrak{P}[-x,-y] = (\mathfrak{P}_w[x] \cap \mathfrak{P}_w[y])[-x,-y]$ is an order. This follows since for any $g_1,\ldots,g_k \in \mathfrak{P}$, the elements g_i, $-x,-y$ are simultaneously positive on a non-empty open set in R^2. (Open sets in R^n are Zariski-dense, relative to the ring of polynomials $R[X_1 \ldots X_n]$, because of the finite Taylor expansion of a polynomial about any point. This holds for any real closed R, since the estimates of 7.2 show that the formal algebraic partial derivatives of a polynomial agree with the "limit" definition of partial derivatives.) Let $\mathfrak{P}' = \mathfrak{P}[-x]$, $\mathfrak{P}'' = \mathfrak{P}'[-y] = \mathfrak{P}[-x,-y]$. It is unclear whether or not $(A,\mathfrak{P}')$ is an RHJ-algebra. In any case, $X(A,\mathfrak{P}') = \{(x,y) \in R^2 | x \leq 0 \leq y\}$. It is then obvious that $(A,\mathfrak{P}'')$ cannot be an RHJ-algebra. In either case, we conclude simple refinements of RHJ-algebras need not be RHJ-algebras.

With this same $(A,\mathfrak{P}) = (R[x,y], \mathfrak{P}_w[x] \cap \mathfrak{P}_w[y])$, we argue that $(A[t], i_*\mathfrak{P})$ cannot be an RHJ-algebra. The point is, one can find an algebraic surface S in three space in the octant $\{(x,y,t) | x,y < 0 < t\}$ which projects on the

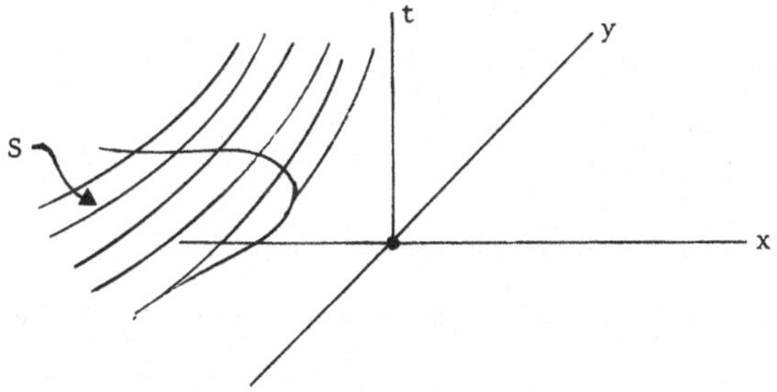

x-y plane to precisely the open third quadrant $\{(x,y) | x,y < 0\}$. Now $X(A[t],i_*\mathfrak{P}) = \{(x,y,t) | x \geq 0 \text{ or } y \geq 0\}$. But the discussion in the paragraph above concerning functions $h(x,y)$ non-negative on $X(A,\mathfrak{P})$ implies the ideal $I(S) \subset A[t]$ is $i_*\mathfrak{P}$ convex. The reason is, if $0 \leq f(x,y,t) \leq g(x,y,t)$ (rel $i_*\mathfrak{P}$), and g vanishes on S, then f must also vanish on at least a non-empty open subset of S, above a region near the (sufficiently) negative x- or y-axis. In general, open subsets of varieties need not be Zariski-dense relative to the coordinate ring of functions $A(S)$, because of "degenerate zeros". But in this case, such an open subset of the hypersurface S will be Zariski-dense, hence $f \in I(S)$.

The ring $(A[t], i_*\mathfrak{P})$ is also easily identified with the fibre sum over R, $(A,\mathfrak{P}) \underset{R}{\otimes} (R[t],\mathfrak{P}_w)$. Thus direct sums of RHJ-algebras need not be RHJ-algebras, in general. In case $(A,\mathfrak{P}_w[g_i])$ and $(B,\mathfrak{P}_w[h_j])$ are RHJ-algebras obtained as finite refinements of weak orders, then $(A,\mathfrak{P}_w[g_i]) \underset{R}{\otimes} (B,\mathfrak{P}_w[h_j]) = (A \underset{R}{\otimes} B, \mathfrak{P}_w[g_i \otimes 1, 1 \otimes h_j])$ will be an RHJ-algebra with $X(A \underset{R}{\otimes} B, \mathfrak{P}_w[g_i \otimes 1, 1 \otimes h_j]) = X(A, \mathfrak{P}_w[g_i]) \times X(B, \mathfrak{P}_w[h_j])$.

If S and T are closed semi-algebraic sets, so is $S \times T$, and $A(S) \underset{R}{\otimes} A(T)$ is a ring of R-valued functions on $S \times T$. Thus $A(S) \underset{R}{\otimes} A(T)$ is a reduced, real R-algebra, in fact, $A(S) \underset{R}{\otimes} A(T) = A(S \times T)$. If $A(S)$ and $A(T)$ are integral domains, so is $A(S) \underset{R}{\otimes} A(T)$. (This is most easily proved geometrically, using the irreducibility of the Zariski closures $\hat{S}$ and $\hat{T}$ of S and T.) Since a homomorphism $A(S) \underset{R}{\otimes} A(T) \to R$ is just a pair of homomorphisms $A(S) \to R$, $A(T) \to R$, it is clear that $X(A(S) \underset{R}{\otimes} A(T), \mathfrak{P}(S) \underset{R}{\otimes} \mathfrak{P}(T)) = S \times T$. However, in general even $\mathfrak{P}(S) \underset{R}{\otimes} \mathfrak{P}(T)$ will be too weak an order and will have convex prime ideals with no zeros in $S \times T$, as in the above example.

Contracted orders are also hard to work with, in general. For example, if the order $\mathfrak{P} = \mathfrak{P}_w[xy^2 - 1] \subset R[x,y]$ is contracted to $R[x]$, one obtains the weak order. Geometrically, it is hard to reconcile $X(R[x,y],\mathfrak{P}) \subset R^2$, which lies in the half plane $x > 0$ with the entire x-axis, which is $X(R[x],\mathfrak{P}_w)$. Presumably, a contraction of an RHJ-algebra need not even be an RHJ-algebra. On the other hand, 8.2.7 states that contractions are very well behaved for semi-integral extensions.

Suppose $Y \subset R^n$ is *any* subset. Then we can define $A(Y) = R[X_1 \ldots X_n]/I(Y)$ and an order $\mathfrak{P}(Y) = \{f \in A(Y) \mid f(y) \geq 0 \text{ all } y \in Y\}$. Of course, $I(Y) = I(\hat{Y})$, where $\hat{Y} = Z(I(Y))$ is the *Zariski* closure of Y in R^n. Also, it is easy to see that $X(A(Y),\mathfrak{P}(Y)) = \overline{Y}$, where $\overline{Y}$ is the *topological* closure of Y in R^n. Namely, any f non-negative on Y is also non-negative on $\overline{Y}$.

If Y is semi-algebraic, then so is $\overline{Y}$ by Tarski-Seidenberg. Ultimately, we will prove that for all closed, semi-algebraic $Y \subset R^n$, $(A(Y),\mathfrak{P}(Y))$ is an RHJ-algebra. So far, we have established this only (a) for closed semi-algebraic sets and (b) for *images* of sets Y with $(A(Y),\mathfrak{P}(Y))$ an RHJ-algebra, under semi-integral extensions $A \subset A(Y)$, inducing $Y = X(A(Y),\mathfrak{P}(Y)) \to X(A, A \cap \mathfrak{P}(Y))$.

If $A \subset B$ is any extension of reduced, finite R-algebras, and $\mathfrak{P} \subset B$ is an order with $(B,\mathfrak{P})$ an RHJ-algebra with $\mathfrak{P} = \mathfrak{P}(X)$, where $X = X(B,\mathfrak{P})$, then consider $i^*\colon X(B,\mathfrak{P}) \to X(A, \mathfrak{P} \cap A)$. Since $i\colon A \to B$ is injective, no function $f \in A$, $f \neq 0$, can vanish on $i^*(X(B,\mathfrak{P})) \subset X(A,\mathfrak{P}_w)$. That is, $i^*(X(B,\mathfrak{P}))$ is Zariski-dense in $X(A,\mathfrak{P}_w)$. Moreover, it is obvious that $f \in \mathfrak{P} \cap A$ if and only if f is non-negative on $i^*(X(B,\mathfrak{P}))$, or, equivalently, on the topological closure $\overline{i^*(X(B,\mathfrak{P})} \subset X(A,\mathfrak{P}_w)$. Thus $X(A, \mathfrak{P} \cap A) = \overline{i^*(X(B,\mathfrak{P}))}$.

Let $P_\alpha \subset B$ be the minimal primes, $B_\alpha = B/P_\alpha$, $A_\alpha = A/i^{-1}(P_\alpha)$. We have inclusions of finite integral domains $A_\alpha \to B_\alpha$ and a diagram of inclusions

$$\begin{array}{ccc} A & \longrightarrow & B \\ \downarrow & & \downarrow \\ \Pi A_\alpha & \longrightarrow & \Pi B_\alpha \end{array}$$

The order $\mathfrak{P} \subset B$ is the contraction of an order on ΠB_α, namely, $\prod_\alpha \mathfrak{P}(X(B_\alpha,\mathfrak{P}_\alpha))$, since $X(B,\mathfrak{P}) = \bigcup_\alpha X(B_\alpha,\mathfrak{P}_\alpha)$, where $\mathfrak{P}_\alpha = \mathfrak{P}/P_\alpha$. (We could also contract $\Pi\mathfrak{P}_\alpha$, and get $\mathfrak{P} \subset B$, but $\mathfrak{P}_\alpha$ could be definitely weaker than $\mathfrak{P}(X(B_\alpha,\mathfrak{P}_\alpha))$.) By the Noether Normalization Lemma, we can obtain B_α as an integral extension of some pure polynomial extension $A_\alpha[X_1 \ldots X_{k_\alpha}]$ of A_α. Thus we can study $A \cap \mathfrak{P}$ by first contracting the $\mathfrak{P}_\alpha(X_\alpha)$, $X_\alpha = X(B_\alpha,\mathfrak{P}_\alpha)$, to $A_\alpha[X_1 \ldots X_{k_\alpha}]$, then to A_α, and then contracting $\Pi(A_\alpha \cap \mathfrak{P}_\alpha(X_\alpha))$ to A. Perhaps surprisingly, the only point at which we might lose RHJ-algebras by contracting is in the pure polynomial extensions $A_\alpha \to A_\alpha[X_1 \ldots X_{k_\alpha}]$. In fact, by induction we could attack this extension by attacking the case of *one* variable $A_\alpha \to A_\alpha[X_1]$, with A_α a finite, real domain, and with an RHJ-algebra order on $A_\alpha[X_1]$, of the form $\mathfrak{P}(Y_1)$ for some subset $Y_1 \subset X(A_\alpha,\mathfrak{P}_w) \times R$. We would then like to conclude that $(A_\alpha, A_\alpha \cap \mathfrak{P}(Y_1))$ was an RHJ-algebra, with $X(A_\alpha, A_\alpha \cap \mathfrak{P}(Y_1)) = \overline{\pi(Y_1)}$, where π: $X(A_\alpha, \mathfrak{P}_w) \times R \to X(A_\alpha, \mathfrak{P}_w)$ is projection. However, it is not clear how generally this holds, and even in the semi-algebraic case, we need the Tarski-Seidenberg theorem to prove it.

We conclude this section by pointing out that we have established an invariant criterion that $(B,\mathfrak{P}) \in (POR)$ is the affine coordinate ring of some closed semi-algebraic set in affine space over R. Namely, B must be a reduced R-algebra of finite type. Also, there must be finitely many $\mathfrak{P}$-convex primes $P_i \subset B$ with $(0) = \cap P_i$, and finite refinements of the weak orders $\mathfrak{P}_i \subset B_i$, $B_i = B/P_i$, with $\mathfrak{P} = B \cap \Pi(\mathfrak{P}_i)_p$, under the inclusion $B \to \Pi B_i$. (Strictly speaking, we must still prove the basic real Nullstellensatz, 8.2.1, and establish Stengle's theorem that $\mathfrak{P}(X(B_i,\mathfrak{P}_i)) = (\mathfrak{P}_i)_p$. We proceed now to these tasks in the next three sections.)

8.3. Real Curves

Let $K = R(x_1 \dots x_n)$ be a real field, finitely generated over our real closed ground field R, with tr. deg.$_R(K) = 1$. Thus K is the field of fractions of the integral domain $A = R[x_1 \dots x_n] = R[X_1 \dots X_n]/P$, where $P \subset R[X_1 \dots X_n]$ is a prime $\mathfrak{P}_w$-convex ideal. We will prove that P has zeros in R^n, that is, there exist homomorphisms over R, $p: A \to R$.

Actually, we will prove something stronger, namely that K can be totally ordered *non-Archimedean* over $R \subset K$, with $A \subset A_R$, where $A_R \subset K$ is the valuation ring of elements finite relative to R. The associated signed place of K, p_R, is then R-valued since a non-trivial place of K over R lowers the transcendence degree over R. (Or, since tr. $\deg_R(K) = 1$, R must be Archimedean closed in K for any order on K non-Archimedean over R. Thus the residue field A_R/Q_R is algebraic over R by results of 7.6.) Thus we obtain p_R: $K \to R, \pm\infty$ and, by restriction, p: $A \to A_R \to R$.

Proposition 8.3.1. Let $K = R(x_1 \dots x_n)$ be a real function field in one variable over a real closed field R. Let $\mathfrak{P} \subset K$ be any total order. Let $x \in K - R$, so that K is algebraic over $R(x)$, and let $\overline{\mathfrak{P}} \subset R(x)$ denote the contracted order $\mathfrak{P} \cap R(x)$. Then (using the notation of 7.5):

Case (i) If $\overline{\mathfrak{P}} = \mathfrak{P}_{\infty}$ [resp. $\mathfrak{P}_{-\infty}$], we can find $b \in R$ such that for all $a > b$ [resp. $a < b$], both orders $\mathfrak{P}_{a,+}$ and $\mathfrak{P}_{a,-}$ on $R(x)$ extend to total orders on K with $A \subset A_R$.

Case (ii) If $\overline{\mathfrak{P}} = \mathfrak{P}_{c,+}$ [resp. $\mathfrak{P}_{c,-}$], $c \in R$, we can find $\varepsilon > 0$ in R such that for all $a \in (c,c+\varepsilon)$ [resp. $a \in (c-\varepsilon,c)$], both orders $\mathfrak{P}_{a,+}$ and $\mathfrak{P}_{a,-}$ on $R(x)$ extend to total orders on K with $A \subset A_R$.

Case (iii) If $\overline{\mathfrak{P}} = \mathfrak{P}_D$ where $D = D_R(x)$, the cut of R defined by the order $\mathfrak{P} \subset K$, is transcendental, then we can find $a', a'' \in R$, $a' < D < a''$, such that for all $a \in (a',a'')$, both orders $\mathfrak{P}_{a,+}$, $\mathfrak{P}_{a,-}$ on $R(x)$ extend to total orders on K with $A \subset A_R$.

Proof: Case (i) occurs only if $D = D_R(x) = \pm\infty$. We can deduce case (i) from case (ii) with $c = 0$, by replacing x by $1/x$.

In cases (ii) and (iii), choose $y \in K$ so that $K = R(x,y)$. Let

$$0 = G(y) = g_0(x)y^m + g_1(x)y^{m-1} + \cdots + g_m(x)$$

$$0 = F_j(x_j) = f_{0,j}(x)x_j^{m_j} + f_{1,j}(x)x_j^{m_j-1} + \cdots + f_{m_j,j}(x)$$

be the minimal polynomials for y and x_j, $1 \leq j \leq n$, with $g_i(x)$, $f_{i,j}(x) \in R[x]$. Consider the Sturm sequence for the irreducible polynomial $G(y)$ over $R(x)$, say $G_0(y) = G(y)$, $G_1(y) = \frac{d}{dy} G_0(y), \ldots, G_k(y)$. The last term is a constant $h(x) \neq 0 \in R(x)$.

Now, by our assumption on K, the polynomial $G(y)$ has real roots over $(R(x), \overline{\mathfrak{P}})$. Suppose $(\alpha(x),\beta(x)) \subset R(x)$ is an interval containing (at least) one of these roots. The terms in the sequences $\{G_0(\alpha(x)),\ldots G_k(\alpha(x)) = h(x)\}$ and $\{G_0(\beta(x)),\ldots,G_k(\beta(x)) = h(x)\}$ have $\overline{\mathfrak{P}}$-signs in $R(x)$. We now appeal to the results of 7.5 to choose $\varepsilon > 0$ in case (ii) [resp. $a' < D < a''$ in case (iii)] so that if $a \in (c, c+\varepsilon)$ or $(c-\varepsilon, c)$, depending on whether $\overline{\mathfrak{P}} = \mathfrak{P}_{c,+}$ or $\mathfrak{P}_{c,-}$ [respectively $a \in (a',a'')$], then the R-signs of all terms $G_i(\alpha(a))$, $G_i(\beta(a))$ agree with the $\overline{\mathfrak{P}}$-signs of the $G_i(\alpha(x))$, $G_i(\beta(x))$. We also make our choices so that $\alpha(a) < \beta(a)$ in R and so as to insure that

a is not a root of any of the numerators or denominators of the rational functions $\alpha(x)$, $\beta(x)$, or the coefficients of the polynomials $G_i(y)$ over $R(x)$. Finally, and this is the most important restriction, we insure that the conditions on a guarantee that a is not a root of any of the leading coefficients $f_{0,j}(x)$ of the minimal polynomials F_j for the elements x_j, $1 \leq j \leq n$.

With all this accomplished, we now start over with either order $\mathfrak{P}_{a,+}$ or $\mathfrak{P}_{a,-}$ on $R(x)$, and consider the Sturm algorithm for real roots of $G(y)$ in the interval $(\alpha(x), \beta(x)) \in R(x)$. If p: $R(x) \to R, \pm\infty$ is the signed place defined by our new order $\mathfrak{P}_{a,+}$ or $\mathfrak{P}_{a,-}$ on $R(x)$, then, by our choices, $p(G_i(\alpha(x))) = p(G_i(\alpha(a)))$ and $p(G_i(\beta(x))) = p(G_i(\beta(a))) \in R$ since $p(x-a) = 0$. The signs of terms which occur in our new Sturm algorithm thus coincide exactly with the signs which occurred in the original $\overline{\mathfrak{P}}$-Sturm algorithm for $G(y)$. Thus $G(y)$ has real roots over $R(x)$ in $(\alpha(x), \beta(x))$ with respect to either order $\mathfrak{P}_{a,+}$ or $\mathfrak{P}_{a,-}$, and this order on $R(x)$ extends to K.

Finally, since $0 = f_{0,j}(x)x_j^{m_j} + \dots + f_{m_j,j}(x) \in K$, $f_{i,j}(x) \in R[x]$ and $f_{0,j}(a) \neq 0$, it follows trivially that $x_j \in A_R \subset K$, hence $A = R[x_1 \dots x_n] \subset A_R$. □

<u>Remark</u>. Because of our careful choice of $a \in R$ in the proof above, additional conclusions can be drawn. Note that the signed place p: $R(x) \to R, \pm\infty$ associated to our order $\mathfrak{P}_{a,+}$ or $\mathfrak{P}_{a,-}$ on $R(x)$ has the property that the sequence $p(G_0(y)), \dots, p(G_k(y)) = p(h(x)) = h(a) \neq 0$, is actually a Sturm sequence over R for the polynomial $G(a,y) = \Sigma g_i(a)y^{m-i}$. This is not completely obvious because construction of a Sturm sequence involves division, so our choice of a avoiding roots of numerators or denominators of the coefficients of the $G_i(y)$ is essential here. In any event, since $h(a) \neq 0$, the polynomial $G(a,y)$ over R has no multiple roots. In fact, with a little further care we could arrange that $G(a,y)$ has the same number of roots in R as $G(y)$ has over $R(x)$ with either of the orders $\mathfrak{P}_{a,+}$, $\mathfrak{P}_{a,-}$, or $\overline{\mathfrak{P}}$, and that the R-roots of $G(a,y)$ are "close" to the $R(x)$-roots of $G(y)$. At these points $(a,b) \in R^2$ on the curve $G(x,y) = 0$, we have $(d/dy)G(a,b) \neq 0$. Thus we have a special type of simple point: the curve $G = 0$ in R^2 crosses the line $x = a$ transversally.

Examples. Consider the curves below:

(a)
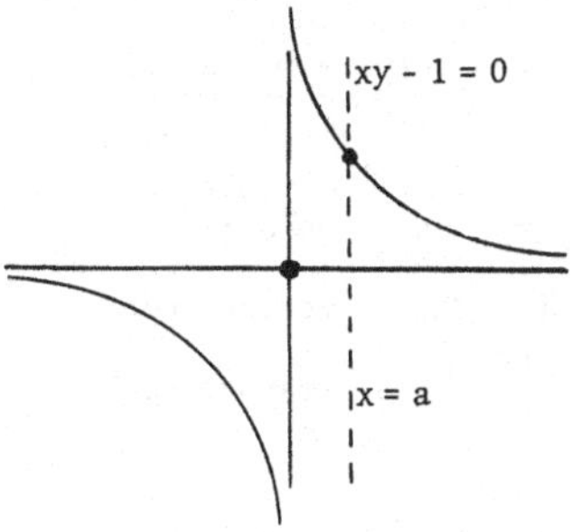

(b)
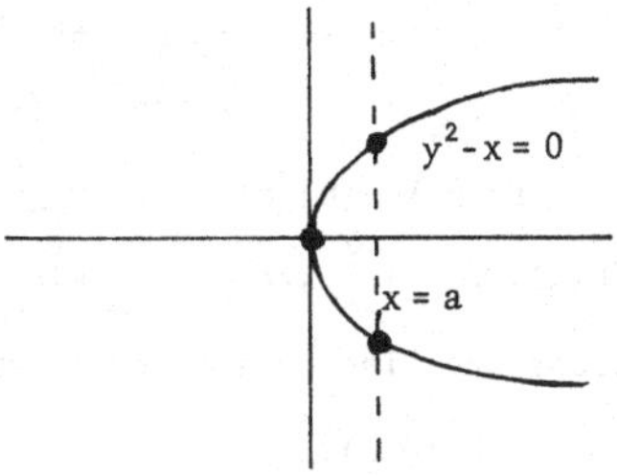

(c)
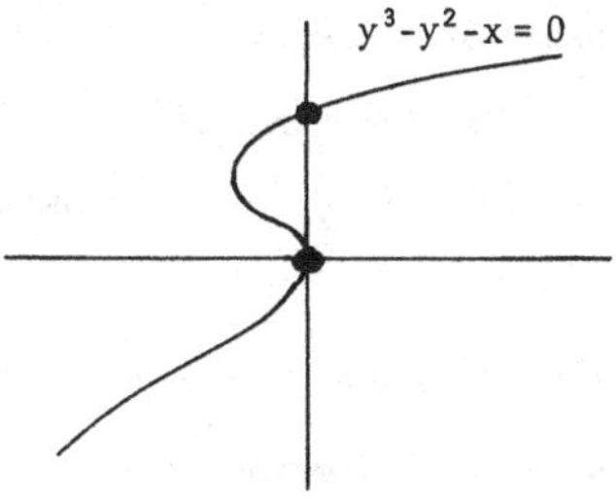

(d)
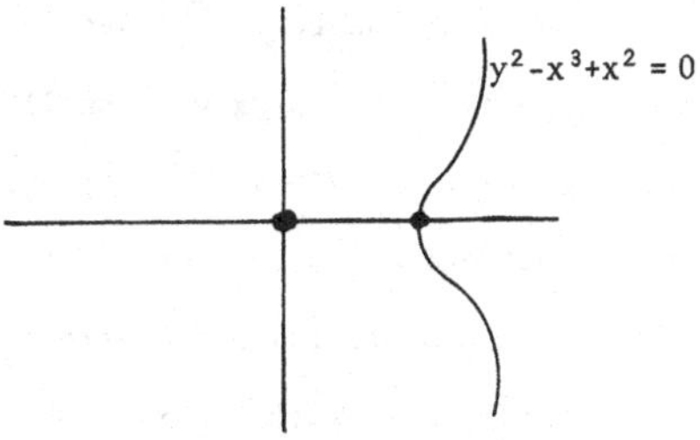

(e)
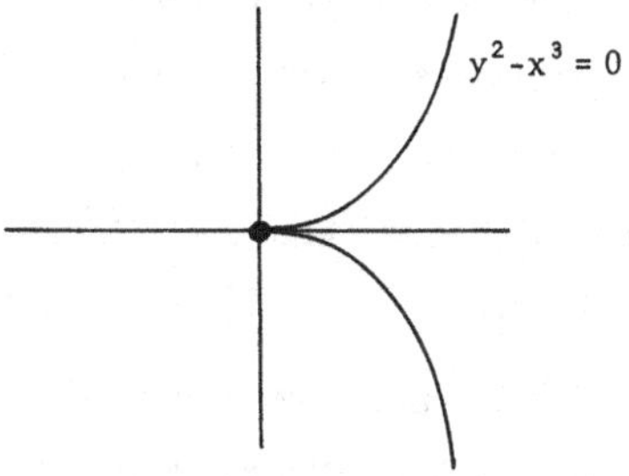

(f)
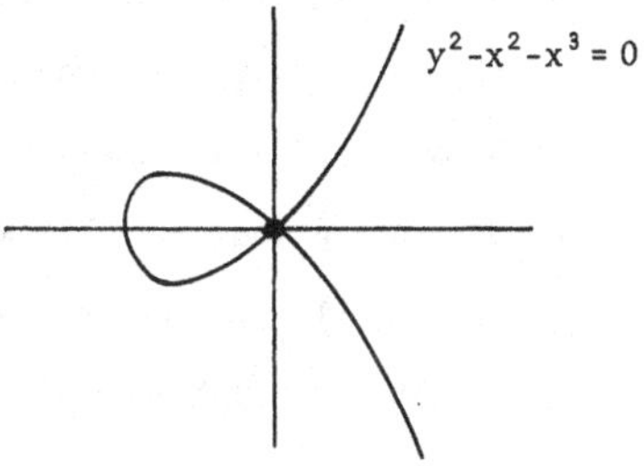

(g)
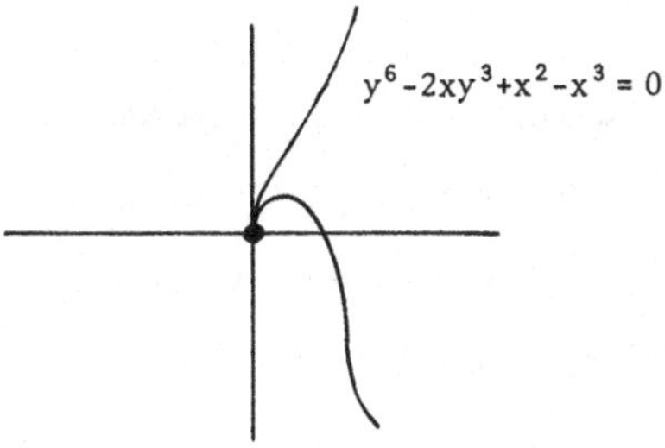

In each case we have an irreducible $G(x,y) = 0$, an affine coordinate ring $A = R[x,y]/(G(x,y))$, and a function field $K = R(x,y)$. We study the behavior near $x = 0$, to illustrate certain aspects of the proof above.

Example (a) illustrates that $R(x,y)$ may admit a signed place $p: R(x,y) \to R, \pm\infty$ with $p(x) = 0$, but with no points $(0,b) \in R^2$ on the curve. If $(R(x), \mathfrak{P}_{0,+})$ is extended to $R(x,y)$, then $p(y) = \infty$ and if $(R(x), \mathfrak{P}_{0,-})$ is extended, $p(y) = -\infty$.

Example (b) has the origin as a simple point, but only the order $(R(x), \mathfrak{P}_{0,+})$ extends to $R(x,y)$ since $x = y^2 > 0$. The curve does not cross the line $x = 0$ transversally, but does cross $x = a$ transversally for $a \in (0,\varepsilon)$.

In example (c), $G(x,y) = y^3 - y^2 - x$ has three real roots over $(R(x), \mathfrak{P}_{0,-})$, but only one real root over $(R(x), \mathfrak{P}_{0,+})$. This behavior is reflected by the specializations $x = a$, small a. If $a < 0$, $y^3 - y^2 - a$ has three real roots, if $a > 0$, $y^3 - y^2 - a$ has one real root, and if $a = 0$, $y^3 - y^2$ has a double root and a simple root.

Example (d) has a zero at the origin, but $R(x,y)$ does not support a total order with $D_R(x) \approx 0$. Since $G(0,0) = 0$, we get a homomorphism $p: A = R[x,y]/(G(x,y)) \to R$, hence $(x,y) \subset A$ is $\mathfrak{P}_w$-convex. However, $(x,y) \subset A$ is not $(\mathfrak{P}_w)_d$-convex. If it were, the signed place extension theorem of 7.7 would imply the existence of a total order on $K = R(x,y)$, with signed place $p_R: K \to R, \pm\infty$ extending $p: A \to R$. Note specifically that $x^2(x-1) = y^2$, hence $x-1 \in (\mathfrak{P}_w)_d$ or $1 \leq x \text{ rel} (\mathfrak{P}_w)_d$. Clearly, then (x,y) is not $(\mathfrak{P}_w)_d$-convex. The conclusion $x-1 \in (\mathfrak{P}_w)_d$ also shows that relative to any total orders on $K = R(x,y)$ we must have $1 \leq x$, which agrees with the picture.

In both examples (e) and (f), $R(x,y)$ supports total orders at the origin, but the origin is a singular point. In example (e), $(R(x), \mathfrak{P}_{0,+})$ extends to $R(x,y)$ in two ways (with $y > 0$ or $y < 0$), but $(R(x), \mathfrak{P}_{0,-})$ does not extend. In example (f), both orders $\mathfrak{P}_{0,+}, \mathfrak{P}_{0,-}$ on $R(x)$ extend in two ways to $R(x,y)$.

In example (g), $(R(x), \mathfrak{P}_{0,+})$ extends to $R(x,y)$ in two ways. In this case, it turns out that there is no element of $R(x)$ between the two real roots of $G(x,y)$ over $(R(x), \mathfrak{P}_{0,+})$. (Geometrically, no graph of $y = f(x)/g(x)$ lies between the two infinitesimal pieces of the curve

$G(x,y) = 0$ at $(0,0)$.) The Sturm algorithm for *any* interval $(\alpha(x), \beta(x))$ in $(R(x), \mathfrak{P}_{0,+})$ will yield either 2 roots or 0 roots for $G(x,y)$ in $(\alpha(x), \beta(x))$, depending on whether or not $0 \in (\alpha(x), \beta(x))$. Of course, when we specialize to small positive values $x = a$, these two roots separate. This example shows how $R(x)$ is not order dense in its real closure (see 7.4.4).

The proof of 8.3.1 leads to a rather nice picture of the real closure of the field $R(x)$ with any order $\mathfrak{P}$. If $\mathfrak{P} = \mathfrak{P}_{\pm\infty}$, we replace x by $1/x$. Also, all the orders $\mathfrak{P}_{a,\pm}$ look the same. Thus we look at only $\mathfrak{P}_{0,+}$ and $\mathfrak{P}_D$ where D is a transcendental cut of R.

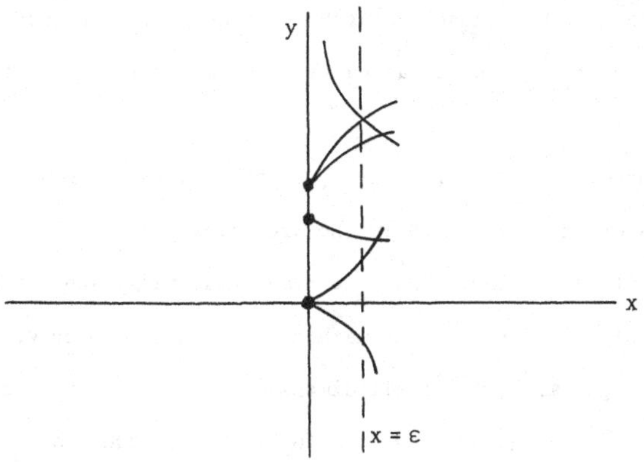

The points in the real closure of $(R(x), \mathfrak{P}_{0,+})$ are roots of irreducible polynomials $G(y)$ over $R(x)$. We can think of these roots as infinitesimal "connected" pieces of plane curves $G(x,y) = 0$, near the y-axis on the positive x-side. One point is greater than another if its graph is above that of the other, infinitesimally near the y-axis. A point is finite relative to R if its graph meets the y-axis, and is infinite relative to R if its graph is asymptotic to the y-axis. The elements which are infinitesimally small relative to R are those graphs through the origin. The addition and multiplication corresponds to addition and multiplication of functions on $(0,\varepsilon) \subset R$.

The case of the transcendental order $\mathfrak{P}_D \subset R(x)$ is slightly simpler. First, the real closure of $(R(x), \mathfrak{P}_D)$ is actually Archimedean over R. We visualize points as infinitesimal pieces of graphs of plane curves

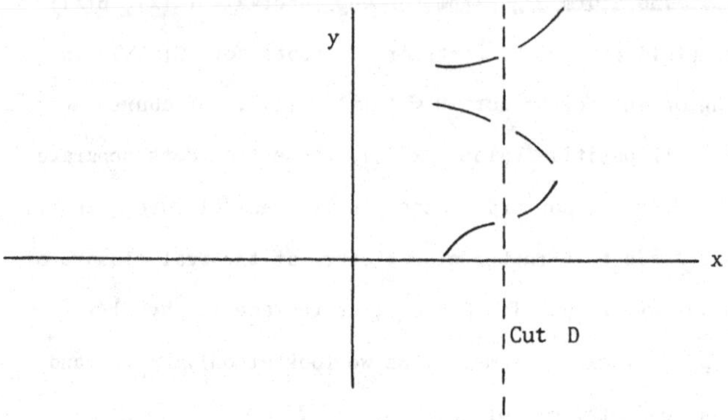

"crossing" the cut at $x \approx D$. (The "values" of the function defined by the graph can be extended to $x = D$. These "values" are necessarily transcendental cuts of R.) In the case of $\mathfrak{P}_D$, R is actually order dense in the real closure of $(R(x), \mathfrak{P}_D)$.

The notion of "connected" piece of a curve, together with rigorous formulations and proofs of the statements in the preceding two paragraphs, would require some effort to develop. Thus, we mean these paragraphs to be intuitive discussions only. After all the machinery of this Chapter VIII is developed, for semi-algebraic sets of all dimensions, it would still be a respectable project to investigate curves carefully. This study would include a detailed analysis of singular points, especially branches at a singular point, and a classification of orders on any function field in one variable, extending the results of 7.5 (which is the special case R(x)).

8.4. Signed Places on Function Fields

In this section we establish the existence of nice signed places on real function fields, in order to study prime convex ideals for various orders on integral domains of finite type over R. We begin with the weak order. Let $P \subset R[X_1 \cdots X_n]$ be a $\mathfrak{P}_W$-convex prime ideal in the polynomial ring, let $A = R[x_1 \cdots x_n] = R[X_1 \ldots X_n]/P$, and let $K = R(x_1 \ldots x_n)$, the fraction field of A. Thus, K is a real function field.

Proposition 8.4.1. (Lang) If $K = R(x_1 \ldots x_n)$ is a real function field with $\mathrm{tr.deg.}_R(K) = r$, then there exists a sequence of signed places over R,

$K \to K_1, \pm\infty \to K_2, \pm\infty \to \cdots \to K_r, \pm\infty$, with each K_i a function field with $\text{tr.deg.}_R(K_i) = r - i$. (Thus, $K_r = R$.) Moreover, if $A_i \subset K$ is the valuation ring of the composite p_i: $K \to K_i, \pm\infty$, then we can arrange that $A \subset A_i$, where $A = R[x_1 \ldots x_n]$. In particular $p_r|_A$: $A \to R$ is a homomorphism, hence $(a_1 \ldots a_n) = (p_r(x_1), \ldots, p_r(x_n)) \in R^n$ is a zero of the prime ideal $P \subset R[X_1 \ldots X_n]$.

Proof: The proof is by induction on r. If $r = 1$, the proposition follows from 8.3.1. If $r > 1$, assume $\{x_1 \ldots x_r\}$ is a transcendence base of K over R, so that K is algebraic over the pure transcendental extension $R(x_1 \ldots x_r)$. Choose any total order on K and let Γ denote the real closure of K, $F \subset \Gamma$ the real closure of $R(x_1 \ldots x_{r-1})$. Let $K' = F(x_r \ldots x_n) \subset \Gamma$. Thus K' is a function field in one variable over the real closed field F. Using 8.3.1 we find p: $K' \to F, \pm\infty$, over F, such that $p(x_i) \in F$ is finite, $r \le i \le n$. Let $p_1 = p|_K$: $K \to F, \pm\infty$. The image of p_1 is a subfield $K_1 \subset F$ with $\text{tr.deg.}_R(K_1) = r-1$ since $p_1(x_j) = x_j$, $1 \le j \le r-1$. Thus K_1 is actually a *function field* over R of transcendence degree $r-1$. (See [68], Vol. II, Chapter VI, §14, Theorem 31.) The proposition now follows by induction applied to K_1. □

Remark. In continuing the proof to obtain $K_1 \to K_2, \pm\infty$, a *new* order on K_1 may need to be chosen, as in the proof of 8.3.1. However, when we ultimately arrive at $K \to K_1, \pm\infty \to \cdots \to R, \pm\infty$, then we order K_{r-1} using $K_{r-1} \to R, \pm\infty$. Now $K_{r-2} \to K_{r-1}, \pm\infty$ is at least a place if the distinction between $\pm\infty$ is dropped. Using Krull's Theorem 7.7.2, we then *reorder* K_{r-2} and recover a signed place $K_{r-2} \to K_{r-1}, \pm\infty$ relative to this "new" order on K_{r-1}. Continuing, we work backwards until all K_i, including $K = K_0$, are reordered so that all $K_i \to K_{i+1}, \pm\infty$ are signed places.

Our proposition has many consequences. The first is the following "weak" weak Nullstellensatz.

Corollary 8.4.2. The maximal $\mathfrak{P}_W$-convex ideals of $R[X_1 \ldots X_n]$ are the ideals $(X_1 - a_1, \ldots, X_n - a_n)$ corresponding to points of R^n.

Proof: In fact, any $\mathfrak{P}_w$-convex ideal I is contained in a prime $\mathfrak{P}_w$-convex ideal P and if $r = \text{tr.deg.}_R(R[X_1 \ldots X_n]/P)$, 8.4.1 shows that there is a chain of prime $\mathfrak{P}_w$-convex ideals $P = P_0 \subsetneqq P_1 \subsetneqq \cdots \subsetneqq P_r = (X_1 - a_1, \ldots, X_n - a_n) \subset R[X_1 \ldots X_n]$. Namely, $P_i = \text{kernel}(p_i|_A : A \to K_i)$, in the notation of 8.4.1. □

Proposition 8.4.3. Let $K = R(x_1 \ldots x_n)$ be a real function field of transcendence degree r over R. Suppose $g_1, \ldots, g_k \in K$, $g_i \neq 0$, are finitely many elements positive in some total order on K. Then there exist signed places over R, $p\colon K \to R, \pm\infty$, with $\text{rank}(p) = r$, $p(x_i), p(g_j) \in R$, and $p(g_j) > 0$.

Proof: Apply 8.4.1 to the real function field $E = R(x_i, y_j, z_j)$ where $y_j^2 = g_j$, $y_j z_j = 1$. □

Remark. An alternate proof of 8.4.3 could be given as follows. If $r = 1$, as in 8.3.1, first any order $\mathfrak{P} \subset K$ is chosen, the replaced by an order extending one of the orders $\mathfrak{P}_{a,\pm} \subset R(x_1) \subset K$, by a careful choice of $a \in R$. This choice can be made exactly as in the proof of 8.3.1, so as to preserve the signs of the g_j and keep the g_j finite, but not infinitesimally small relative to R. This technique is now combined with the inductive proof of 8.4.1 to give 8.4.3.

Using either method of proof of 8.4.3, the conclusion can be strengthened to include an approximation statement. Precisely, the original order on K defines a signed place $p_R\colon K \to \Delta, \pm\infty$, where Δ is Archimedean over R. The new order on K corresponds to an R-valued signed place $p\colon K \to R, \pm\infty$. The conclusion of 8.4.3 can then be extended to say that the $p(g_j) \in R$ approximate the $p_R(g_j) \in \Delta, \pm\infty$ as closely as desired. (If $p_R(g_j) = \pm\infty$, this just means $p(g_j)$ can be made large.)

The next corollary constitutes a "weak" strong Nullstellensatz.

Corollary 8.4.4. Let $P \subset R[X_1 \ldots X_n]$ be a prime $\mathfrak{P}_w$-convex ideal, $g \notin P$. Then there exist zeros of P, $a \in R^n$, with $g(a) \neq 0$.

Proof: $g \neq 0 \in R(x_1 \dots x_n) = K$, the fraction field of $R[X_1 \dots X_n]/P$. Choose a total order on K and apply 8.4.3 to the single element g or $-g$, whichever is positive. □

Remark. Corollary 8.4.4 is exactly the basic Nullstellensatz stated previously as Proposition 8.2.1. Thus all the consequences of 8.2.1 are now established. For example, if $I \subset R[X_1 \dots X_n]$ is any ideal, $Z(I) = \{a \in R^n | h(a) = 0 \text{ all } h \in I\}$, then $\sqrt{H(I,\mathfrak{P}_w)} = \{f \in R[X_1 \dots X_n] | f(a) = 0 \text{ all } a \in Z(I)\}$. If we combine this result with the characterization 2.2.4 of $\sqrt{H(I,\mathfrak{P}_w)}$, we conclude that the ideal of functions which vanish on $Z(I)$ consists of exactly those f such that $f^{2s} + \Sigma p_i^2 = \Sigma q_j^2 h_j^2$, for some $s \geq 1$, $p_i, q_j \in R[X_1 \dots X_n]$, $h_j \in I$. This result, or similar versions, is known as the Dubois-Risler Nullstellensatz in the recent literature, [14], [16], [17]. We point out that our proposition 8.4.3 as well as Corollary 8.4.4 date back to the early work of Artin, at least in essentially equivalent forms. The other ingredient of the Dubois-Risler Nullstellensatz is then the characterization of the radical of a hull for general partially ordered rings, given in 2.2.4.

In fact, little extra work is now required to characterize those functions $f \in R[X_1 \dots X_n]$ which vanish on a basic closed semi-algebraic set of the form $Z(I) \cap W\{g_j\} \subset R^n$. First, assume $J \subset R[X_1 \dots X_n]$ is a radical ideal so that $\mathfrak{P} = \mathfrak{P}_w[g_j]$ is an order on $A = R[X_1 \dots X_n]/J$ and assume $J \subset I$. Then $(A,\mathfrak{P})$ is an RHJ-algebra, $X(A,\mathfrak{P}) = Z(J) \cap W\{g_j\} \supset Z(I) \cap W\{g_j\}$ and the functions $f \in A$ which vanish on $Z(I) \cap W\{g_j\}$ are exactly the functions $f \in \sqrt{H(I,\mathfrak{P})} \subset A$. In terms of polynomials $f \in R[X_1 \dots X_n]$, we conclude f vanishes on $Z(I) \cap W\{g_j\}$ if and only if there is a formula

$$f^{2s} + \Sigma h^2_{i_1 \dots i_r} g_{i_1} \cdots g_{i_r} = \Sigma (k^2_{j_1 \dots j_q} g_{j_1} \cdots g_{j_q}) y^2_{j_1 \dots j_q} + z$$

for some $s \geq 1$ and polynomials h,k,y,z, with $y \in I$ and $z \in J$. (We allow the "empty product" 1 of the g_j here.)

Secondly, it is not hard to see that the smallest such ideal J is exactly $I(W\{g_j\})$. Certainly, $\mathfrak{P}_w[g_j]$ is an order on $R[X_1 \dots X_n]/I(W\{g_j\})$.

If J is any radical ideal with this property and $h \in I(W\{g_j\})$, then h vanishes on $X(R[X_1 \ldots X_n]/J, \mathfrak{P}_w[g_j])$, hence $h = 0 \in R[X_1 \ldots X_n]/J$. Moreover, if $J = I(W\{g_j\})$, then $Z(I) \cap W\{g_j\} = Z(I + J) \cap W\{g_j\}$, and $J \subset I + J$.

The two paragraphs above reduce the determination of $I(Z(I) \cap W\{g_j\})$ to the determination of $I(W\{g_j\})$. If there exists $x \in R^n$ with $g_j(x) > 0$, all j, then, of course, $I(W\{g_j\}) = 0$. Even if $W\{g_j\}$ has no interior in R^n, in a given case we may know an ideal $J \subset I$ with $\mathfrak{P}_w[g_j]$ an order on $R[X_1 \ldots X_n]/J$. It would then be unnecessary to actually determine $I(W\{g_j\})$.

Next we state a weaker form of 8.4.3 which is as useful for most purposes.

Corollary 8.4.5 (Artin) Let $A = R[x_1 \ldots x_n]$ be a real integral domain $\mathfrak{P}_w[g_1 \ldots g_k]$ a finitely generated refinement of the weak order on A, $g_i \neq 0$. Then there is a homomorphism $p\colon A \to R$ with $p(g_i) > 0$, $1 \leq i \leq k$.

Proof: Extend $\mathfrak{P}_w[g_i]$ to a total order on A and apply 8.4.3. □

In particular, this result applies to the polynomial ring $R[X_1 \ldots X_n]$. In fact, we deduce the following, also due to Artin. (Part (b) is Hilbert's 17th problem.)

Corollary 8.4.6.

(a) If $g_i \in R[X_1 \ldots X_n]$ are nonzero polynomials, $1 \leq i \leq k$, then $\mathfrak{P}_w[g_i]$ is an order on $R[X_1 \ldots X_n]$ if and only if there exists $a \in R^n$ with $g_i(a) > 0$, $1 \leq i \leq k$.

(b) The polynomials $f \in R[X_1 \ldots X_n]$ nowhere negative as functions on R^n coincide with elements of the derived order

$$(\mathfrak{P}_w)_d = \{f \mid h^2 f = \Sigma h_j^2, \text{ some } h, h_j \in R[X_1 \ldots X_n], h \neq 0\}.$$

Proof: The "only if" statement in (a) is 8.4.5. The "if" statement is true because the set $U\{g_i\} = \{a \mid g_i(a) > 0, 1 \leq i \leq k\}$ is Zariski dense if nonempty.

Similarly, a function $f \in (\mathfrak{P}_w)_d$ is clearly nowhere negative since if $f(a) < 0$ and $h^2 f = \Sigma h_j^2$, then h would vanish identically in some ball around a. Conversely, if $f \notin (\mathfrak{P}_w)_d$, then $\mathfrak{P}_w[-f]$ is an order on $R[X_1 \dots X_n]$ and (a) applies. □

Remark 8.4.7. The "if" statement of 8.4.6(a) is not true for arbitrary finite real domains $A = R[X_1 \dots X_n]/P$. That is, the existence of a homomorphism $p: A \to R$ with $p(g_i) > 0$ does not imply $\mathfrak{P}_w[g_i]$ is an order on A. The point $a = (p(X_1), \dots, p(X_n))$ could be a "degenerate" zero of P as in Example (d) of 8.3. The argument of 8.4.6(a) shows that if $U\{g_i\} \cap X(A, \mathfrak{P}_w)$ is Zariski dense in $X(A, \mathfrak{P}_w)$, then $\mathfrak{P}_w[g_i] \subset A$ is an order. Here $X(A, \mathfrak{P}_w) = Z(P) \subset R^n$, the zeros of P, and $U\{g_i\} = \{b \in R^n | g_i(b) > 0\}$. Conversely, if $\mathfrak{P}_w[g_i] \subset A$ is an order and $h \neq 0 \in A$, then either $\mathfrak{P}_w[g_i, h]$ or $\mathfrak{P}_w[g_i, -h]$ is an order on A, since A is a domain. In either case, 8.4.5 guarantees that h does not vanish identically on $U\{g_i\} \cap X(A, \mathfrak{P}_w)$. Thus $U\{g_i\}$ is Zariski dense in $X(A, \mathfrak{P}_w)$.

Remark 8.4.8. Consider the orders $\mathfrak{P}_w \subset (\mathfrak{P}_w)_s \subset (\mathfrak{P}_w)_p \subset (\mathfrak{P}_w)_m = (\mathfrak{P}_w)_d = \mathfrak{P}(R^n)$ on the polynomial ring $R[X_1 \dots X_n]$. In the next section we will prove a theorem of Stengle which implies $(\mathfrak{P}_w)_p = \mathfrak{P}(R^n)$. Thus, there are only three distinct orders here, $\mathfrak{P}_w \subset (\mathfrak{P}_w)_s \subset \mathfrak{P}(R^n)$. These inclusions are definitely proper. For example, Hilbert [1] gave an example of a strictly positive polynomial in two variables, $f(X_1, X_2)$, which is not a sum of squares. Our proof that $(\mathfrak{P}_w)_p = \mathfrak{P}(R^n)$ will show, in fact, that any strictly positive polynomial belongs to $(\mathfrak{P}_w)_s$. Thus $\mathfrak{P}_w \neq (\mathfrak{P}_w)_s$, if $n = 2$. On the other hand, by definition, $(\mathfrak{P}_w)_s = \{f | (1 + \Sigma h_i^2) f = \Sigma g_j^2\}$. Immediately, the homogeneous term of f of lowest degree must be a sum of squares. But now if, say, Hilbert's example $f(X_1, X_2)$ is made homogeneous, $F(X_0, X_1, X_2)$, so that $f(X_1, X_2) = F(1, X_1, X_2)$, then $F(X_0, X_1, X_2)$ cannot be a sum of squares and hence cannot belong to $(\mathfrak{P}_w)_s$. Thus $(\mathfrak{P}_w)_s \neq \mathfrak{P}(R^n)$ if $n = 3$.

The last result in this section is a signed place perturbation theorem for function fields. The statement and proof are similar to 8.4.1, but instead of trying to make elements finite relative to a discrete rank r signed place, we assume certain elements infinitesimally small relative to one signed place, then try to keep them that way relative to a discrete rank r signed place. This result will be the basis of our study of derived orders in 8.6 and dimension theory in 8.10. It provides a tool for studying the geometry of a semi-algebraic set near a point, or more generally near a semi-algebraic subset.

Proposition 8.4.9. Let $A = R[x_1 \ldots x_n]$ be a finite integral domain over R with $\text{tr.deg.}_R(A) = r$. Let K be the fraction field of A and suppose p': $K \to \Delta, \pm\infty$ is a signed place over R, with Δ Archimedean over R and $p'(x_i) = 0$. (Necessarily $p' = p_R$, relative to the order on K induced by p'. This follows easily from results of 7.6.) Then there exists a discrete rank r signed place p: $K \to R, \pm\infty$ with $p(x_i) = 0$.

Proof: If $r = 1$, there is nothing to prove since we assume $p'(x_i) = 0$, hence p' is already discrete rank 1 and R-valued. If $r \geq 2$, we prove the proposition by induction.

By the Noether Normalization Lemma we can find r linear combinations of the x_i, say $u_1,\ldots,u_r$, such that A is integral over $R[u_1 \ldots u_r]$. Certainly $p'(u_i) = 0$. Thus we may as well assume that A is already integral over $R[x_1 \ldots x_r]$.

Let $f_j(x_{r+j}) = \Sigma a_{ij}(x_1 \ldots x_r)x^i_{r+j} = 0$ be the minimal polynomial for x_{r+j}, $j > 0$, with coefficients in $R[x_1 \ldots x_r]$. The $f_j(x_{r+j})$ are monic polynomials, hence applying p' we get that $0 \in R$ is a root of the *nontrivial* polynomial $\Sigma a_{ij}(0 \ldots 0)T^i \in R[T]$. Choose $0 < \varepsilon \in R$ such that ε is less than the absolute value of all other roots $\{\alpha_k\}$ of the $\Sigma a_{ij}(0 \ldots 0)T^i$ over R. Let L be the algebraic extension of K obtained by adjoining $\sqrt{\varepsilon^2 - x^2_{r+j}}$, $j > 0$, and replace A by $R[x_i, y_j]$, where $y_j = \sqrt{\varepsilon^2 - x^2_{r+j}} - \varepsilon$. The order on K defined by p' extends to L. In fact, the place p' extends to L, namely, take $p' = p_R$, $R \subset L$. The residue field is Archimedean over R, hence replacing K by L does not change our hypotheses. Also, any discrete rank r signed place p: $L \to R, \pm\infty$ restricts to a discrete rank r signed place of K. Thus we may assume $K = L$.

With this assumption, given any signed place $p\colon K \to R, \pm\infty$ over R with $p(x_i) = 0$, $1 \leq i \leq r$, then necessarily also $p(x_{r+j}) = 0$, $j > 0$. The reason is that $0 \in R$ is the only root of the $\Sigma a_{ij}(0\cdots0)T^i$ in the interval $(-\varepsilon, \varepsilon)$ and $p(x_{r+j}) \in (-\varepsilon, \varepsilon)$ since now $x_{r+j}^2 < \varepsilon^2$ in K.

Let Γ be the real closure of K relative to the order defined by p'. Let $F \subset \Gamma$ be the real closure of the subfield $R(x_1 \ldots x_{r-1})$ of K. Write $K = R(x_1 \ldots x_r, y)$, with y algebraic over $R(x_1 \ldots x_r)$. Then $F(x_r, y)$ is a function field in one variable x_r over F, totally ordered as a subfield of Γ. With respect to this total order, the x_i, $1 \leq i \leq r$, are infinitesimally small relative to the subfield R. We may assume the x_i positive, be replacing x_i by $-x_i$ if necessary.

The order or $F(x_r, y)$ induces an order on the simple transcendental extension $F(x_r)$ of F. This order is completely determined by the cut $D_F(x_r) = D$ of F, as in 7.5.1. We have $0 < D$, that is, $0 \in D$ if D is regarded as a subset of F. Now, D may be algebraic or transcendental. If D is algebraic, we will not yet perturb the order on $F(x_r, y)$. If D is transcendental, we appeal to Proposition 8.3.1 to choose an element $\beta \in F$, $0 < \beta < D$, so that if $F(x_r)$ is *reordered* over F with either order $\mathfrak{P}_{\beta,\pm}$, as in 7.5.1, then this new order extends to $F(x_r, y)$. Since $0 < \beta < D$, we still have x_r infinitesimally small relative to $R \subset F$. Of course, the x_i, $i < r$, are also still infinitesimally small relative to R since F has not changed.

We now consider the signed place p_F on $F(x_r, y)$, with respect to our (possibly) new order. We have forced p_F to be non-trivial, thus p_F is F-valued. In fact, we have forced the restriction $p_F\colon R(x_1 \ldots x_r, y) = K \to F, \pm\infty$ to be non-trivial. (Specifically, $p_F(g(x_1 \ldots x_r)) = 0$ where $g(x_1 \ldots x_{r-1}, \beta) = 0$ is the minimal polynomial for $\beta \in F$ over $R[x_1 \ldots x_{r-1}]$.) But p_F is still trivial on $R(x_1 \ldots x_{r-1})$, hence $p_F\colon K \to F, \pm\infty$ is a discrete rank 1 signed place.

The image $p_F(K) = K' \subset F$ is thus a function field of transcendence degree $r-1$ over R. In fact, the proof of this result in, say, [68, vol.II, Chap. VI, §14, Theorem 31] shows more. Let $A' = R[x_i, z_j]$ denote the integral closure of $A = R[x_i]$ in K. Then K' is the field of fractions of A'/P', where $P' \subset A'$ is some minimal prime ideal. With the order on K' induced by the inclusion $K' \subset F$, the z_j are finite relative to R, since z_j is integral over A. Thus we can write $A'/P' = R[x_i', z_j'] \subset K'$, where $x_i' = x_i \pmod{P'}$ and $z_j' = z_j - p_R(z_j) \pmod{P'}$. By induction, applied to the signed place p_R: $K' \to \Delta', \pm\infty$, Δ' Archimedean over R, there is a discrete rank $r-1$ signed place q: $K' \to R, \pm\infty$, with $q(x_i') = 0$.

We now compose p_F and q, $K \to K', \pm\infty \to R, \pm\infty$, and reorder K by altering p_F on infinte elements, as in 7.7.2, to take into account the fact that the order on K' induced by p_F is not the same as that induced by q. We end up with our desired rank r signed place p: $K \to R, \pm\infty$. □

Remark. Here is an extension of 8.4.9 which is useful in studying geometry near a point. In the notation of 8.4.9, suppose given finitely many elements $g_j \in A$, $g_j \neq 0$, which are positive in the order on K induced by p': $K \to \Delta, \pm\infty$. Then we may add to the conclusion the statement that the g_j are also positive in the order induced by the discrete rank r signed place p: $K \to K_1, \pm\infty \to \ldots \to K_{r-1}, \pm\infty \to R, \pm\infty$, and moreover, we may arrange that $p_{r-1}(g_j) \neq 0$, where p_{r-1}: $K \to K_{r-1}, \pm\infty$ is the composition of the first $r-1$ maps in the chain.

If $g_j \notin (x_1 \ldots x_n) \subset A = R[x_1 \ldots x_n]$, then already $p(g_j) \neq 0$. There are various ways to keep the g_j positive relative to our new order on K. For example, using the technique of the proof of 8.2.10, one can adjoin certain $\sqrt{g_j}$ to K, then replace A by $R[x_1 \ldots x_n, \sqrt{g_j} - \varepsilon_j]$, where $p'(g_j) = \varepsilon_j^2 \in R$, $0 < \varepsilon_j$. (See also the proof of Proposition 8.6.5 in a later section for this technique.) This technique works whether or not $g_j \in (x_1 \ldots x_n)$. Thus we can concentrate on the case $g_j \in (x_1 \ldots x_n)$, and just try to arrange that $p_{r-1}(g_j) \neq 0$. Setting $g = \Pi g_j$, we may assume there is only a single g. Now, we just add the dummy generator g to our formula for $A = R[x_1 \ldots x_n, g]$. Then we might as well rename the generators so that g becomes x_1, the first element of the transcendence base $\{x_1 \ldots x_r\}$. If one inspects the proof above, one finds $g = x_1 \in R(x_1 \ldots x_{r-1}) \subset F$, so that at the crucial inductive step constructing p_F: $F(x_r, y) \to F, \pm\infty$, we obviously have $p_F(g) = g \neq 0$. Our extension of 8.4.9 is thus proved by induction. (Note that, again, if $r = 1$ there is nothing to be proved.)

8.5. Characterization of Non-Negative Functions

In this section we prove the theorem of Stengle [22], characterizing the functions nowhere negative on a closed semi-algebraic set.

Begin with a reduced, real R-algebra A of finite type, and a finite refinement of the weak order $\mathfrak{P} = \mathfrak{P}_w[g_j]$.

Proposition 8.5.1 (Stengle). An element $f \in A$ is nowhere negative on $X(A,\mathfrak{P})$ if and only if there is an equation $(f^{2n} + p)f = q$, some $n \geq 0$, $p,q \in \mathfrak{P}$.

Proof. The "if" statement is trivial. We prove the "only if" statement by applying the Nullstellensätze of 8.2 and 8.4 to the ring $A' = A[t]$ of polynomials in one indeterminate over A.

Let $\mathfrak{P}' = \mathfrak{P}_w[g_j] \subset A'$. We know $\mathfrak{P}'$ is an order on A' since any order on A extends to an order on A' by 6.1. A homomorphism $x'\colon A' \to R$ is simply a homomorphism $x\colon A \to R$ together with a value $x'(t) \in R$. It follows that $X(A',\mathfrak{P}')$ is identified with $X(A,\mathfrak{P}) \times R$. (This we also discussed in 8.2.)

Suppose $f \in A$ is nonnegative on $X(A,\mathfrak{P})$. Then $f \in A \subset A'$ vanishes on all zeros of $t^2 + f$ in $X(A',\mathfrak{P}')$. Thus, by the Nullstellensatz, $f \in \sqrt{H(t^2 + f,\mathfrak{P}')} \subset A'$. From 2.2, we obtain an equation

$$(*) \qquad f^{2n}(x) + P(x,t) = Q(x,t)(t^2 + f(x))$$

for some $n \geq 1$, $P(x,t) \in \mathfrak{P}'$. We are regarding elements of A,A' as functions on $X(A,\mathfrak{P})$, $X(A',\mathfrak{P}')$, respectively.

Now any element $G(x,t) \in A'$ can be *uniquely* written $G(x,t) = G_0(x,t^2) + tG_1(x,t^2)$. Applying this decomposition to (*), we can replace $P(x,t)$, $Q(x,t)$ by $P_0(x,t^2)$, $Q_0(x,t^2)$, respectively. But any identity $F(x,t^2) = 0 \in A'$ implies $F(x,t) = 0 \in A'$ and thus $F(x,a) = 0 \in A$, any $a \in A$. Thus we obtain a relation in A,

$$(**) \qquad f^{2n}(x) + P_0(x, -f(x)) = 0.$$

Now, we investigate $P_0(x, -f(x))$ more closely. Since $P(x,t) \in \mathfrak{P}'$, we can find a formula

$$P(x,t) = \Sigma P_J^2(x,t) g_J(x), \quad g_J \in \mathfrak{P} \subset A.$$

Specifically, the g_J are finite products of the g_j, where $\mathfrak{P} = \mathfrak{P}_w[g_j] \subset A$. Thus

$$P_0(x,t^2) = (P_{J,0}^2(x,t^2) + t^2 P_{J,1}^2(x,t^2)) g_J(x)$$

by writing $P_J(x,t) = P_{J,0}(x,t^2) + t P_{J,1}(x,t^2)$, expanding, and using uniqueness of the decomposition $P(x,t) = P_0(x,t^2) + t P_1(x,t^2)$. Replacing t^2 by t and t by $-f(x)$, we find

$$P_0((x, -f(x)) = (\Sigma P_{J,0}^2(x, -f) g_J(x)) - f(x)(\Sigma P_{J,1}^2(x, -f) g_J(x)).$$

Thus (**) gives an equation $f^{2n}(x) + p(x) = q(x)f(x)$ with $p(x), q(x) \in \mathfrak{P}$. We obtain 8.5.1 by multiplying this last by $f(x)$. □

Corollary 8.5.2. Let A be a reduced R-algebra of finite type, $\mathfrak{P} = \mathfrak{P}_w[g_j]$ a finite refinement of the weak order on A, $f \in A$ a function nowhere negative on $X(A,\mathfrak{P})$. Suppose $g \in A$ vanishes on $Z(f) \cap X(A,\mathfrak{P})$, so that $g \in \sqrt{H(\{f\}, \mathfrak{P})} \subset A$. Then, there is an inequality $0 \leq g^{2n} \leq qf$ (rel $\mathfrak{P}$) for some $q \in \mathfrak{P}$.

Proof. This is a consequence of the proof of 8.5.1. We introduce $A' = A[t]$ again, and $\mathfrak{P}' = \mathfrak{P}_w[g_k] \subset A'$. Then $g \in \sqrt{H(\{t^2 + f\}, \mathfrak{P}')} \subset A'$. Going through the proof of 8.5.1 yields an equation $g^{2n} + p = qf$, $p, q \in \mathfrak{P}$, which is exactly our assertion. □

Remark. Corollary 8.5.2 should be compared with 2.2.3. We have replaced the hypothesis $f \in \mathfrak{P}$ of 2.2.3 by the weaker hypothesis f non-negative on $X(A,\mathfrak{P})$, and have established the same conclusion in both 8.5.2 and 2.2.3.

Corollary 8.5.3. Let A be a reduced R-algebra of finite type, $\mathfrak{P} = \mathfrak{P}_w[g_j]$ a finite refinement of the weak order on A, and $f \in A$ a function *strictly* positive on $X(A,\mathfrak{P})$. Then there is an equation $1+p = (1+q)f$, some $p,q \in \mathfrak{P}$.

Proof. Corollary 8.5.2 gives equations $(1+q')f = p'$, $p',q' \in \mathfrak{P}$ by setting $g = 1$. But now p' is also strictly positive on $X(A,\mathfrak{P})$. Applying 8.5.2 again gives $1+p'' = q''p'$, $p'',q'' \in \mathfrak{P}$. Now check that $(1+q')(1+q'')f = 1 + p' + p''$. □

Note that the equation $1+p = (1+q)f$ can be written $f = 1+p/1+q \in \mathfrak{P}_{S(1)} \subset A_{S(1)}$.

Let A be a reduced, real R-algebra of finite type, $\mathfrak{P} = \mathfrak{P}_w[g_j]$ a finite refinement of the weak order on A. We have now established that $(A,\mathfrak{P})$ is an RHJ-algebra and that the functions $f \in A$ nonnegative on $X = X(A,\mathfrak{P})$ coincide with $\mathfrak{P}_p = \{f \in A | (f^{2n} + p)f = q,\ \text{some}\ n \geq 1,\ p,q \in \mathfrak{P}\}$. Suppose $I \subset A$ is a radical $\mathfrak{P}$-convex ideal. Then I is $\mathfrak{P}_p$-convex by 8.2.2, and the orders $(\mathfrak{P}/I)_p$, $\mathfrak{P}_p/I$, $(\mathfrak{P}_p/I)_p \subset A/I$ are defined. By 8.5.1, $(\mathfrak{P}/I)_p \subset A/I$ consists of all functions $f \in A/I$ non-negative on $Z(I) \subset X$. On the other hand, $\mathfrak{P}_p/I$ consists of the restrictions to $Z(I)$ of the functions $f \in A$ non-negative on all of X. In general, $\mathfrak{P}_p/I \subset (\mathfrak{P}/I)_p$ will be a proper inclusion. For example, if $A = R[X,Y]$, $I = (Y^2 - X^3)$, then X is nowhere negative on $Z(I) \subset R^2$, but no polynomial $f(X,Y)$ congruent to X modulo $(Y^2 - X^3)$ can be non-negative on all of R^2. (Consider $\partial f/\partial X$ at the origin.) We do have $(\mathfrak{P}_p/I)_p = (\mathfrak{P}/I)_p$, since, first, $\mathfrak{P}/I \subset \mathfrak{P}_p/I$, which gives $(\mathfrak{P}/I)_p \subset (\mathfrak{P}_p/I)_p$, and, secondly, any $f \in (\mathfrak{P}_p/I)_p$ is easily seen to be non-negative on $Z(I) \subset X$, which gives $(\mathfrak{P}_p/I)_p \subset (\mathfrak{P}/I)_p$ by Stengle's theorem.

If A is a reduced real R-algebra of finite type and $\mathfrak{P} = \mathfrak{P}_w[g_i]$ is a finite refinement of the weak order, then the order $\mathfrak{P}_s \subset A$, consisting of all $f \in A$ with $(1+p)f = q$, some $p,q \in \mathfrak{P}$, is *not* a geometric invariant of the ring of functions A on the set $X(A,\mathfrak{P})$. For example, if $A = R[t]$, $\mathfrak{P} = \mathfrak{P}_w[t^3]$, then $t \notin \mathfrak{P}_s$.

On the other hand, if f is *strictly* positive on $X(A,\mathfrak{P})$, then 8.5.3

guarantees that $f \in \mathfrak{P}_s$. Also, recall that $\mathfrak{P}_s \subset A$ is defined as the contraction to A of the order $\mathfrak{P}_{S(1)} \subset A_{S(1)}$. Now, $A_{S(1)}$ can be invariantly characterized as the ring of functions on $X = X(A,\mathfrak{P})$ obtained by inverting all $f \in A$ with no zeros on X. This is immediate from 8.5.3, or directly from the Nullstellensatz, since if f is nowhere zero on X, then $1 \in H(\{f\}, \mathfrak{P})$ and by 2.2.4, $1+p = fg$, some $p \in \mathfrak{P}$, $g \in A$. The ring $A_{S(1)}$ is perhaps a more natural geometric invariant of X than the ring A. If we replace the order $\mathfrak{P}$ by $\mathfrak{P}(X)$, then $\mathfrak{P}(X)_{S(1)} \subset A(X)_{S(1)}$ is invariantly characterized as the functions nowhere negative on X. Recall from Chapter V that $(A(X)_{S(1)}, \mathfrak{P}(X)_{S(1)})$ is the ring of global section of the structure sheaf associated to $(A(X), \mathfrak{P}(X))$. Stengle's theorem has a nice interpretation in terms of the structure sheaf associated to $(A,\mathfrak{P})$. Namely, $f \in A$ is nowhere negative on X if and only if $f \in \mathfrak{P}_{S(f)} \subset A_{S(f)}$. That is, f should be "positive" in the ring of sections over the basic open set $D(f) \subset X$.

Stengle's theorem also generalizes Artin's result 8.4.6(b), in the case $A = R[X_1 \ldots X_n]$, $\mathfrak{P} = \mathfrak{P}_w$. Namely, it is trivial from 8.5.1 that if f is nowhere negative on R^n, then $f \in (\mathfrak{P}_w)_d$. Let $(\mathfrak{P}_w : f) = \{h^2 \in R[X_1 \ldots X_n] | h^2 f = \Sigma h_i^2$, some $h_i\}$. It would perhaps be interesting to characterize the zeros $Z(\mathfrak{P}_w : f) = Z(\sqrt{H((\mathfrak{P}_w : f), \mathfrak{P}_w)}) \subset R^n$. By 2.2.3, if $Z(\mathfrak{P}_w : f) = \emptyset$, then $f \in (\mathfrak{P}_w)_s$, and conversely. By Stengle's theorem $Z(\mathfrak{P}_w : f) \subset Z(f)$. It is also easy to see that $Z(\mathfrak{P}_w : f)$ must include all zeros x of f such that the homogeneous term of lowest degree in the Taylor expansion of f about x is *not* a sum of squares.

8.6. Derived Orders

Let A be a reduced, real R-algebra of finite type, $\mathfrak{P} = \mathfrak{P}_w[g_j]$ a finite refinement of the weak order $X = X(A,\mathfrak{P})$. In previous sections we have identified X with the maximal convex ideals of $(A,\mathfrak{P})$ and established that the functions nowhere negative on X coincide with $\mathfrak{P}_p = \{f \in A | (f^{2n}+p)f = q$, some $n \geq 1$, $p,q \in \mathfrak{P}\}$. In this section we study the derived order $\mathfrak{P}_d$ of 3.12. Also, if $I \subset A$ is a radical $\mathfrak{P}$-convex ideal, we investigate briefly the orders $(\mathfrak{P}_p/I)_d$ and, when defined, the orders $\mathfrak{P}_d/I$ and $(\mathfrak{P}_d/I)_d$. All of these orders are geometric invariants of the rings of functions A, A/I on $X(A,\mathfrak{P})$, $X(A/I, \mathfrak{P}/I)$, respectively.

Recall that the derived order $\mathfrak{P}_d$ consists of those $f \in A$ with $pf = q$, some $p,q \in \mathfrak{P}$ and p not a zero divisor. Our study of $\mathfrak{P}_d$, $\mathfrak{P} = \mathfrak{P}_w[g_j]$ will include, of course, the orders $(\mathfrak{P}/I)_d$. Also, we will then have a characterization of $\mathfrak{P}_d/I$, when defined. Note that since $(\cap \mathfrak{P}_i)_d = \cap(\mathfrak{P}_i)_d$ by 3.12, we will also have described $\mathfrak{P}_d$ for a finite intersection of orders of type $\mathfrak{P}_w[g_j]$.

An equation $pf = q$, $p,q \in \mathfrak{P}$, p not a zero divisor, does not imply f nowhere negative on $X = X(A,\mathfrak{P})$. The function p could vanish on the set $U(-f) \subset X$ where f is negative. We need to distinguish between "degenerate" and "non-degenerate" points of our semi-algebraic sets.

In general, let $(B,\mathfrak{P})$ be any RHJ-algebra, $X = X(B,\mathfrak{P})$. We say $x \in X$ is *degenerate* if for some $f,h \in B$, h not a zero divisor, we have $x \in U(f)$ and $h(y) = 0$, all $y \in U(f)$. (We will write $h(U(f)) = 0$.) Note that the definition of degenerate point depends only on X and B, not on the particular order $\mathfrak{P}$ with $X = X(B,\mathfrak{P})$. We denote the set of *non-degenerate* points of X by X_d. The subscript refers to "derived" or "dense" for reasons which will appear shortly.

Given any point $x \in X$ we can associate an ideal of degeneracy $I_x \subset B$ as follows. For each open U, $x \in U$, let $I(U) = \{h \in B | h(U) = 0\}$. Clearly $U \subset U'$ implies $I(U') \subset I(U)$. Since B is Noetherian, the ideal $I_x = \bigcup_{x \in U} I(U)$ is finitely generated, say by $h_1,\ldots,h_k$, with $x \in U(f_i)$, $h_i(U(f_i)) = 0$. Now, if $U(f) \subset U\{f_i\}$ is chosen by 8.1.1, then $h_i \in I(U(f))$, hence $I_x = I(U(f))$. Choosing an affine embedding $X \subset R^n$, we can interpret $U(f) \subset X$ as a small open ball in R^n, intersected with X.

If B is an integral domain, it is clear that $x \in X_d$ if and only if $I_x = (0)$, which just says that any open neighborhood of x is Zariski dense in X. For general B, $x \in X_d$ if and only if $I_x \not\subset \cup P_i$, where $P_i \subset B$ are the minimal primes. This is equivalent to $I_x \subset P_j$, some j, by a well-known argument in commutative algebra. Recall that the P_i are convex for any order $\mathfrak{P} \subset B$. We claim that $I_x \subset P_j$ if and only if $x \in Z(P_j)$ and $I_x = (0) \subset B/P_j$. From 8.2, $Z(P_j) = X(B/P_j, \mathfrak{P}/p_j)$ and $X(B,\mathfrak{P}) = \bigcup_i X(B/P_i, \mathfrak{P}/P_i)$.

The "if" part of the claim is easy, since if some $h \notin P_j$ vanished on a neighborhood $U(f)$ of $x \in X$, then $\bar{h} \in B/P_j$ is non-zero, but vanishes on $U(f) \cap Z(P_j)$, contradicting $I_x = (0) \subset B/P_j$. Conversely, suppose $I_x \subset P_j$, $f \in P_j$, $f(x) > 0$. Then any element $h \in \bigcap_{i \neq j} P_i - P_j$ vanishes on $U(f)$ since $hf = 0 \in B$. Thus $x \in Z(P_j)$. Finally, if $h_1 \notin P_j$ vanished on a neighborhood $U(f) \cap Z(P_j)$, choose $h_2 \in \bigcap_{i \neq j} P_i - P_j$, and consider $h = h_1 h_2$. We have $h \notin P_j$, $h(Z(P_i)) = 0$, $i \neq h$, hence $h(U(f)) = 0$, since $U(f) = \bigcup_i (Z(P_i) \cap U(f))$. The arguments in these last two paragraphs prove the following.

<u>Proposition 8.6.1</u>. $X(B,\mathfrak{P})_d = \bigcup_{P_i \text{ minimal}} X(B/P_i, \mathfrak{P}/P_i)_d$. □

<u>Proposition 8.6.2</u>. Suppose $P_i \subset B$ is a finite collection of primes with $(0) = \cap P_i$, and suppose $\mathfrak{P}_i \subset B_i = B/P_i$ are orders. If $\mathfrak{P} = B \cap \Pi \mathfrak{P}_i$, where $B \to \Pi B_i$ is the natural inclusion, then $X(B,\mathfrak{P})_d = \bigcup_{P_j \text{ minimal}} X(B_j, \mathfrak{P}_j)_d$.

<u>Proof</u>: The starting point is 8.2.9, which asserts that $X(B, \mathfrak{P}) = \bigcup_i X(B_i, \mathfrak{P}_i)$. We know $x \in X(B, \mathfrak{P})_d$ if and only if $I_x \subset P_j$, some *minimal* P_j. If $x \in X(B_j, \mathfrak{P}_j)_d$, then $I_x \subset P_j$, just as in the proof of 8.6.1. If $I_x \subset P_j$, but $x \notin X(B_j, \mathfrak{P}_j)$, let $f \in B$, $f(x) < 0$, $f(\text{mod } P_j) \in \mathfrak{P}_j \subset B_j$, then $x \in U(-f) \subset X(B, \mathfrak{P}) - X(B_j, \mathfrak{P}_j) \subset \bigcup_{i \neq j} X(B_i, \mathfrak{P}_i)$ and, again, any $h \in \bigcap_{i \neq j} P_i - P_j$ vanishes on $U(-f)$. Thus $x \in X(B_j, \mathfrak{P}_j)$. The proof that $x \in X(B_j, \mathfrak{P}_j)_d$ is just like the last step of the proof of 8.6.1, with the $X(B_i, \mathfrak{P}_i)$ replacing the $Z(P_i)$. □

Proposition 8.6.1 seems more natural than 8.6.2, since it applies to any RHJ-algebra. But 8.6.2 is better since it applies more directly to the situation of 8.2.11, where we showed how any closed semi-algebraic set is of the form $X(B, \mathfrak{P})$.

We next show how non-degenerate points behave with respect to intersections of orders.

<u>Proposition 8.6.3</u>. Suppose $(B, \mathfrak{P}_i)$ are RHJ-algebras, $\mathfrak{P} = \cap \mathfrak{P}_i$, $i \leq i \leq k$, and suppose $P \subset B$ is a $\mathfrak{P}$-convex prime ideal. Let $\mathfrak{P}_{i'}$ denote

those $\mathfrak{P}_i$ so that P is $\mathfrak{P}_{i'}$ convex. (Thus if $P = (0)$, we get all the $\mathfrak{P}_i$.) Then $X(B/P, \mathfrak{P}/P)_d = \bigcup_{i'} X(B/P, \mathfrak{P}_{i'}/P)_d$.

Proof: First note that by 2.7 we know some $\mathfrak{P}_{i'}$ exist. If $x \in X(B/P, \mathfrak{P}_{i'}/P)_d$, but some $f \in B$, $f \notin P$, vanished on a neighborhood U of x in $X(B/P, \mathfrak{P}/P)$, then f also vanishes on $U \cap X(B/P, \mathfrak{P}_{i'}/P)$. This contradiction shows $\bigcup_{i'} X(B/P, \mathfrak{P}_{i'}/P)_d \subset X(B/P, \mathfrak{P}/P)_d$.

Conversely, if $x \in X(B/P, \mathfrak{P}/P)_d$, assume $x \notin X(B/P, \mathfrak{P}_{i'}/P)_d$ all i'. Let $f_{i'} \notin P$ vanish on a neighborhood of x, $U(h_{i'}) \cap X(B/P, \mathfrak{P}_{i'}/P)$. Since P is not $\mathfrak{P}_{i''}$-convex, where the $\mathfrak{P}_{i''}$ are the other orders among the $\mathfrak{P}_i$, it follows that P is *properly* contained in the hull $H(P, \mathfrak{P}_{i''}) \subset B$. Thus let $f_{i''} \notin P$ vanish identically on $Z(P) \cap X(B, \mathfrak{P}_{i''}) = X(B/P, \mathfrak{P}/P) \cap X(B, \mathfrak{P}_{i''})$. From 8.1.1, choose $x \in U(h) \in \cap U(h_{i'})$ and let $f = \Pi f_{i'} \cdot \Pi f_{i''} \notin P$. Then f vanishes on the neighborhood $U(h)$ of x in $X(B/P, \mathfrak{P}/P)$ since

$$U(h) = \bigcup_{i'} [U(h) \cap X(B, \mathfrak{P}_{i'}) \cap Z(P)] \cup \bigcup_{i''} [U(h) \cap X(B, \mathfrak{P}_{i''}) \cap Z(P)]. \qquad \square$$

Remark 8.6.4. The study of derived orders of the type we are interested in has already been reduced to the case of integral domains in Proposition 3.12.1. Thus, if B is a ring, $P_i \subset B$ a finite collection of primes with $(0) = \cap P_i$ and we have either (1) an order $\mathfrak{P} \subset B$ with all P_i convex, or (2) orders $\mathfrak{P}_i \subset B_i = B/P_i$ and $\mathfrak{P} = B \cap \Pi \mathfrak{P}_i$, then $\mathfrak{P}_d = B \cap \Pi(\mathfrak{P}/P_i)_d$ in case (1) and $\mathfrak{P}_d = B \cap \Pi(\mathfrak{P}_i)_d$ in case (2).

The next result is the central result of this section, and gives the relation between non-degenerate points and derived orders in a crucial special case.

Proposition 8.6.5. Let A be a real, finitely generated integral domain over R. Let $\mathfrak{P} = \mathfrak{P}_w[g_i]$ be a finite refinement of the weak order on A, $g_i \neq 0$. Then $(A, \mathfrak{P}_d)$ is an RHJ-algebra, with $X(A, \mathfrak{P}_d) = X(A, \mathfrak{P})_d$. Moreover, $\mathfrak{P}_d = \{f \in A | f(x) \geq 0 \text{ all } x \in X(A, \mathfrak{P})_d\}$. Finally, $X(A, \mathfrak{P})_d$ also coincides with the following subsets of $X = X(A, \mathfrak{P}_w)$, (the irreducible real algebraic variety associated to A):

(i) $\{x \in X \mid \text{exists total orders } \bar{\mathfrak{P}} \supset \mathfrak{P} \text{ on } A \text{ with } x\ \bar{\mathfrak{P}}\text{-convex}\}$

(ii) $\{x \in X \mid \text{for all } g \in A \text{ with } g(x) > 0,\ \mathfrak{P}[g] \text{ is an order on } A\}$.

Before giving the proof, we discuss some applications. First note that the Proposition gives us a class of RHJ-algebras $(A,\mathfrak{P}_d)$ for which it is not at all obvious that $X(A,\mathfrak{P}_d)$ is a closed semi-algebraic set.

Corollary 8.6.6. If B is any reduced real R-algebra of finite type, $\mathfrak{P} = \mathfrak{P}_w[g_i]$ a finite refinement of the weak order, $X = X(B,\mathfrak{P})$, then $\mathfrak{P}_d = \{f \in B \mid f(x) \geq 0 \text{ all } x \in X_d\}$.

Proof: Let $P_j \subset B$ be the minimal primes. By 8.6.1, $X_d = \cup X(B/P_j, \mathfrak{P}/P_j)_d$, and by Remark 8.6.4, $\mathfrak{P}_d = B \cap \Pi(\mathfrak{P}/P_j)_d$. But $\mathfrak{P}/P_j = \mathfrak{P}_w[g_i \mid g_i \notin P_j]$, and by 8.6.5, $f \in (\mathfrak{P}/P_j)_d$ if and only if f is non-negative on $X(B/P_j, \mathfrak{P}/P_j)_d$. Thus $f \in \mathfrak{P}_d$ if and only if f is non-negative on X_d. □

Next, let S be any closed semi-algebraic set, and represent $S = \cup S_i$, with $S_i = Z\{f_{ik}\} \cap W\{g_{ik}\} \subset R^n$, as in 8.1, where f_{ik}, g_{jk} are polynomials in n-variables. Then $I(S) = \cap I(S_i) = \bigcap_i \bigcap_\alpha P_{i\alpha}$, where $P_{i\alpha}$ are the minimal primes above $I(S_i)$. We also have the order $\mathfrak{P}_{i\alpha} = \mathfrak{P}_w[g_{ik} \notin P_{i\alpha}]$ on $A_{i\alpha} = R[X_1 \ldots X_n]/P_{i\alpha}$. From 8.2.11, we have the inclusion $A(S) \to \Pi A_{i\alpha}$ and we have $S = X(A(S), \mathfrak{P})$, where $\mathfrak{P} = A(S) \cap \Pi \mathfrak{P}_{i\alpha}$. There may be repetition among the $P_{i\alpha}$, but a given prime occurs at most once among the $P_{i\alpha}$, fixed i. We can collect similar terms, and denote by P_α the various primes which occur. On each ring $A_\alpha = [R[X_1 \ldots X_n]/P_\alpha$, put the order $\mathfrak{P}_\alpha = \cap \mathfrak{P}_{i\alpha}$, the intersection taken over those indices with $P_{i\alpha} = P_\alpha$. We still have inclusion $A(S) \to \Pi A_\alpha$ and $\mathfrak{P} = A(S) \cap \Pi \mathfrak{P}_\alpha$. Denote by P_β the minimal primes of $A(S)$. Necessarily, they all occur among the P_α. From Remark 8.6.4, $\mathfrak{P}_d = A(S) \cap \Pi(\mathfrak{P}_\beta)_d$. Moreover, $\mathfrak{P}_\beta = \cap \mathfrak{P}_{i\beta}$, so from 3.12, $(\mathfrak{P}_\beta)_d = \cap(\mathfrak{P}_{i\beta})_d$. Each $\mathfrak{P}_{i\beta}$ is a finite refinement of the weak order on the integral domain A_β, namely, $\mathfrak{P}_{i\beta} = \mathfrak{P}_w[g_{ik} \mid g_{ik} \notin P_\beta]$. Now 8.6.5 describes $(\mathfrak{P}_{i\beta})_d$ as the functions in A_β nowhere negative on $X(A_\beta, \mathfrak{P}_{i\beta})_d$. Finally, from 8.6.2, $S_d = X(A(S), \mathfrak{P})_d =$

$\bigcup_{P_\beta \text{ minimal}} X(A_\beta, \mathfrak{P}_\beta)_d$. From 8.6.3 (with $P = (0)$), we conclude $S_d = \bigcup_{P_\beta \text{ minimal}} \bigcup_{P_{i\beta} = P_\beta} X(A_\beta, \mathfrak{P}_{i\beta})_d$, and moreover, $\mathfrak{P}_d \subset A(S)$ consists precisely of functions f nowhere negative on S_d. We summarize these arguments in the following.

<u>Corollary 8.6.7</u>. Let S be a closed semi-algebraic set, $\mathfrak{P} \subset A(S)$ a specific order of the form considered in 8.2.11 with $S = X(A(S), \mathfrak{P})$. Then for any order $\mathfrak{P}'$ with $\mathfrak{P} \subset \mathfrak{P}' \subset \mathfrak{P}(S)$, we have that $(A(S), \mathfrak{P}'_d)$ is an RHJ-algebra and $S_d = X(A(S), \mathfrak{P}'_d)$.

<u>Proof</u>: The argument above gives the result for $\mathfrak{P}' = \mathfrak{P}$. For any such $\mathfrak{P}'$, $\mathfrak{P}_d \subset \mathfrak{P}'_d \subset \mathfrak{P}(S)_d$. But if $f \in \mathfrak{P}(S)_d$, trivially f is nowhere negative on S_d. Thus $\mathfrak{P}(S)_d \subset \mathfrak{P}(S_d) = \mathfrak{P}_d$. □

We remark that if $\mathfrak{P}'' \subset \mathfrak{P}$ is a *weaker* order with $S = X(A(S), \mathfrak{P}'')$, then it is not clear when $\mathfrak{P}''_d = \mathfrak{P}(S_d)$. In fact, we do not assert that if $(A,\mathfrak{P})$ is any RHJ-algebra, then necessarily $(A,\mathfrak{P}_d)$ is also an RHJ-algebra. Our arguments definitely use properties of orders of specific types.

Next let $(A,\mathfrak{P})$ be an RHJ-algebra with $\mathfrak{P} = \mathfrak{P}_w[g_i]$ a finite refinement of the weak order. Let $I \subset A$ be a $\mathfrak{P}$-convex, radical ideal $X = X(A,\mathfrak{P})$.

<u>Corollary 8.6.8</u>

(a) I is $\mathfrak{P}_d$-convex if and only if each minimal prime $P_i \subset A$ over I is $\mathfrak{P}_d$-convex if and only if $P_i = I(Z(P_i) \cap X_d)$ if and only if $I = I(Z(I) \cap X_d)$.

(b) If I is $\mathfrak{P}_d$-convex, then $(A/I, \mathfrak{P}_d/I)$ is an RHJ-algebra with $X(A/I, \mathfrak{P}_d/I) = Z(I) \cap X_d$. Moreover, $\mathfrak{P}_d/I \subset A/I$ consists of the restrictions to $Z(I) \cap X_d$ of functions $f \in A$ nowhere negative on X_d.

(c) For any radical $\mathfrak{P}$-convex $I \subset A$, $(A/I, (\mathfrak{P}/I)_d)$ is an RHJ-algebra, with $X(A/I, (\mathfrak{P}/I)_d) = X(A/I, \mathfrak{P}/I)_d$. Moreover, $(\mathfrak{P}/I)_d$ consists of the functions $f \in A/I$ nowhere negative on $X(A/I, \mathfrak{P}/I)_d$.

(d) $(\mathfrak{P}_p)_d = \mathfrak{P}_d \subset A$, hence for any $\mathfrak{P}$-convex radical $I \subset A$, $(\mathfrak{P}/I)_d = (\mathfrak{P}_p/I)_d = ((\mathfrak{P}/I)_p)_d$.

Proof: (a), (b), (c) are routine restatements of various results. The first statement in (d) holds, since by 8.5.1 and 8.6.5 $\mathfrak{P}_p \subset \mathfrak{P}_d$, and we always have $(\mathfrak{P}_d)_d = \mathfrak{P}_d$. The last part of (d) holds because $\mathfrak{P}/I \subset A/I$ is still a finite refinement of the weak order, so $(\mathfrak{P}/I)_d = ((\mathfrak{P}/I)_p)_d$. Also, by 8.5.1, $\mathfrak{P}_p/I \subset (\mathfrak{P}/I)_p$, so we have $(\mathfrak{P}/I)_d \subset (\mathfrak{P}_p/I)_d \subset ((\mathfrak{P}/I)_p)_d$. □

An interesting order not covered by 8.6.8 is $(\mathfrak{P}_d/I)_d \subset A/I$, when $I \subset A$ is $\mathfrak{P}_d$-convex. We state the following without proof.

Proposition 8.6.9. $(A/I, (\mathfrak{P}_d/I)_d)$ is an RHJ-algebra, with $X(A/I, (\mathfrak{P}_d/I)_d) = (Z(I) \cap X_d)_d$. Moreover, $f \in (\mathfrak{P}_d/I)_d$ if and only if f is non-negative on $(Z(I) \cap X_d)_d$. □

Using techniques above, one can reduce to the case A an integral domain and $I = P \subset A$ a prime $\mathfrak{P}_d$-convex ideal. The last two assertions are not too hard. But the RHJ-property seems to be a fairly strenuous extension of 8.6.5. We will provide the necessary ingredient for this extension in 8.10.

We now return to the proof of Proposition 8.6.5. We follow the notation in the statement of that proposition.

Proof of 8.6.5: First, if $f \in \mathfrak{P}_d$, say $h^2 f = p$, $p \in \mathfrak{P}$, $h \neq 0$, then we must have f non-negative on $X(A,\mathfrak{P})_d$. Otherwise, h would vanish on a neighborhood of a point of $X(A,\mathfrak{P})_d$, contradicting the definition of $X(A, \mathfrak{P})_d$. Thus, points of $X(A, \mathfrak{P})_d$ are $\mathfrak{P}_d$-convex. By 7.7.3, the $\mathfrak{P}_d$-convex points are exactly the points x for which x is convex for some total order refinement of $\mathfrak{P}$ on A. If $x \notin X(A, \mathfrak{P})_d$, choose $f,h \in A$, $h \neq 0$, $f(x) > 0$ and $h(U(f)) = 0$. Then $h^2 f \leq 0$ on $X(A, \mathfrak{P})$. By 8.4.3, f is therefore negative in any total refinement of $\mathfrak{P}$, that is, $-f \in \mathfrak{P}_d$. Since $-f(x) < 0$, x is not $\mathfrak{P}_d$-convex. We have now proved $X(A,\mathfrak{P}_d) = X(A,\mathfrak{P})_d$, and have established the characterization (i) of this set.

Next, suppose $f \notin \mathfrak{P}_d$. Then $\mathfrak{P}[-f]$ is an order on A. By 8.4.3, we find a signed place $p\colon K \to R, \pm\infty$, with $p(A) \subset R$, $p(g_i) > 0$, $p(-f) > 0$. Here, K is the fraction field of A. But then p gives a point $x \in X(A,\mathfrak{P})$ which is convex for the total order refinement of $\mathfrak{P}$ defined by p. Thus x

is $\mathfrak{P}_d$-convex and $f(x) < 0$. This proves that $\mathfrak{P}_d = \{f \in A \mid f(x) \geq 0$, all $x \in X(A,\mathfrak{P})_d\}$, since $X(A,\mathfrak{P})_d = X(A,\mathfrak{P}_d)$.

Third, if $x \in X(A,\mathfrak{P})_d$ and $g(x) > 0$, then g is positive in all total orders for which x is convex. Since some such order exists, refining $\mathfrak{P}$, we see that $\mathfrak{P}[g]$ is an order on A. Conversely, note that by Remark 8.4.7, whenever $\mathfrak{P}_w[f_i] \subset A$ is an order $U\{f_i\} \subset X(A,\mathfrak{P}_w)$ is Zariski dense. Now if $x \notin X(A,\mathfrak{P})_d$, we can find $h,g \in A$, $h \neq 0$, $g(x) > 0$, and $h(X(A,\mathfrak{P}) \cap U(g)) = 0$. In particular, $U\{g_i,g\}$ is not Zariski dense in $X(A,\mathfrak{P}_w)$ hence $\mathfrak{P}[g]$ is not an order. This establishes characterization (ii) of $X(A,\mathfrak{P})_d$.

Finally, we come to the hard part of the theorem. We must show that if $P \subset A$ is a $\mathfrak{P}_d$-convex prime ideal, $g \notin P$, then there exists $x \in Z(P) \cap X(A,\mathfrak{P})_d$, with $g(x) \neq 0$. Note that the set of degenerate points of $X(A,\mathfrak{P})$ is obviously open, hence $X(A,\mathfrak{P})_d$ is closed. In a later section we will give another characterization of $X(A,\mathfrak{P})_d$, which will imply by Tarski-Seidenberg that $X(A,\mathfrak{P})_d$ is a semi-algebraic subset. (This is not clear yet since the definition of $X(A,\mathfrak{P})_d$ contains functions in A as quantified variables. If we knew only finitely many distinct ideals of degeneracy $I_x \subset A$ occur, for $x \in X(A,\mathfrak{P})$, we could conclude that $X(A,\mathfrak{P})_d$ is semi-algebraic without the new characterization.) But even granting all this, we would still only know that $X(A,\mathfrak{P})_d$ was closed, semi-algebraic. If unproved Proposition 8.1.2 were available, then we would have a proof that $(A,\mathfrak{P}_d)$ is an RHJ-algebra by the constructions in 8.2. However, we must make do without 8.1.2 and give a direct proof of this part of 8.6.5.

Assuming $P \subset A$ is $\mathfrak{P}_d$ convex, we can find a total order $\mathfrak{P}'$ on A, refining $\mathfrak{P}$, with P still convex, by 7.7.3. By adjoining $\sqrt{g_i}$ to A, we can construct a *domain* B, integral over A, with a total order extending $\mathfrak{P}'$ on A. (We use the method described in the proof of 8.2.10 and adjoin no more of the $\sqrt{g_i}$ than we need to insure that the weak order on B contracts to $\mathfrak{P} = \mathfrak{P}_w[g_i]$ on A.) By the going up theorem 6.4.2, we can find a convex prime $Q \subset B$ for this total order on B, lying over $P \subset A$. The reason for passing to B is because now *any* total order on B will refine $\mathfrak{P}_w[g_i]$ on A.

Let K be the fraction field of B, with the total order induced by that on B. Write $B = R[x_1 \ldots x_n]$, and suppose $\bar{x}_i = x_i \pmod{Q}$, $1 \leq i \leq s$, gives a transcendence base for B/Q over R. Let $K \subset \Gamma$ be a real closure of K and let E be the real closure of $R(x_1 \dots x_s)$ in Γ. The fraction field of B/Q is $R(\bar{x}_1 \dots \bar{x}_n)$, with $\bar{x}_{s+1}, \dots, \bar{x}_n$ algebraic over the pure transcendental extension $R(\bar{x}_1 \dots \bar{x}_s)$. Consider $B = R[x_1 \dots x_n] \subset E(x_{s+1} \dots x_n) = F \subset \Gamma$, with convex ideal $Q \subset B$ for our total order $\mathfrak{P}'$. (Originally, $\mathfrak{P}'$ was an order on A. We extended it to B, K and Γ, and we continue to call the extension $\mathfrak{P}'$.)

By the signed place existence theorem 7.7.4, we can find a signed place $p'\colon F \to \Delta, \pm\infty$, inducing the order $\mathfrak{P}'$, finite on B with center $Q \subset B$ and with Δ Archimedean over $R(\bar{x}_1 \dots \bar{x}_n)$. Since p' is trivial on $R(x_1 \dots x_s)$, p' is also trivial on the algebraic extension E, and we identify E with $p'(E)$. Since Δ is Archimedean over $R(\bar{x}_1 \dots \bar{x}_s)$, it is also Archimedean over E. Since the $\bar{x}_{s+j}$ are algebraic over $R(\bar{x}_1 \dots \bar{x}_s)$, $j > 0$, there are polynomials f_j with coefficients in $R[x_1 \dots x_s]$ and with $f_j(x_{s+j}) \in Q$. It follows that $p'(x_{s+j}) \in E$. We can then write $p'(x_i) = \zeta_i \in E$, all $1 \le i \le n$. Thus we have $p'(x_i - \zeta_i) = 0$ and also, of course, if $Q = (h_j) \subset R[x_1 \dots x_n]$, then $p'(h_j) = 0$.

Certainly, the function field $F = E(x_{s+1} \dots x_n)$ over E is the fraction field of $E[x_i - \zeta_i, h_j]$. Let $r = \text{tr.deg}_E(F)$. (Thus $r+s = \text{tr.deg}_R(B)$.) By the signed place perturbation theorem 8.4.9, there exists a discrete rank r signed place over E, $p\colon F \to E, \pm\infty$ (inducing a new order on F), with $p(x_i - \zeta_i) = 0$, and $p(h_j) = 0$. Certainly, p is finite on $E[x_{s+1} \dots x_n]$. Let $\tilde{Q}_1 \subset \tilde{Q}_2 \subset \dots \subset \tilde{Q}_r$ denote the chain of maximal ideals in the valuation rings $\tilde{B}_1 \supset \tilde{B}_2 \supset \dots \supset \tilde{B}_r \supset E[x_{s+1} \dots x_n]$ in F, associated to p. The $\tilde{Q}_i \subset \tilde{B}_i$ are all convex for the total order defined by p. Let $Q_i = \tilde{Q}_i \cap B$, so that $Q_1 \subset Q_2 \subset \dots \subset Q_r$. Also, $Q \subset Q_r$ since $p(h_j) = 0$. Each $\tilde{B}_i/\tilde{Q}_i = F_i$ is a function field over E with $\text{tr.deg}_E(F_i) = r - i$. This follows by repeated application of [68, vol. II, Chapter VI, §14, Theorem 31]. From the diagram below we deduce that $\text{tr.deg}_R(B/Q_i) = s+r-i$. (The integers indicate transcendence degrees.)

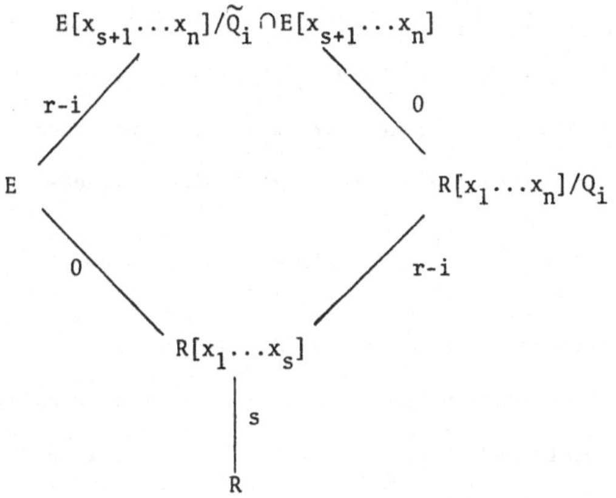

In particular, $\text{tr.deg}_R(B/Q_r) = \text{tr.deg}_R(B/Q) = s$, and since $Q \subset Q_r$, we have $Q = Q_r$. Note this implies that if $g \notin Q$, $g \in B$, then $p(g) \neq 0$.

We now restrict the signed place $p\colon F \to E, \pm\infty$ to $K \subset F$. (Recall $K = R(x_1 \ldots x_n)$ is the fraction field of B.) We then have a discrete rank r signed place $p\colon K \to E, \pm\infty$ over the ground field $R \subset K$, finite on $B = R[x_1 \ldots x_n]$, with $\text{kernel}(p|_B) = Q \subset B$, hence with $\text{tr.deg}_R(p(B)) = s$. Again, the image field $L = p(K) \subset E$ is a function field over R, with $\text{tr.deg}_R(L) = s$. If $g \in B$, $g \notin Q$, then by 8.4.3 we can find a discrete rank s signed place $q\colon L \to R, \pm\infty$, finite on $B/Q \subset L$, with $q(g) \neq 0 \in R$. We then compose p and q, $K \to L, \pm\infty \to R, \pm\infty$, and change p on infinite elements as in 7.7.2 to get a new order on K and a discrete rank $r+s$ signed place $q \circ p\colon K \to R, \pm\infty$.

We have thus constructed a strictly increasing chain of convex prime ideals $Q_1 \subset Q_2 \subset \cdots \subset Q_r = Q \subset Q_{r+1} \subset \cdots \subset Q_{r+s} \subset B$, with $B/Q_{r+s} = R$, $g \notin Q_{r+s}$. We now contract to our original ring $A \subset B$ of Proposition 8.6.5. The (new) total order on B still refines $\mathfrak{P} = \mathfrak{P}_w[g_i]$ on A by construction of B. The ideals $P_i = Q_i \cap A$ are then all $\mathfrak{P}_d$=convex and form a strictly increasing chain $P_1 \subset P_2 \subset \ldots \subset P_r = P \subset P_{r+1} \subset \cdots \subset P_{r+s} \subset A$, with $A/P_{r+s} = R$. If $g \in A$, $g \notin P$, then $g \in B$, $g \notin Q$, and we may assume $g \notin Q_{r+s}$, hence $g \notin P_{r+s}$. Thus, P_{r+s} corresponds to a point $x \in Z(P) \cap X(A,\mathfrak{P}_d)$, with $g(x) \neq 0$, completing the proof of 8.6.5. □

For later reference we state a corollary of this proof of 8.6.5.

Proposition 8.6.10. Let A be a real finite integral domain, $\mathfrak{P} = \mathfrak{P}_w[g_j] \subset A$ a finite refinement of the weak order, $P \subset A$ a $\mathfrak{P}_d$-convex prime ideal. Let $r = \text{tr.deg}_R(A) - \text{tr.deg}_R(A/P)$, and let K denote the fraction field of A. Then there is a sequence of discrete, rank 1 signed places

$$K \xrightarrow{p_1} K_1, \pm\infty \xrightarrow{p_2} \cdots \xrightarrow{p_r} K_r, \pm\infty,$$

inducing a total order on K refining $\mathfrak{P}$, all finite on A, with $P = \text{kernel}(p|_A)$, where p is the composition $p\colon K \to K_r, \pm\infty$. In particular, we obtain a chain of prime ideals $(0) \subsetneqq P_1 \subsetneqq \cdots \subsetneqq P_r = P \subset A$, all convex for a total order refinement of $\mathfrak{P}$. □

8.7. A Preliminary Inverse Function Theorem

In order to make sense of differential topology over an arbitrary real closed field, it is imperative to investigate purely algebraic versions of the inverse function theorem. The result proved in this section is a rather special case of a better algebraic inverse function theorem, but is strong enough to provide a good picture of a real algebraic variety near an algebraic simple point. This application will be given in the next section. This, in turn, will be used to stratify arbitrary closed, semi-algebraic sets and prove that any such is the maximal convex ideal spectrum of an RHJ-algebra. Also crucial for this discussion will be the work on derived orders in the previous section. (We refer again to 8.1 for the distinction between closed, semi-algebraic sets and closed semi-algebraic sets.)

Before stating the main theorem of this section, we digress a bit in order to put in perspective a consequence of the Tarski-Seidenberg theorem which we seem to require at this point. The result is that if $f(x_1 \ldots x_n)$ is a polynomial and $B = \{(x_1 \ldots x_n) \in R^n | a_i \leq x_i \leq b_i\}$ is a closed bounded rectangle in R^n (or a closed ball), then the set of values $\{t \in R | t = f(x)$, some $x \in B\}$ is a semi-algebraic subset of the line R.

This is trivially a consequence of the Tarski-Seidenberg theorem (see the Appendix), but it does not seem any easier to prove directly than the

Tarski-Seidenberg theorem itself. We can reformulate the result as follows. Let $F = \{(x_1 \ldots x_n, y) \in R^n \times R = R^{n+1} | (x_1 \ldots x_n) \in B,\ y = f(x_1 \ldots x_n)\}$. Thus, F is the graph of f over B, and is obviously a closed semi-algebraic set in R^{n+1}. The values of f on B are obtained by projecting F onto R, via the map $y\colon\ R^n \times R \to R$.

As discussed in 8.1, the Tarski-Seidenberg theorem easily can be used to prove certain sets are semi-algebraic, but it does not give much information on whether sets are closed or open, without further work. On the other hand, our going-up theorem for semi-integral extensions provides a nice tool for concluding that certain sets are closed. This was formalized in Proposition 8.2.7, where we showed that if $A \subset B$ is a semi-integral extension of finitely generated R-algebras relative to an RHJ-order $\mathfrak{P} \subset B$, then $\text{image}(X(B,\mathfrak{P})) = X(A,\mathfrak{P} \cap A)$, under the projection $X(B,\mathfrak{P}_w) \to X(A,\mathfrak{P}_w)$. Of course, $X(A,\mathfrak{P} \cap A)$ is always closed. If $X(B,\mathfrak{P})$ is semi-algebraic, Tarski-Seidenberg implies $\text{image}(X(B,\mathfrak{P}))$ is also semi-algebraic.

As an example, we can deduce that a polynomial $f(x_1 \ldots x_n)$ assumes a *maximum* value on any bounded closed semi-algebraic set $S \subset R^n$. Consider the graph of f over S, say $F \subset R^n \times R = R^{n+1}$. Then F is a bounded closed semi-algebraic set and the RHJ-algebra $(A(F),\mathfrak{P}(F))$ has $X(A(F),\mathfrak{P}(F)) = F$. Moreover, $(A(F),\mathfrak{P}(F))$ is semi-integral over R, hence also over $R[y] \subset A(F)$. We project F onto the y-axis and get a bounded closed, (comma) semi-algebraic subset of the line. But in dimension one, the distinction between closed, semi-algebraic and closed semi-algebraic obviously is unnecessary, both notions simply corresponding to finitely many closed intervals (including single points and closed rays). Thus, bounded, closed, semi-algebraic implies a maximum element in dimension one.

Now, as another way of applying the Taski-Seidenberg theorem, one could draw the same conclusion about maximum values directly from the fact that it is true in the case of the real numbers. But this is a "transcendental proof", whereas we have just given a "purely algebraic proof".

We now state an inverse function theorem.

Proposition 8.7.1. Suppose $Y_1 \ldots Y_n \in R[X_1 \ldots X_n]$. We regard the Y_i

as functions on R^n and we regard the n-tuple $Y = (Y_1 \ldots Y_n)$ as a map $R^n \to R^n$. Assume $Y(0) = 0$ and assume that the derivative matrix $((\partial Y_i/\partial X_j)(0))$ is non-singular. Then:

(a) There exists $\varepsilon > 0$ such that the map Y restricted to the closed ball $B(0,\varepsilon) \subset R^n$ of radius ε at the origin is injective.

(b) Given $\varepsilon > 0$, there exists $\delta > 0$ such that $B(0,\delta) \subset Y(B(0,\varepsilon))$.

(c) From (a), (b), near the origin the coordinate functions X_i are functions of the Y_j. These functions are in fact algebraic functions in the sense that for suitable polynomials f_i, we have $f_i(Y_1 \ldots Y_n, X_i) \equiv 0$, $1 \leq i \leq n$.

(d) Let $A(x) = ((\partial Y_i/\partial X_j)(x))$, the derivative of Y at $x \in R^n$. Then sufficiently near the origin the inverse function $X = X(y)$ is differentiable with derivative A^{-1}, in the sense that if $y_0 = Y(x_0)$,

$$\lim_{y \to y_0} \frac{X(y) - X(y_0) - A^{-1}(x_0)(y - y_0)}{\|y - y_0\|} = 0$$

Proof: The point to be made at the outset is that once we have the maximum value property of polynomials on closed bounded semi-algebraic sets (which we have just established purely algebraically) one can write out word for word one of the standard proofs of the inverse function theorem for real numbers entirely in elementary algebraic terms.

Part (a), the local injectivity of Y is very easy since it is simply necessary to find $\varepsilon > 0$ and a constant $c > 0$ such that for all $x, x' \in B(0,\varepsilon)$, we have

$$\|x - x'\| \leq c\|Y(x) - Y(x')\| .$$

By a linear change of coordinates, we may assume that $((\partial Y_i/\partial X_j)(0)) = A(0)$ is the identity matrix, that is,

$$Y_i = X_i + (\text{terms of degree} \geq 2).$$

Now using the estimates of 7.2 we can, in fact, find $\varepsilon_1 > 0$ so that for $x, x' \in B(0,\varepsilon_1)$ we have

$$(*) \qquad \frac{1}{2}\|Y(x) - Y(x')\| \leq \|x - x'\| \leq 2\|Y(x) - Y(x')\| .$$

(Note that the first inequality will guarantee that the inverse function $X = X(y)$ is continuous near 0.)

We next prove (b), the local surjectivity of Y. Directly from (*), if $\|x\| < \varepsilon < \varepsilon_1$, then $\frac{\varepsilon}{2} < \|Y(x)\|$. Let us now assume $\varepsilon_2 < \varepsilon_1$ so small that $A(x)$ is non-singular for all $x \in B(0,\varepsilon_2)$. If $\|y\| < \frac{\varepsilon_2}{4}$, consider the function $P(x) = \sum_{i=1}^{n} (Y_i(x) - y_i)^2 = \|Y(x) - y\|^2$ on $B(0,\varepsilon_2)$. If $\|x\| = \varepsilon_2$, then $\|Y(x)\| > \frac{\varepsilon}{2}$, hence $P(x) \geq (\varepsilon_2/4)^2 > \|y\|^2 = P(0)$. Thus the function $P(x)$ does not assume a minimum value on the boundary of $B(0,\varepsilon_2)$. If $x_0 \in B(0,\varepsilon_2)$ is an *interior* point at which $P(x_0)$ is minimum, then

$$0 = (\partial P/\partial X_j)(x_0) = \Sigma 2(Y_i(x_0) - y_i)((\partial Y_i/\partial X_j)(x_0))$$

for all $1 \leq j \leq n$. That is, $A(x_0)(Y(x_0) - y_0) = 0$. Since $A(x_0)$ is non-singular, we deduce $Y(x_0) = y_0$, as desired.

Since $A(0) = ((\partial Y_i/\partial X_j)(0))$ is nonsingular, it is easy to see that $R(Y_1 \dots Y_n)$ has transcendence degree n over R. Thus $R(Y_1 \dots Y_n) \subset R(X_1 \dots X_n)$ is an algebraic extension. This proves (c).

Finally, to prove (d), apply the linear isomorphism $A(x_0)$ to the limit statement of (d), transforming (d) to

$$\lim_{y \to y_0} \frac{A_0(X(y) - X(y_0)) - (y - y_0)}{\|y - y_0\|} = \lim_{y \to y_0} \frac{A_0(x - x_0) - (Y(x) - Y(x_0))}{\|x - x_0\|} \frac{\|x - x_0\|}{\|y - y_0\|} = 0,$$

which *does* hold by continuity of $X(y)$ and differentiability of $Y(x)$. □

We discuss further the differentiability of the inverse function $X = X(y)$. Once we have (d), it is easy to see that the partial derivatives

$$(\partial X_j/\partial Y_i)(y_0) = \lim_{\varepsilon \to 0} \frac{X_j(y_1, \dots, y_i + \varepsilon, \dots y_n) - X_j(y_1 \dots y_n)}{\varepsilon}$$

exist and, in fact, that the matrix $((\partial X_j/\partial Y_i)(y_0))$ coincides with $A(x_0)^{-1}$, where $y_0 = Y(x_0)$. Thus, Cramer's rule expresses the $(\partial X_j/\partial Y_i)(y_0)$ as specific rational algebraic functions of the polynomials $(\partial Y_i/\partial X_j)(x_0)$, with denominator $\det(A(x_0))$.

Continuing this discussion, it can be shown that the $X_j(y)$ are actually infinitely differentiable functions of y. The higher derivatives are formally computable in terms of $x_0 = X(y_0)$ as follows.

First, rewrite the polynomials $Y_i - y_i = Y_i(X_1 \ldots X_n) - Y_i(x_1 \ldots x_n)$ as polynomials in $X_j - x_j$, where $x_0 = (x_1 \ldots x_n)$, $y_0 = (y_1 \ldots y_n)$. The constant terms vanish and the linear terms will be $((\partial Y_i/\partial X_j)(x_0))(X - x_0)$. Then formally solve for the $X_j - x_j$ as power series in the $Y_i - y_i$. The coefficient of $(Y_1 - y_1)^{i_1} \cdot \ldots \cdot (Y_n - y_n)^{i_n}$ will then be $\frac{1}{(i_1)! \cdot \ldots \cdot (i_n)!} (\partial^I X_j/\partial Y^I)(y_0)$, $I = (i_1 \ldots i_n)$.

Alternatively, the total differential dY can be regarded as a polynomial function $dY\colon R^n \times R^n \to R^n \times R^n$, linear on the second factor at each point of the first factor, and, moreover, dY has non-singular differential at $(0,0)$. The second derivatives $(\partial^2 Y_i/\partial X_j \partial X_k)$ then occur as part of the first derivative of dY. Applying the general discussion of first derivatives to dY shows that the inverse function of Y is twice differentable, and leads to a computation of the $(\partial^2 X_i/\partial Y_j \partial Y_k)$.

Here is perhaps a more algebraic approach to the derivatives $(\partial X_j/\partial Y_i)$. The *derivation* $\partial/\partial Y_i\colon R[Y_1 \ldots Y_n] \to R(Y_1 \ldots Y_n) \subset R(X_1 \ldots X_n)$ extends uniquely to a derivation $D_i\colon R[X_1 \ldots X_n] \to R(X_1 \ldots X_n)$. Specifically, if $f_j(Y, X_j) = a_0(Y)X_j^m + a_1(Y)X_j^{m-1} + \ldots + a_m(Y) = 0$ is the minimal polynomial for X_j over $R[Y_1 \ldots Y_n]$, then we must have

$$
\begin{aligned}
0 &= D_i(f_j(Y, X_j)) \\
&= ((\partial a_0/\partial Y_i)X_j^m + \ldots + (\partial a_m/\partial Y_i)) + ((\partial f_j/\partial X_j)(Y, X_j))D_i(X_j),
\end{aligned}
$$

and this last equation can be solved for $D_i(X_j)$.

Using the inverse function theorem 8.7.1, we can easily establish the following implicit function theorems.

Proposition 8.7.2. Suppose $Y_1, \ldots, Y_k \in R[X_1 \ldots X_n]$, $k < n$, $Y_i(0) = 0$, and suppose the vectors $dY_i(0) = ((\partial Y_i/\partial X_1)(0),\ldots,(\partial Y_i/\partial X_n)(0))$, $1 \leq i \leq k$, are linearly independent. Reordering the variables if necessary, assume that $\{dY_i(0),\ 1 \leq i \leq k;\ dX_j(0),\ k < j \leq n\}$ span R^n. Then, sufficiently near $0 \in R^k$, the equations $Y_i(x) = c_i$, $1 \leq i \leq k$, $c_i \in R$, $x = (x_1\ldots x_k, x_{k+1}\ldots x_n)$ uniquely define the coordinates $x_1\ldots x_k$ as functions of $x_{k+1}\ldots x_n$.

Proof: Define $F\colon R^n \to R^n$ by $F(x) = (Y_1(x),\ldots,Y_k(x),X_{k+1}(x),\ldots,X_n(x))$. Then F is locally 1-1 and onto by 8.7.1. The result then follows easily.

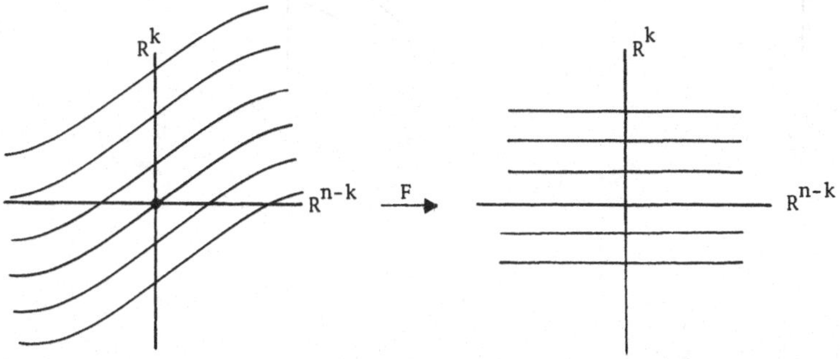

Surfaces $Y_i(x) = c_i$, $1 \leq i \leq k$ Surfaces $X_i = c_i$, $1 \leq i \leq k$. □

Remark. Proposition 8.7.1 and the discussion of derivatives above also gives the tangent plane of the surfaces $Y_i(x) = c_i$. For example, if $(\partial Y_i/\partial X_j)(0) = 0$, $j > k$, then the tangent plane to $Y_i(x) = 0$, $1 \leq i \leq k$, at the origin is the coordinate plane $\{x_1 = \cdots = x_k = 0\} = R^{n-k} \times \{0\}$. In general, the tangent plane of $Y_i(x) = 0$ at the origin is the annihilator with respect to the usual scalar product on R^n of the k-vectors $(dY_i)(0)$, $1 \leq i \leq k$, $dY_i = (\partial Y_i/\partial X_1,\ldots,\partial Y_i/\partial X_n)$.

Proposition 8.7.3. Suppose $Y_1\ldots Y_k \in R[X_1\ldots X_n]$, $k > n$, $Y_i(0) = (0)$, and suppose the vectors $dY_i(0)$, $1 \leq i \leq k$ span R^n. Reordering the variables if necessary, assume that $dY_1(0),\ldots,dY_n(0)$ span R^n. Then after smooth algebraic change of coordinates the map $Y\colon R^n \to R^k$ looks like the standard inclusion, mapping $(x_1\ldots x_n)$ to $(x_1\ldots x_n, 0\ldots 0)$, for $\|x\|$ sufficiently small.

Proof: Define $G(x_1 \dots x_n, x_{n+1} \dots x_k) = (Y_1(x_1 \dots x_n), \dots, Y_k(x_1 \dots x_n)) + (0 \dots 0, x_{n+1} \dots x_k)$, so that $G\colon R^k \to R^k$ is locally 1-1 and onto near 0 by 8.7.1.

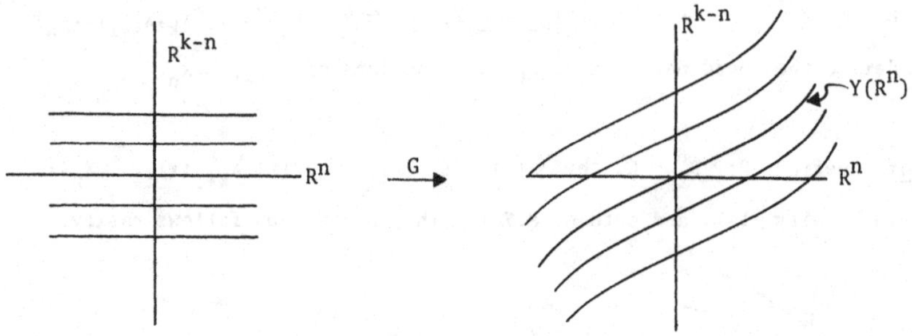

□

8.8. Algebraic Simple Points, Dimension, Codimension, and Rank

We continue our study of the polynomial ring $R[X] = R[X_1 \dots X_n]$ over a real closed field R. If $f \in R[X]$, we have $df = (\partial f/\partial X_1, \dots, \partial f/\partial X_n)$, an n-vector of functions. If $P \subset R[X]$ is a prime ideal, define the *dimension of* P to be $\dim(P) = \text{tr.deg}_R(R[X]/P)$. Thus $\dim(P)$ is an invariant of the finite integral domain $A = R[X]/P$. Suppose $P = (f_1 \dots f_s)$. Define the *codimension of* P to be the maximum number of A-linearly independent vectors $df_i \in A^n$. It is routine to check that $\operatorname{codim}(P)$ is independent of the choice of generators of P, using the formula $d(hf) = f\,dh + h\,df = h\,df \in A^n$ if $f \in P$.

A standard result in commutative algebra is that $\dim(P) + \operatorname{codim}(P) = n$. In fact, $\dim(P)$ can also be characterized as the dimension over K of the space of R derivations $D\colon A \to K$, where K is the fraction field of A. This coincides with the space of R-derivations $D\colon R[X] \to K$ which vanish on P. Such a D is determined by $DX = (DX_1, \dots, DX_n) \in K^n$, subject to precisely the linear conditions $0 = Df = df \cdot DX$, if $f = f(X) \in P$. In other words, we have $\operatorname{codim}(P)$ linear conditions on DX, so $n = \dim(P) + \operatorname{codim}(P)$. We will use the fact that if $P \subsetneq Q$, then $\dim(Q) < \dim(P)$, hence $\operatorname{codim}(P) < \operatorname{codim}(Q)$.

It is also clear that codim(P) is the rank over A of the matrix $(\partial f_i/\partial X_j)$, where $P = (f_1 \ldots f_s)$. Let $\nabla \in R[X]$ be the determinant of some $r \times r$ submatrix of maximal rank over A. Thus $\nabla \notin P$, that is, $\nabla \neq 0 \in A$. Let $Z(P) \subset R^n$ denote the zeros of P. If $x \in Z(P)$, define the *rank of* P *at* x to be $\text{rank}_x(P) = \text{rank}_R(\frac{\partial f_i}{\partial X_j}(x))$. Obviously, $\text{rank}_x(P) \leq \text{rank}_A(\frac{\partial f_i}{\partial X_j}) =$ codim(P). From the Nullstellensatz 8.4.3 we obtain the following.

Proposition 8.8.1. Suppose $P \subset R[X]$ is a prime and $\mathfrak{P} = \mathfrak{P}_w[g_i]$ is an order on $A = R[X]/P$, $g_i \neq 0 \in A$, $1 \leq i \leq k$. Then there exist zeros $x \in Z(P)$ with $g_i(x) > 0$ and $\nabla(x) \neq 0$. In particular, $\text{rank}_x(P) = \text{codim}(P)$. □

The points $x \in X(A,\mathfrak{P}_w)$ with $\text{rank}_x(P) = \text{codim}(P)$ will be called *algebraic simple points* of $X(A,\mathfrak{P}_w)$. This definition refers to a particular affine embedding of $X(A,\mathfrak{P}_w)$. However, from commutative algebra, algebraic simple points can be invariantly characterized as follows. Let A_x be the local ring associated to the point x and let $m_x \subset A_x$ be the maximal ideal. Then x is a simple point if and only if $\dim_R(m_x/m_x^2) = \dim(A) = \text{tr.deg}_R(A)$.

If we refine the order on A, say to $\mathfrak{P} = \mathfrak{P}_w[g_i]$, we are not especially interested in all the algebraic simple points belonging to $X(A,\mathfrak{P})$. The figures below illustrate what might occur. In both cases $X(A,\mathfrak{P})$, where $\mathfrak{P} = \mathfrak{P}_w[g]$, contains degenerate points which are algebraic simple points of

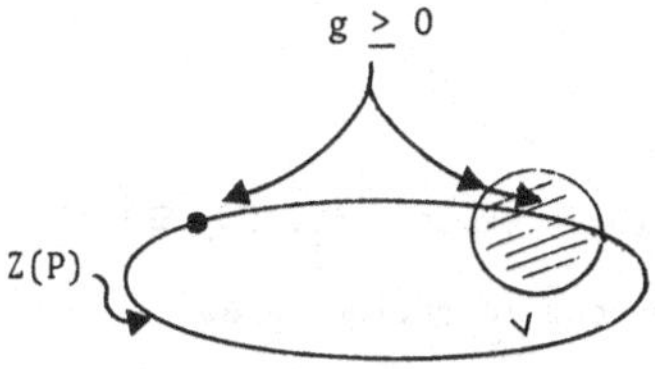

Figure (a)

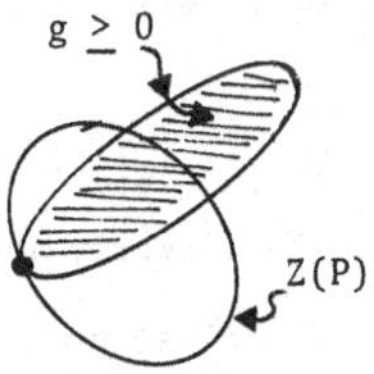

Figure (b)

$X(A,\mathfrak{P}_w)$. We will define the algebraic simple points of $X(A,\mathfrak{P})$ to be the non-degenerate algebraic simple points, that is, simple points of $X(A,\mathfrak{P}_w)$ belonging to $X(A,\mathfrak{P}_d)$. Propositions 8.8.1 and 8.6.5 show that these simple points are, in fact, dense in $X(A,\mathfrak{P}_d)$. Namely, if $x \in X(A,\mathfrak{P}_d) = X(A,\mathfrak{P})_d$ and $g_i(x) > 0$, then $\mathfrak{P}[g_i]$ is an order on A and 8.8.1 guarantees simple points exist in the neighborhood $x \in U\{g_i\} \subset X(A,\mathfrak{P})_d$.

Now, however, we should check that an algebraic simple point of $X(A,\mathfrak{P}_w)$ is automatically non-degenerate. This requires a somewhat careful argument. In fact, we will base the proof on the inverse function theorem. The following is the central result of this section.

Proposition 8.8.2. Let $g, f_1, \ldots, f_r \in R[X]$, $x \in R^n$, with $f_i(x) = 0$ $df_i(x)$ linearly independent, $g(x) > 0$, and assume the neighborhood $U(g)$ of x is sufficiently small. Then the ideal $I(Z\{f_i\} \cap U(g)) = P$ is prime. In fact, P is the unique minimal prime of the ideal $(f_1 \ldots f_r)$, with $x \in Z(P)$. Moreover, P is convex for any order $\mathfrak{P} = \mathfrak{P}_w[g_j] \subset R[X]$, with $g_j(x) > 0$.

Before indicating the proof of 8.8.2, we give applications. In practice, the "small" neighborhoods $U(g)$ of x can be thought of as small balls centered at x. The first part of the proposition says that $A = R[X]/P$ is a ring of functions on *any* neighborhood of x in the variety $Z\{f_i\}$ defined by the vanishing of the f_i, $1 \leq i \leq r$. If $g_j(x) > 0$, then choose $U(g) \subset U\{g_i\}$. The g_j are positive in the order $\mathfrak{P}(Z\{f_i\} \cap U(g)) \subset A$, hence P is convex for $\mathfrak{P}_w[g_j]$. This proves the last part of the proposition.

Corollary 8.8.3. If A is a real finite integral domain over R and $x \in X = X(A,\mathfrak{P}_w)$ is a simple point, then $x \in X_d$.

Proof: Write $A = R[X_1 \ldots X_n]/P$. By assumption, there exists $f_1 \ldots f_r \in P$, $r = \operatorname{codim}(P)$, with $df_i(x)$ linearly independent. Rank considerations show that P is necessarily a minimal prime of $(f_1 \ldots f_r)$, and since $x \in Z(P)$, we are in the situation of 8.8.2. The zeros of P *near* x thus coincide with the zeros of the f_i near x and P is exactly the ideal vanishing on these zeros. Thus every neighborhood of x is Zariski dense in X, as desired. □

Corollary 8.8.4. If A is a real finite integral domain over R, $\mathfrak{P} = \mathfrak{P}_w[g_i]$ a finite refinement of the weak order on A, $g_i \neq 0$, then the non-degenerate set $X(A,\mathfrak{P})_d$ is exactly the closure of the set of algebraic simple points $x \in X(A,\mathfrak{P}_w)$ with $g_i(x) > 0$.

Proof: We know from 8.8.1 and 8.6.5 that any neighborhood of a non-degenerate point contains such simple points. Conversely, 8.8.2 guarantees that a function which vanishes on a neighborhood of such a simple point is already 0 in A. □

Corollary 8.8.5. With the same assumptions as in 8.8.4, the set $X(A,\mathfrak{P})_d$ is a closed, semi-algebraic set.

Proof: If $A = R[X]/P$, $P = (f_i)$, then the simple points of $X(A,\mathfrak{P}_w)$ are the points in the open, semi-algebraic set $\{x \mid \text{rank}((\partial f_i/\partial X_j)(x)) = \text{codim } P\}$. (In fact, by looking at the possible subdeterminants, this is even an open semi-algebraic set.) We intersect with $U\{g_i\}$, then take closure. Tarski-Seidenberg guarantees that the closure of a semi-algebraic set is semi-algebraic. □

Proof of 8.8.2. First, part of the proposition is a purely algebraic result. Denote by A_x the local ring obtained by localizing $R[X]$ at x and dividing by the ideal generated by the f_i. The assumption $df_i(x)$ independent implies that the graded ring associated to the maximal ideal $m_x \subset A_x$ is a polynomial ring, in particular, a domain. Therefore, A_x is a domain, or equivalently, $(f_1 \ldots f_r)$ is a prime ideal in $R[X]_x$. In turn, this says precisely that x is a zero of precisely one minimal prime P of $(f_1 \ldots f_r)$ in $R[X]$. Geometrically, $Z(P)$ and $Z(f_1 \ldots f_r)$ coincide in a *Zariski open* neighborhood of $x \in R^n$. For details of this argument, see [63, Chapter 11], or texts on algebraic geometry.

Now we must look more closely at small semi-algebraic neighborhoods of x. We must show that if h vanishes on $Z\{f_i\} \cap U(g)$, $g(x) > 0$, then $h \in P$. Throughout the argument we may assume $U(g)$ as small as desired. Thus, we assume $df_1(y) \ldots df_r(y)$ are independent, $y \in U(g)$, and we assume the zeros of P and $(f_1 \ldots f_r)$ coincide in $U(g)$.

Suppose otherwise, that is, $P \subsetneq I(Z\{f_i\} \cap U(g)) = \cap P_j$, where the P_j

are convex prime ideals, strictly containing P. We know $x \in Z(P_j)$, some j, say $x \in Z(P_0)$. Then $\operatorname{codim}(P_0) > \operatorname{codim}(P)$ and by 8.8.1 we can find $f_0 \in P_0$, $y \in Z(P_0) \cap U(g) - \bigcup_{j \neq 0} Z(P_j)$, such that $df_0(y), df_1(y), \ldots, df_r(y)$ are linearly independent. Since $Z\{f_i\} \cap U(g) = \bigcup_j (Z(P_j) \cap U(g))$ and $y \notin Z(P_j)$, $j \neq 0$, we know that near y the zeros of $(f_1 \ldots f_r)$ coincide with the zeros of P_0. Thus, $f_0 \in P_0$ vanishes identically on $Z\{f_i\}$, *near* y, yet $df_0(y)$ is independent of $df_1(y), \ldots, df_r(y)$. (Alternatively, if we choose $f'_0 \in \bigcap_{j \neq 0} P_j$, $f'_0(y) = 1$, then we can replace f_0 by $f_0 f'_0$ without changing $df_0(y)$. Now, $f_0 f'_0 \in \bigcap P_j = I(Z\{f_i\} \cap U(g))$, which certainly vanishes on $Z\{f_i\}$ near y, yet still has differential independent of $df_1(y), \ldots, df_r(y)$.) From this, we will derive our contradiction, by appeal to the implicit function theorem 8.7.2.

By translation and linear change of coordinates, we may assume $y = 0 \in R^n$ and $f_i = x_i + (\text{terms of degree} \geq 2) \in R[X]$, $1 \leq i \leq r$. The implicit function theorem states that near 0 the surface $f_1 = \cdots = f_r = 0$ is given by the graph of a map $R^{n-r} \to R^r$ in $R^n = R^r \times R^{n-r}$, $\{y_i \ldots y_r, x_{r+1} \ldots x_n\}$, where $y_i = y_i(x_{r+1} \ldots x_n)$ are smooth algebraic functions of the last $n-r$ variables. Moreover, the tangent plane of the surface is the plane $x_1 = \ldots = x_r = 0$, or equivalently, $(\partial y_i/\partial x_j)(0)$, $1 \leq i \leq r$, $r+1 \leq j \leq n$. If $f_0 \in R[X]$ is a function with $df_0(0)$ independent of $df_i(0) = dx_i(0)$, $1 \leq i \leq r$, we must have $(\partial f_0/\partial x_j)(0) \neq 0$, some $j \geq r+1$, say $j = r+1$. But then a simple calculation of derivatives of f_0 along the curve $(y_1(t,0\ldots0), \ldots, y_r(t,0\ldots0), t,0\ldots0)$ shows that f_0 cannot vanish identically on the surface $f_1 = \ldots = f_r = 0$. □

The last part of Proposition 8.8.2 states roughly that a prime P is convex if it has enough zeros. Precisely, one can say that if $P \subset R[X]$ is prime, $\mathfrak{P} = \mathfrak{P}_w[g_i] \subset R[X]$ a finite refinement of the weak order, $g_i \notin P$, then P is $\mathfrak{P}$-convex if P has algebraic simple zeros $x \in R^n$ with $g_i(x) > 0$. We will now establish another criterion for convexity with this same flavor. The result can be found in [21]. It could be derived from the implicit function theorem and existence of simple zeros, but we will follow the proof of [21], which avoids the implicit function theorem.

Proposition 8.8.6. Let $A = R[x_1 \ldots x_n] = R[X_1 \ldots X_n]/P$ be a finite integral domain over R, $U(g) \subset R^n$ a non-empty open set, $g \notin P$. Let $x_1 \ldots x_d \in A$ be a transcendence base over R and let $\pi\colon Z(P) \to R^d$ be the projection defined by $R[x_1 \ldots x_d] \to A$. Then P is convex for $\mathfrak{P}_w[g] \subset R[X]$ (equivalently, $\mathfrak{P}_w[g] \subset A$ is an order), if and only if $\pi(Z(P) \cap U(g))$ contains a disc $B(y,\varepsilon) = \{z \in R^d \mid \|z-y\| \leq \varepsilon\}$, some $y \in R^d$, $\varepsilon > 0 \in R$.

Proof: First assume $\pi(Z(P) \cap U(g))$ contains a disc. Now, P is $\mathfrak{P}_w[g]$-convex if and only if $P = I(Z(P) \cap U(g))$. Suppose not, that is, $P \subsetneq I(Z(P) \cap U(g)) = \cap P_i$, where the P_i are $\mathfrak{P}_w[g]$-convex primes. Then $P \subsetneq P_i$ and $\text{tr.deg}_R(R[x]/P_i) < d$. Choose $f_i \in R[x_1 \ldots x_d]$ in the kernel of $R[x_1 \ldots x_d] \to A \to A/P_i$, $f_i \neq 0$, and let $f = \Pi f_i$. Then f vanishes on $\pi(Z(P) \cap U(g))$, which contains a disc, hence $f = 0$. This contradiction shows P is $\mathfrak{P}_w[g]$-convex after all.

Conversely, if $\mathfrak{P}_w[g] \subset A$ is an order, let $t \in A$ be a primitive element for the field extension $R(x_1 \ldots x_d) \subset K$, where K is the fraction field of A. Let $f(T) = \Sigma_{i=0}^m a_i(x_1 \ldots x_d)T^i$ be the minimal polynomial for t over $R[x_1 \ldots x_d]$. We can write $x_j = \Sigma_{i=0}^{m-1} (b_{ij}(x_1 \ldots x_d)/c_{ij}(x_1 \ldots x_d))t^i \in K$, $d+1 \leq j \leq n$. By the Nullstellensatz we can find $y = (y_1 \ldots y_n) \in Z(P) \cap U(g)$ such that $a_m(y_1 \ldots y_d) \neq 0$, $c_{ij}(y_1 \ldots y_d) \neq 0$, and $\partial f/\partial t(y_1 \ldots y_d) = \Sigma_{i=1}^m i \cdot a_i(y_1 \ldots y_d)t^{i-1} \neq 0$. Thus $t(y_1 \ldots y_d) \in R$ is a simple root of the polynomial $f(y_1, \ldots, y_d, T) \in R[T]$.

Suppose $(z_1 \ldots z_d)$ is very near $(y_1 \ldots y_d)$ in R^d. Then $f(z_1 \ldots z_d, T)$ will have a root τ near $t(y_1 \ldots y_n)$, by a simple estimate argument showing $f(z_1 \ldots z_d, T)$ changes sign near $T = t(y_1 \ldots y_n)$. If $d+1 \leq j \leq n$, define $z_j = x_j(z_1 \ldots z_d, \tau)$ by the rational formula above and consider $z = (z_1 \ldots z_n) \in R^n$. Watching our estimates carefully, we may assume $z \in U(g) \subset R^n$. We are finished if we prove $z \in Z(P) \subset R^n$.

Now, the kernel of $R[x_1 \ldots x_d, T] \to R[x_1 \ldots x_d, t] = B \subset A$ is precisely $f(x_1 \ldots x_d, T)$. Define $\rho\colon R[x_1 \ldots x_d, T] \to R$ by $\rho(x_i) = z_i$, $\rho(T) = \tau$. Then $f(z_1 \ldots z_d, \tau) = 0$ implies ρ factors through $\rho\colon B \to R$. Since $c_{ij}(y_1 \ldots y_d) \neq 0$, a little further care with estimates will insure $c_{ij}(z_1 \ldots z_d) \neq 0$ and thus

ρ extends to $\rho: A \to R$, $\rho(x_1 \ldots x_n) = (z_1 \ldots z_n)$. Thus $z \in Z(P)$ as desired. □

8.9. Stratification of Semi-Algebraic Sets

In this section we give procedures for stratifying semi-algebraic sets, that is to say, for breaking arbitrary semi-algebraic sets up into "simple" pieces. We should emphasize that we deal only with the most elementary aspects of stratification questions. For example, we do not consider the Whitney-Thom regularity conditions, nor questions of equisingularity. Actually, our primary goal in this section is to show that if $F \subset R^n$ is any closed, semi-algebraic set, $I(F) = \{f \in R[X_1 \ldots X_n] | f(x) = 0, \text{ all } x \in F\}$, $A(F) = R[X_1 \ldots X_n]/I(F)$, $\mathfrak{P}(F) = \{g \in A(F) | g(x) \geq 0, \text{ all } x \in F\}$, then $(A(F), \mathfrak{P}(F))$ is an RHJ-algebra (with, of course, $X(A(F), \mathfrak{P}(F)) = F$.) Stratification considerations arise naturally in the proof of this result.

Let $E \subset R^n$ be any semi-algebraic set, and represent E as $E = \cup E_i$, $E_i = Z\{f_{ij}\} \cap U\{g_{ik}\}, f_{ij}, g_{ik} \in R[X_1 \ldots X_n]$. Let $\overline{E} \subset R^n$ be the closure of E. Then $\overline{E} = \cup \overline{E}_i$ and $I(E) = I(\overline{E}) = \cap I(\overline{E}_i)$ Write $I(E_i) = I(\overline{E}_i) = \cap P_{i\alpha}$, where the $P_{i\alpha}$ are the minimal primes of $I(E_i)$. Then $f_{ij} \in P_{i\alpha}$, all j, α, and $g_{ik} \notin P_{i\alpha}$. This last holds since if, say, $g_{io} \in P_{i\alpha}$, we could choose $h_{io} \in \bigcap_{\beta \neq \alpha} P_{i\beta} - P_{i\alpha}$. Then $g_{io} h_{io} \in I(E_i)$, but g_{io} is *strictly* positive on E_i, hence $h_{io} \in I(E_i)$, contradiction.

Let $A_{i\alpha} = R[X_1 \ldots X_n]/P_{i\alpha}$, $\mathfrak{P}_{i\alpha} = \mathfrak{P}_w[g_{ik}] \subset A_{i\alpha}$. Now, we definitely do not have $X(A_{i\alpha}, \mathfrak{P}_{i\alpha}) \subset \overline{E}_i$, in general, because the g_{ik} could have degenerate zeros far away from $U\{g_{ik}\}$. However, we do have the following.

Proposition 8.9.1. Let $S_{i\alpha} = X(A_{i\alpha}, \mathfrak{P}_{i\alpha})_d$, the non-degenerate points of $X(A_{i\alpha}, \mathfrak{P}_{i\alpha})$. Then $S_{i\alpha} \subset \overline{E}_i$. Moreover, $I(\overline{E}_i) \subsetneq I(\overline{E}_i - S_{i\alpha})$. In fact, $I(\overline{E}_i - S_{i\alpha}) \not\subset P_{i\alpha}$.

Proof: We use 8.8.4, which characterizes $S_{i\alpha}$ as the closure of the set of algebraic simple zeros x of $P_{i\alpha}$, with $g_{ik}(x) > 0$. Call this last set $V_{i\alpha}$, so $S_{i\alpha} = \overline{V}_{i\alpha}$. Since $f_{ij} \in P_{i\alpha}$, we have $Z(P_{i\alpha}) \subset Z\{f_{ij}\}$ and thus clearly $V_{i\alpha} \subset E_i = Z\{f_{ij}\} \cap U\{g_{ik}\}$. Thus $S_{i\alpha} = \overline{V}_{i\alpha} \subset \overline{E}_i$.

Let $P_{i\alpha} = (h_j) \subset R[X_1 \ldots X_n]$. Let $\{A_\rho\}$ denote the set of $r \times r$ submatrices of $(\partial h_j/\partial X_k)$, and let $\Delta_{i\alpha} = \sum_\rho \det(A_\rho)^2$, where $r = \operatorname{codim}(P_{i\alpha})$. Let $g_i = \prod_k g_{ik}$. Then $g_i \cdot \Delta_{i\alpha} \neq 0 \in A_{i\alpha}$, that is, $g_i \cdot \Delta_{i\alpha} \notin P_{i\alpha}$. Moreover, it is clear that $V_{i\alpha} = X(A_{i\alpha}, \mathfrak{P}_{i\alpha}) \cap U(g_i \cdot \Delta_{i\alpha})$ where $V_{i\alpha}$ is the set of algebraic simple zeros $x \in Z(P_{i\alpha})$ with $g_{ik}(x) > 0$, as above. Since $g_i \cdot \Delta_{i\alpha} \in \mathfrak{P}_{i\alpha}$, it is obvious that $g_i \cdot \Delta_{i\alpha}$ vanishes on $\overline{E}_i - V_{i\alpha}$, hence on $\overline{E}_i - S_{i\alpha}$. □

We are now in Fat City. We have $I(E) = I(\overline{E}) = \bigcap_{i,\alpha} P_{i\alpha}$. Define the *dimension of* E to be $\max_{i,\alpha} \dim(P_{i\alpha})$. This is the maximum dimension of the minimal primes of $I(E)$, which all must occur among the $P_{i\alpha}$. We consider the inclusion $A(E) = A(\overline{E}) \to \prod_{i,\alpha} A_{i\alpha}$ and the order $A(\overline{E}) \cap \Pi(\mathfrak{P}_{i\alpha})_d$. Let $S = \bigcup_{i,\alpha} S_{i\alpha} \subset \bigcup \overline{E}_i = \overline{E}$. Then, by 8.6.5, each $(A_{i\alpha}, (\mathfrak{P}_{i\alpha})_d)$ is an RHJ-algebra, with $X(A_{i\alpha}, (\mathfrak{P}_{i\alpha})_d) = S_{i\alpha}$ and $(\mathfrak{P}_{i\alpha})_d = \mathfrak{P}(S_{i\alpha})$. By 8.2.9, $(A(\overline{E}), A(\overline{E}) \cap \Pi(\mathfrak{P}_{i\alpha})_d)$ is an RHJ-algebra, with $X(A(\overline{E}), A(\overline{E}) \cap (\mathfrak{P}_{i\alpha})_d) = S$. Obviously, $A(\overline{E}) \cap \Pi(\mathfrak{P}_{i\alpha})_d = \mathfrak{P}(S)$, the functions in $A(\overline{E})$ nowhere negative on S.

Consider $\overline{E} - S$. By Tarski-Seidenberg $\overline{E} - S$ is a semi-algebraic set. Clearly, $\overline{E} - S = \bigcup_i \overline{E}_i - \bigcup_{i,\alpha} S_{i\alpha} \subset \bigcup_i (\overline{E}_i - \bigcup_\alpha S_{i\alpha})$, thus $\dim(\overline{E} - S) \leq \max \dim(\overline{E}_i - \bigcup_\alpha S_{i\alpha})$. But $I(\overline{E}_i) \subsetneq I(\overline{E}_i - \bigcup_\alpha S_{i\alpha})$. In fact, by 8.9.1, each minimal prime of $I(\overline{E}_i - \bigcup_\alpha S_{i\alpha})$ must *properly* contain some $P_{i\alpha}$. Thus, we have strict inequality $\dim(\overline{E}_i - \bigcup_\alpha S_{i\alpha}) < \dim(\overline{E}_i)$, hence $\dim(\overline{E} - S) < \max \dim(\overline{E}_i) = \dim(\overline{E})$. We can now repeat the whole process above, beginning with the semi-algebraic set $\overline{E} - S$. After finitely many steps we succeed in writing $\overline{E} \subset R^n$ as a finite union of sets of the form $X(A, \mathfrak{P})_d \subset R^n$, where $A = R[X_1 \ldots X_n]/P$, P prime, and where $\mathfrak{P} \subset A$ is a finite refinement of the weak order. As consequences, we have proved the following two results.

Proposition 8.9.2. Let $F \subset R^n$ be a closed, semi-algebraic set. Then $(A(F), \mathfrak{P}(F))$ is an RHJ-algebra. □

Proposition 8.9.3. Let $(A, \mathfrak{P}) \in$ (PORNN), with A and R-algebra of finite type. In order that $(A, \mathfrak{P}) \cong (A(F), \mathfrak{P}(F))$ for some closed, semi-algebraic set F, it is necessary and sufficient that there exist (1) finitely many primes

$P_i \subset A$ with $(0) = \cap P_i$ and (2) finite refinements of the weak order $\mathfrak{P}_i \subset A_i = A/P_i$, such that $\mathfrak{P} = A \cap \Pi(\mathfrak{P}_i)_d$, under the inclusion $A \to \Pi A_i$. □

Proposition 8.9.3 should be compared with the results of 8.2, especially 8.2.11, 8.2.12, and the last paragraph of 8.2. Working with derived orders $\mathfrak{P}_d$ instead of $\mathfrak{P}_p$, we can now deal with all closed, semi-algebraic sets, not just the closed semi-algebraic sets.

If we combine the Tarski-Seidenberg theorem, the going-up theorem for semi-integral extensions, and 8.9.2 we can prove the following.

Proposition 8.9.4. Let φ^*: $R[Y_1 \ldots Y_m] \to R[X_1 \ldots X_n]$ be a homomorphism, inducing φ: $R^n \to R^m$. Suppose $S \subset R^n$ is a closed, bounded, semi-algebraic set. Then $\varphi(S) \subset R^m$ is a closed, bounded, semi-algebraic set.

Proof: The image $\varphi(S)$ is semi-algebraic by Tarski-Seidenberg and bounded by simple estimate arguments. The homomorphism φ^* induces an inclusion $A(\varphi(S)) \to A(S)$, where $A(S) = R[X_1 \ldots X_n]/I(S)$ and $A(\varphi(S)) = R[Y_1 \ldots Y_m]/I(\varphi(S))$. Moreover, $\mathfrak{P}(\varphi(S)) = A(\varphi(S)) \cap \mathfrak{P}(S)$. Since $A(S)$ is semi-integral over R, $A(S)$ is certainly semi-integral over $A(\varphi(S))$. By 8.9.2, $(A(S), \mathfrak{P}(S))$ is an RHJ-algebra, hence by 8.2.7, so is $(A(\varphi(S)), \mathfrak{P}(\varphi(S)))$, and $\varphi(S) = X(A(\varphi(S)), \mathfrak{P}(\varphi(S)))$. Thus, $\varphi(S) \subset R^m$ is closed. □

Remark. As an immediate corollary of 8.9.4, we get that any polynomial function on a closed, bounded, semi-algebraic set assumes maximum values. This generalizes the discussion at the beginning of 8.7, where we proved this result for bounded, closed semi-algebraic sets.

In many ways the non-degenerate sets $X(A, \mathfrak{P})_d$, where A is a finite integral domain over R, $\mathfrak{P} = \mathfrak{P}_w[g_j] \subset A$ a finite refinement of the weak order, are more natural "building blocks" for semi-algebraic sets than the $X(A, \mathfrak{P})$. Although $X(A, \mathfrak{P})_d$ is not a manifold, as we have seen it is the *closure* of a d-manifold, where $d = \text{tr.deg}_R(A)$. Specifically, it is the closure of the set V of algebraic simple points x of $X(A, \mathfrak{P}_w)$ with $g_j(x) > 0$, and our implicit function theorem of 8.7 guarantees that V is an algebraic d-manifold, that is, locally like R^d. Thus, $X(A, \mathfrak{P})_d$ is a sort of closed "d-manifold with boundary and singularities". Secondly, these orders $\mathfrak{P}_d \subset A$ are exactly the contractions to A of finite refinements of the weak order on the field of fractions K of A. Thus we can view $X(A,\mathfrak{P}_d)$ as an "affine model" for $(K,\mathfrak{P})$, $\mathfrak{P} = \mathfrak{P}_w[g_j] \subset K$.

If E is a semi-algebraic set, the procedure above for obtaining $\overline{E} = \cup X(A_i,\mathfrak{P}_i)_d$, where $P_i \subset A(E)$ are primes, $A_i = A(E)/P_i$, and $\mathfrak{P}_i \subset A_i$ a finite refinement of the weak order, definitely depends on a specific presentation of $E = \cup E_i$, $E_i = Z\{f_{ij}\} \cap U\{g_{ik}\}$. It is also not really a stratification of E, since the pieces $X(A_i,\mathfrak{P}_i)_d$ can overlap, and, in fact, can overlap on more than their "boundaries", as in Figure (a). Even if they overlap only on boundaries, we may end up with a very unnatural decomposition of $\overline{E}$, as in Figure (b).

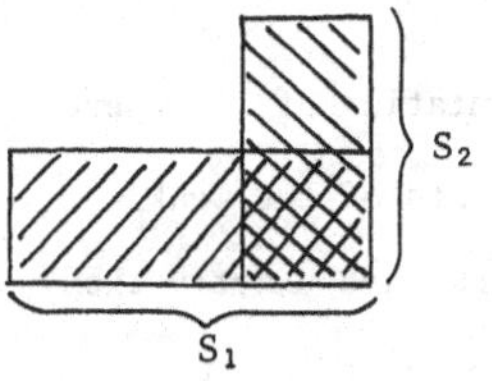

Figure (a)

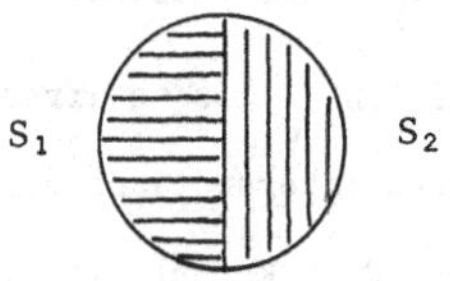

Figure (b)

We can rectify these problems somewhat by replacing the orders $(\mathfrak{P}_i)_d \subset A_i$ by finite intersections of such orders. Specifically, let $\{P_\alpha\}$ index the *distinct* primes which occur among the P_i. We assume our decomposition $\overline{E} = \cup X(A_i,\mathfrak{P}_i)_d$ is irredundant, in the sense that no term can be omitted. On $A_\alpha = A(E)/P_\alpha$, we impose the order $\bigcap_{P_i = P_\alpha} \mathfrak{P}_i = \mathfrak{P}_\alpha$. Then we have $(\mathfrak{P}_\alpha)_d = \cap(\mathfrak{P}_i)_d$ and $X(A_\alpha,\mathfrak{P}_\alpha)_d = \bigcup_{P_i = P_\alpha} X(A_i,\mathfrak{P}_i)_d$. Now, we claim that the RHJ-algebras $(A_\alpha,(\mathfrak{P}_\alpha)_d)$ are *geometric invariants* of $\overline{E}$, $\overline{E} = \cup X(A_\alpha,\mathfrak{P}_\alpha)_d$, $(A(\overline{E}),\mathfrak{P}(\overline{E})) = (A(\overline{E}),A(\overline{E}) \cap \Pi(\mathfrak{P}_\alpha)_d)$, and $\dim(X(A_\alpha,\mathfrak{P}_\alpha)_d \cap X(A_\beta,\mathfrak{P}_\beta)_d) < \dim X(A_\alpha,\mathfrak{P}_\alpha)_d$ if $\alpha \neq \beta$. The proof is not hard. The primes P_γ of maximal dimension are certainly minimal primes of $A(E)$, hence are characterized as the minimal primes of $A(E)$ of maximal dimension. Moreover, $X(A_\gamma,\mathfrak{P}_\gamma)_d = (\overline{E} \cap X(A_\gamma,\mathfrak{P}_w))_d$, which is a geometric invariant. Now $\overline{E} - \bigcup_\gamma X(A_\gamma,\mathfrak{P}_\gamma)_d$ has strictly lower dimension than $\overline{E}$. Thus, by induction, the whole decomposition of $\overline{E}$ is invariant. This discussion indicates that very nice basic building blocks for semi-algebraic sets are affine models of orders on function fields $\mathfrak{P} \subset K$, where $\mathfrak{P}$ is a finite intersection of finite refinements of the weak order.

We can also use the ideals of degeneracy $I_x \subset A(\overline{E})$, $x \in \overline{E}$, to study stratifications. By definition, $I_x = \underset{x \in U}{\varinjlim} I(\overline{E} \cap U)$, where U parametrizes smaller and smaller open neighborhoods of x. Of course, the limit stabilizes $I_x = I(\overline{E} \cap U_o)$ if U_o is small enough, say a small ball centered at $x \in \overline{E} \subset R^n$. Write $I_x = I(\overline{E} \cap U_o) = \cap P_{ix}$, where the $P_{ix} \subset A(\overline{E})$ are minimal primes of I_x.

Proposition 8.9.5.

(a) Only finitely many distinct prime ideals P_{ix} occur, as x varies over $\overline{E}$. Each $P_{ix} = I_y$, for suitable y near x.

(b) If $\overline{E} = \cup X(A_\alpha, \mathfrak{P}_\alpha)_d$ is an irredundant representation of $\overline{E}$, where the $\mathfrak{P}_\alpha \subset A_\alpha$ are finite intersections of finite refinements of the weak order, then the primes $P_\alpha = \text{kernel}(A(\overline{E}) \to A_\alpha)$ which occur are exactly the prime ideals which occur as ideals of degeneracy, I_y prime, $y \in \overline{E}$.

(c) The subset $X(A_\alpha, \mathfrak{P}_\alpha)_d \subset \overline{E}$ is characterized as those $x \in \overline{E}$ such that P_α is a minimal prime of I_x.

Proof: The proof consists of reviewing the various results of 8.6, 8.7, 8.8, and this section. For example, since each P_{ix} is $I(U_o \cap \overline{E})$-convex, we know that arbitrarily near x in $\overline{E}$ there will exist algebraic simple zeros of P_{ix}. In fact, we can find $y \in \overline{E} \cap U_o \cap Z(P_{ix})$ such that $y \notin Z(P_{jx})$, $P_{jx} \neq P_{ix}$, and an open set $y \in U_1$ such that $\overline{E} \cap U_1 = Z(P_{ix}) \cap U_1$ consists entirely of simple zeros of P_{ix}. Then $I_y = P_{ix}$.

The finiteness of the set of all P_{ix}, $x \in \overline{E}$, follows from (b) and (c). These two statements can be proved readily with all the machinery at hand. For example, suppose $I_y = P$ is prime, and assume $P = I(\overline{E} \cap U)$, $y \in U$. Consider the $X(A_\beta, \mathfrak{P}_\beta)_d$ which intersect $\overline{E} \cap U$. Since P vanishes on $\overline{E} \cap U$, and since $U \cap X(A_\beta, \mathfrak{P}_\beta)_d$ is Zariski dense in $X(A_\beta, \mathfrak{P}_w)$, we have $P \subset P_\beta$.

But also, some $X(A_\gamma, \mathfrak{P}_\gamma)_d$ will contain a whole neighborhood of simple zeros of P in $\bar{E} \cap U$, so $P_\gamma \subset P$, hence $P_\gamma = P$. We leave the rest of the details of (b) and (c) to the reader. □

The ideals of degeneracy can be used to define local notions of rank and dimension. Namely, if $I_x = \cap P_{ix}$, define

$$\dim_x(E) = \max \dim(P_{ix})$$

$$\operatorname{rank}_x(E) = \operatorname{rank}_x(I_x) .$$

($\operatorname{Rank}_x(I) = \operatorname{rank}\{dh_j(x)\}$, where $I = (h_j)$, $x \in Z(I)$.) We can define stratifications of E based on rank and dimension. Note that for any integer k, $\{x \in E | \dim_x(E) \geq k\}$ is a closed subset of E and $\{x \in E | \operatorname{rank}_x(E) \geq k\}$ is an open subset of E. We do not quite want to begin a stratification with all points of maximal rank, since this set will have (possibly rather singular) boundary points due to inequalities defining E. However, we can define $E^{(0)}$ to be the "interior points of maximal rank". These are obtained as follows. Take the irreducible components $X(A_\alpha, \mathfrak{P}_\alpha)_d \subset \bar{E}$ of *least* dimension. Then take only those points y which are simple points, lying on a unique $X(A_\alpha, \mathfrak{P}_\alpha)_d$, and for which an entire neighborhood of y in $Z(P_\alpha)$ belongs to E. We will have $I(E - E^{(0)}) \supsetneq I(E)$, so the process can be iterated, but will terminate. Using the implicit function theorem of 8.7, $E^{(0)}$ is a manifold.

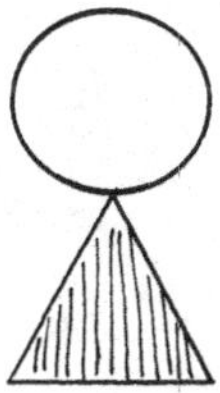

In the figure above, $E^{(0)}$ consists of the circle minus the vertex of the triangle. Then at the next stage $E^{(1)}$ we get the *interior* points of the triangle. Next, $E^{(2)}$ consists of the interiors of the edges of the triangle and finally $E^{(3)}$ is the set of vertices.

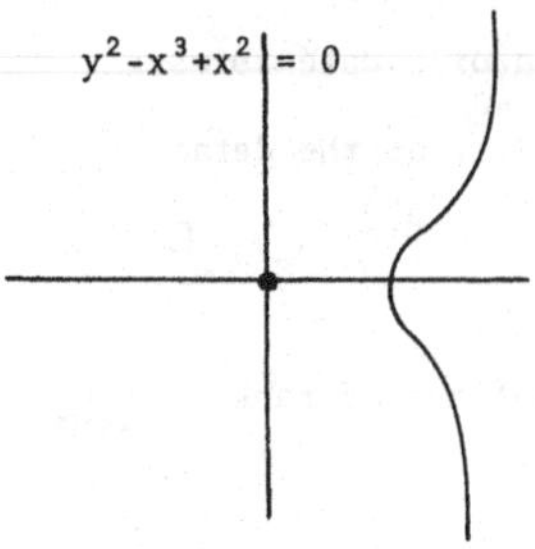

Figure (a)

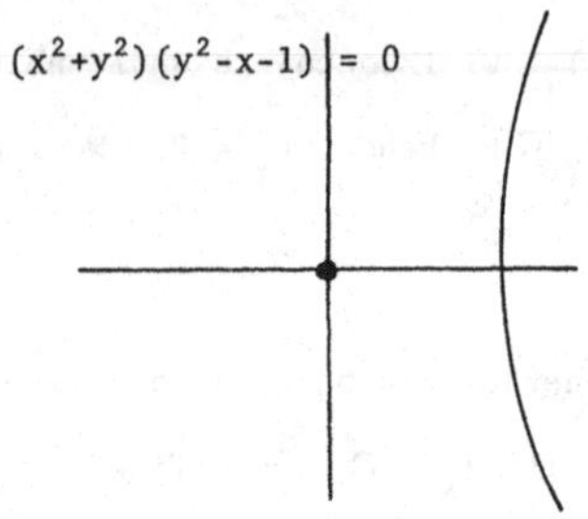

Figure (b)

We point out that by defining rank *locally*, our stratification by rank is not the same as that of Whitney [44] for algebraic varieties $Z(I) \subset R^n$. In our stratification, the origin has rank 2 in both varieties above, hence is the first stratum. On the other hand, in Whitney's stratification, the origin has rank 0 in Figure (a) and the curve has rank 1, whereas in Figure (b), the origin has rank 2 and the curve rank 1. (In Figure (b), the polynomials $x(y^2 - x - 1)$ and $y(y^2 - x - 1)$ vanish on the variety, but have independent differentials at the origin.)

We can also find a manifold stratification of E by dimension, $E \supset E_{(0)} \supset E_{(1)} \supset \cdots$. We begin with the pieces $X(A_\alpha, \mathfrak{P}_\alpha)_d$ of $\overline{E}$ of greatest dimension, and take for $E_{(0)}$ the "interior" simple points, lying on a unique $X(A_\alpha, \mathfrak{P}_\alpha)_d$. Then $\dim(E - E_{(0)}) < \dim(E)$, and the stratification continues inductively.

8.10. Krull Dimension

Let A be an integral domain, $\mathfrak{P} \subset A$ an order. By a *weak $\mathfrak{P}$-chain* of prime ideals we mean a strictly increasing sequence of $\mathfrak{P}$-convex primes $(0) \subsetneq P_1 \subsetneq \cdots \subsetneq P_r \subset A$. The *length* of the chain is r. We define the *weak Krull dimension*, $\dim_w(A,\mathfrak{P})$, to be the maximum length of such a chain.

In complete generality, this notion is probably uninteresting. Even if we make the drastically simplifying assumption that A is finitely generated over a ground field R, the order $\mathfrak{P}$ can make things complicated. For example, if R[T] is ordered with T infinitely large relative to R, then R[T] is a semi-field, with weak dimension zero. This pathology is caused by lack of finiteness conditions on the order.

Suppose, then, that we begin with a finite real domain A over R, R real closed, and an order $\mathfrak{P} = \mathfrak{P}_w[g_j]$ obtained by finitely extending the weak order. If $Q \subsetneq P$ are prime ideals, then $\text{tr.deg}_R(A/P) < \text{tr.deg}_R(A/Q)$. Thus $\dim_w(A,\mathfrak{P}) \leq r = \text{tr.deg}(A)$. But from 8.4.3 we have chains of $\mathfrak{P}$-convex primes of length r, thus $\dim_w(A,\mathfrak{P}) = \text{tr.deg}(A)$. In fact, if $Q \subset A$ is any $\mathfrak{P}$-convex prime, we can apply 8.4.3 to $(A/Q, \mathfrak{P}/Q)$ and obtain chains of $\mathfrak{P}$-convex primes above Q, $Q \subsetneq P_1 \subsetneq \cdots \subsetneq P_s \subset A$, where $s = \text{tr.deg}(A/Q)$.

By analogy with classical Krull dimension theory, at least two other questions come to mind. First, given such a $Q \subset A$, can one find a chain of $\mathfrak{P}$-convex primes below Q, $(0) \subsetneq Q_1 \subsetneq \cdots \subsetneq Q_{r-s} = Q$? More generally, given $Q \subset P \subset A$, both $\mathfrak{P}$-convex, can one find a chain of $\mathfrak{P}$-convex primes between Q and P, $Q = P_0 \subsetneq P_1 \subsetneq \cdots \subsetneq P_t = P$, where $t = \text{tr.deg}(A/Q) - \text{tr.deg}(A/P)$?

The answer turns out to be "yes", but before discussing this further, we want to argue that there are other, perhaps more natural, notions of dimension. Again, we begin with $(A,\mathfrak{P})$, $\mathfrak{P} = \mathfrak{P}_w[g_j]$ a finite refinement of the weak order on the finite domain A. Define a *strong $\mathfrak{P}$-chain* of prime ideals to be a strictly increasing chain $(0) \subsetneq P_1 \subsetneq \cdots \subsetneq P_r$, such that for some total order refinement $\overline{\mathfrak{P}} \supset \mathfrak{P}$, all P_i are $\overline{\mathfrak{P}}$-convex. Note that a weak $\mathfrak{P}$-chain corresponds to a sequence of (non-injective) epimorphisms of domains $(A,\mathfrak{P}) \to (A/P_1, \mathfrak{P}/P_1) \to \cdots \to (A/P_r, \mathfrak{P}/P_r)$. From Proposition 7.7.9, a strong $\mathfrak{P}$-chain requires much more restrictive convexity hypotheses on the primes P_i, and by 7.7.10, a strong $\mathfrak{P}$-chain corresponds to a sequence of signed places $K \to K_1, \pm\infty \to \cdots \to K_r, \pm\infty$ over R, finite on A, with $P_i = \text{kernel}(A \to K_i)$, where K is the fraction field of A. We define the *strong Krull dimension* $\dim_s(A,\mathfrak{P})$ to be the maximum length of a strong $\mathfrak{P}$-chain. Of course, 8.4.3 actually produces strong $\mathfrak{P}$-chains in A, hence we still have $\dim_s(A,\mathfrak{P}) = \text{tr.deg}(A)$. But what is more important, we have available the signed place perturbation theorem 8.4.9 and its consequence Proposition 8.6.10 which gives more delicate information immediately.

Proposition 8.10.1. Let $Q \subset A$ be a $\mathfrak{P}_d$-convex prime $\mathfrak{P} = \mathfrak{P}_w[g_j]$. Then there exists a strong $\mathfrak{P}$-chain $(0) \subsetneq P_1 \subsetneq \cdots \subsetneq P_s = Q \subsetneq \cdots \subsetneq P_r \subset A$, where $s = \text{tr.deg}(A) - \text{tr.deg}(A/Q) = \text{codim}(Q)$.

Proof: The chain below Q comes from 8.6.10, which gives signed places $K \to K_1, \pm\infty \to \cdots \to K_s, \pm\infty$. Then K_s is still a function field, to which we apply 8.4.3, and extend the strong $\mathfrak{P}$-chain above Q.

Actually, a little more is required if some $g_j \in Q$ in order to conclude that the total order one finally obtains on K actually extends $\mathfrak{P}$. Specifically, one constructs an integral extension $A \subset B$ by adjoining as many $\sqrt{g_j}$ as necessary in order that the weak order on B contracts to $\mathfrak{P} \subset A$. The prime Q lifts to a convex prime of B, and we apply 8.6.10 and 8.4.3 to the fraction field of B, then restrict to K. This argument was actually used in the proof of 8.6.10. □

The reason strong dimension is a more natural concept than weak dimension is that strong dimension is a *local* concept, giving geometric information about a semi-algebraic set infinitesimally near a point. Specifically, suppose the prime Q of 8.10.1 is a point $x \in X(A,\mathfrak{P})$. Then $\mathfrak{P}_d$-convexity of Q says x is a non-degenerate point $x \in X(A,\mathfrak{P})_d$. Moreover, if $g(x) > 0$, $g \in A$, then any $\mathfrak{P}_d$-convex prime P is $\mathfrak{P}[g]$-convex. Thus in any neighborhood U of x in $X(A,\mathfrak{P})_d$, the prime ideals P_i of the strong $\mathfrak{P}$-chain of 8.10.1 have lots of zeros. Specifically, $P_i = I(Z(P_i) \cap U)$, we can find simple zeros of P_i in U, and so on. The strong chain of primes going down from Q correspond to a chain of subvarieties going up from the point x, of dimension $1,2,\ldots,r$. In other words, *infinitesimally near* x, we can move about on the semi-algebraic set $X(A,\mathfrak{P})$ with r-degrees of freedom.

By way of contrast, we reconsider weak dimension. The result which allows the "desired" conclusion is the following.

Unproved Proposition 8.10.2. Suppose $Q \subset A$ is $\mathfrak{P}$-convex, $\mathfrak{P} = \mathfrak{P}_w[g_j]$. Then there exists a $\mathfrak{P}_d$-convex prime Q', with $Q' \subset Q$ and $\text{tr.deg}(A/Q') - \text{tr.deg}(A/Q) = 1$.

Corollary 8.10.3. Suppose $P_{i_1} \subsetneq \cdots \subsetneq P_{i_j} \subset A$ is any $\mathfrak{P}$-convex chain in A. Then there exists a $\mathfrak{P}$-convex refinement $(0) \subsetneq P_1 \subsetneq \cdots \subsetneq P_r \subset A$ of maximal length $r = \text{tr.deg}(A)$.

Proof: It suffices to insert an appropriate chain between any two $\mathfrak{P}$-convex primes $P \subset Q$. Passing to $(\bar{A},\bar{\mathfrak{P}})$, where $\bar{A} = A/P$, $\bar{\mathfrak{P}} = \mathfrak{P}/P = \mathfrak{P}_w[\bar{g}_j]$, we may assume $P = (0)$. Apply 8.10.2 to find a $\mathfrak{P}_d$-convex prime $Q' \subset Q$ of dimension one greater than that of Q. Then apply 8.6.10 to go down from Q'. □

A proof of Proposition 8.10.2 can be extracted from [21]. Here is a rough outline of the geometry involved. By extending the ground field from R to an appropriate transcendental extension, we may assume that Q is zero dimensional, that is, a *point*. Now, we know the variety $X(A,\mathfrak{P})$ has lots of points. For example, if $\text{tr.deg}(A) = r$, near a simple point $X = X(A,\mathfrak{P})$ looks like affine r-space smoothly embedded in some higher dimensional affine space R^n. We consider *sections* of X by n-r+1-planes through the point Q. The implicit function theorem guarantees that many of these sections will be 1-dimensional semi-algebraic sets. (Simply take the n-r+1 plane in general position with respect to the tangent r-plane of X at a simple point.) The difficulty is, the point Q may be an isolated point of such a section Y, and Y may not be algebraically irreducible. However, the method of [21]

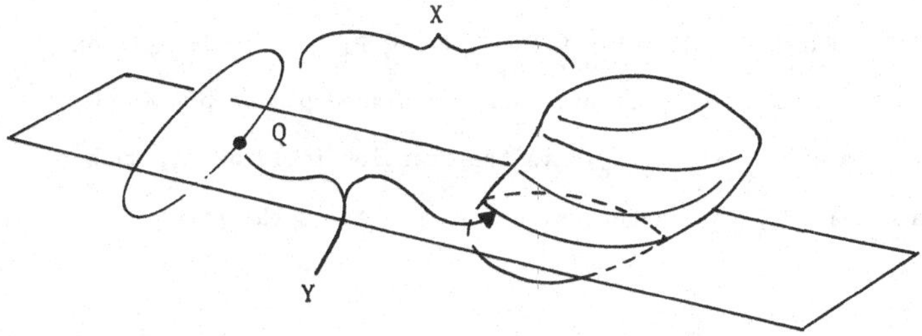

is essentially to argue that generically these sections Y *are* algebraically irreducible 1-dimensional sets through Q, although Q may indeed be a degenerate point. Such a set corresponds to the desired prime $Q' \subset Q$. This

Q' is $\mathfrak{P}_d$-convex since, by construction, Q' has sufficiently many zeros in X_d.

Note the proof of this dimension theorem 8.10.2 uses completely different concepts than those required for the study of strong dimension. This is because 8.10.2 really is not a local geometric result at all, but a global property of semi-algebraic sets. Near a degenerate point Q, the variety $X(A,\mathfrak{P})$ will not be described by r-independent parameters; although globally we can pass a curve, then a surface containing the curve, and so on, up to the r-fold itself, through the point Q. Also, we point out that the commutative algebra analogue of the dimension theorem 8.10.2 can be proved using integral extensions and various going-up and going-down theorems for prime ideals. In our real setting this method seems to break down because of the special hypotheses needed in 6.4, especially in Proposition 6.4.2(b).

The analog of 8.10.3 for strong chains does *not* follow routinely from 8.10.1. The reason is that given $P \subset Q \subset A$ with P a $\mathfrak{P}_d$-convex prime of A and Q/P a $(\mathfrak{P}_d/P)_d$-convex prime of A/P, we cannot immediately construct strong $\mathfrak{P}$-chains between P and Q by passing to $(A/P, (\mathfrak{P}_d/P)_d)$. The finiteness condition on the order is lost by this process. What is needed instead is a more vigorous version of the signed place perturbation theorem 8.4.9, which was the basis of 8.10.1.

Proposition 8.10.4. Suppose $A = R[x_1 \ldots x_n]$ is a finite integral domain over R, and $\mathfrak{P}' \subset A$ is a total order such that all x_i are infinitesimally small relative to R. Suppose $(0) = P_{i_0} \subsetneqq P_{i_1} \subsetneqq \cdots \subsetneqq P_{i_s}$ are prime $\mathfrak{P}'$-convex ideals of A. Then there exists a discrete rank r signed place $p\colon\ K \to R, \pm\infty$, where $r = \text{tr.deg}(A)$ and $K = R(x_1 \ldots x_n)$ is the fraction field of A, such that $p(x_i) = 0$ and all P_{i_j} are $\mathfrak{P}$-convex, where $\mathfrak{P} \subset A$ is the total order on A induced by p.

Proof: The proof is by induction on r, there being nothing to prove if r is quite small. Let $r_j = \text{tr.deg}(A/P_{i_j})$, so that $r = r_0 > r_1 > \cdots > r_s$. As in the proof of 8.4.9, we may assume A is integral over $R[x_1 \ldots x_r]$, and we may also assume that the $\{\bar{x}_k = x_k (\text{mod } P_{i_j}),\ 1 \le k \le r_j\}$, give a transcendence base for A/P_{i_j}, all j, by simply rearranging $x_1 \ldots x_r$ if necessary.

If $r_0 - r_1 = 1$, then the place $p'_{R(x_1 \ldots x_{r_1})}: K \to \Delta, \pm\infty$ defined by the total order $\mathfrak{P}'$ on K and the subfield $R(x_1 \ldots x_{r_1}) \subset K$ is a discrete rank 1 place, with image a function field K' over R with $\text{tr.deg}(K') = r-1$. Moreover, if $A' \subset K$ is the integral closure of A in K, we can write $A' = R[x_i, y_j]$, with x_i and y_j infinitesimally small relative to R, we can lift all the $\mathfrak{P}'$-convex primes P_{i_j} of A to $\mathfrak{P}'$-convex primes P'_{i_j} of A', and they necessarily form a chain in A' by 7.7.8. Also, K' is the fraction field of a quotient of A' (in fact, of A'/P'_{i_1}), hence by induction, we can find a discrete rank $r-1$ signed place $q: K' \to R, \pm\infty$, preserving the convexity of the P'_{i_j}, $j \geq 2$. Composition and reordering K then gives the desired $p: K \to K', \pm\infty \to R, \pm\infty$. Thus we may assume $r_0 - r_1 \geq 2$.

Write $A/P_{i_j} = R[\bar{x}_1 \ldots \bar{x}_{r_j}][\bar{x}_{r_j+1}] \cdots [\bar{x}_n]$, and for $k > r_j$, let $f_{j,k} = f_{j,k}(x_k) \in P_{i_j}$ represent the minimal polynomial for $\bar{x}_k$ over $R[\bar{x}_1 \ldots \bar{x}_{k-1}]$. If $S_j \subset A$ is the multiplicative set of non-zero elements of $R[x_1 \ldots x_{r_j}]$, then $\{f_{j,k}\}_{k > r_j}$ generate the prime $P_{i_j}A_{S_j}$ in A_{S_j}.

If $k > r = r_0$, then $f_{j,k}(x_k)$ is integral over $R[x_1 \ldots x_r]$. Let $0 = \Sigma a_{ijk}(x_1 \ldots x_r) f^i_{j,k}$ be a monic polynomial equation for $f_{j,k}$, with coefficients in $R[x_1 \ldots x_r]$. We have our total order $\mathfrak{P}' \subset K$. Let Γ denote a real closure of K, and let $F_j \subset \Gamma$ be the real closure of $R(x_1 \ldots x_{r_j})$, so that $F_s \subset F_{s-1} \subset \cdots \subset F_1$. Since all x_i are infinitesimally small relative to R, the x_i are certainly finite relative to $R(x_1 \ldots x_{r_j})$. In fact, $P_{i_j} \subset A$ is the center in A of the place $p'_j = p_{R(x_1 \ldots x_{r_j})}$. (This follows easily from Proposition 7.7.4 and results in 7.6.) Thus elements of P_{i_j} are infinitesimally small relative to $R(x_1 \ldots x_{r_j})$. It follows that the $p'_j(x_i)$ are all algebraic over $R(x_1 \ldots x_{r_j})$, in the appropriate residue field, and we can identify $p'_j(x_i)$ with some $\zeta_{j,i} \in F_j$. Applying the place p'_j to our integral dependence relation for $f_{j,k}$ in K, $k > r$, we get that 0 is a root of the non-trivial polynomials $\Sigma a_{ijk}(\zeta_1 \ldots \zeta_r) T^i$ over F_j. Also, if for $k > r$ we let $0 = \Sigma b_{ik}(x_1 \ldots x_r) x^i_k$ be an integral dependence relation for x_k over $R[x_1 \ldots x_r]$, then 0 is also a root of the non-trivial polynomials $\Sigma b_{ik}(0 \ldots 0) T^i$ over R.

Let $0 < \varepsilon \in R$ be less than all other roots of the $\Sigma b_{ik}(0\ldots0)T^i$ over R and let $0 < \varepsilon_j \in F_j$ be less than all other roots of the $\Sigma a_{ijk}(\zeta_1\ldots\zeta_r)T^i$ over F_j. Since $x_1 \in F_j$, we may also assume $\varepsilon_j^2 < x_1^2$. We now extend K to a field L by adjoining the elements $\zeta_{j,i}$, $\sqrt{\varepsilon^2 - x_k^2}$, and $\sqrt{\varepsilon_j^2 - f_{j,k}^2}$, where $1 \leq j \leq s$, $1 \leq i \leq n$, and $r < k \leq n$. (Note these elements belong to Γ so L is real.) We replace A by the ring $B = [x_i, x_i - \zeta_{j,i}, \sqrt{\varepsilon^2 - x_k^2} - \varepsilon, \sqrt{\varepsilon_j^2 - f_{j,k}^2}]$. Since the $f_{j,k}$ and $x_i - \zeta_{j,i}$ are infinitesimally small relative to F_j, hence also relative to R, and since $\varepsilon_j^2 < x_1^2$, all these generators of B are infinitesimally small relative to R.

The places p_j' are still defined on L since $R(x_1\ldots x_{r_j}) \subset L$. The center of p_j' on B is a prime Q_{i_j} which contracts to $P_{i_j} \subset A$. Thus all our hypotheses hold for B. Since any discrete rank r place $p\colon L \to R, \pm\infty$ restricts to a discrete rank r place of K, we may as well assume $A = B$, $K = L$. The advantage in this: if we now construct any total order $\mathfrak{P} \subset K$ such that the $\{x_1\ldots x_r\}$ are infinitesimally small relative to R and the $\{f_{j,k} | r_j < k \leq r\}$ are infinitesimally small relative to $R(x_1\ldots x_{r_j})$, then necessarily all x_i are infinitesimally small relative to R and all $f_{j,k}$ are infinitesimally small relative to $R(x_1\ldots x_{r_j})$. As we observed, $\{f_{j,k}\}_{k > r_j}$ generate $P_{i_j} \cdot A_{S_j}$, where $S_j = R[x_1\ldots x_{r_j}]$. Thus if p_j is the place $p_{R(x_1\ldots x_{r_j})}$ on K relative to this (new) order $\mathfrak{P}$, we still have that the center of p_j on A is $P_{i_j} \subset A$.

Let $E \subset \Gamma$ be the real closure of $R(x_1\ldots x_{r-1})$. Since $r - r_1 \geq 2$, we have $F_1 \subsetneq E$. Consider the function field $E(x_r, x_{r+1}\ldots x_n)$ of transcendence degree 1 over E, which can be written $E(x_r, y)$ or $E(f_{1,r}, z)$, where, recall, $f_{1,r} = f_{1,r}(x_r) \in P_{i_1}$ represents a minimal polynomial for $\bar{x}_r$ over $R[\bar{x}_1\ldots\bar{x}_{r-1}]$, $\bar{x}_i = x_i (\text{mod } P_{i_1})$. We consider the cut $D = D_E(x_r)$ of E defined by x_r. If D is algebraic, the place p_E is non-trivial, hence $p_{R(x_1\ldots x_{r-1})}$ is discrete rank 1 on K. We are then finished by induction, as in the second paragraph of this proof. We propose to reorder $E(x_r, y)$ over E so that $D_E(x_r)$ is algebraic and so that all our hypotheses on $K \subset E(x_r, y)$ are preserved in the reordering.

It is better to work with the transcendental element $f_{1,r}(x_r) \in E(f_{1,r}, z)$

$= E(x_r,y)$, since $f_{1,r}$ is infinitesimally small relative to the (large) field $R(x_1 \ldots x_{r_1}) \subset E$. Thus let $D' = D_E(F_{1,r})$. We may assume $0 < D'$, by changing sign of $f_{1,r}$ if necessary. We appeal to 8.3.1 to choose $\beta \in E$, $0 < \beta < D'$, so that if $E(f_{1,r})$ is reordered over E with $D_E(f_{1,r}) = \beta$, then this new order extends to $E(f_{1,r},z)$. Of course, β is still infinitesimally small relative to $R(x_1 \ldots x_{r_1})$.

We now consider our hypotheses on the elements x_i, $i \leq r$ and $f_{j,k}$, $r_j < k \leq r$. Since the order on E is unchanged, we need only worry about x_r and the $f_{j,r}$, $1 < j \leq s$. But $f_{1,r} = f_{1,r}(x_r) = \Sigma c_i(x_1 \ldots x_{r-1})x_r^i$, say, and if we apply the place p_{f_1}, we get $0 = \Sigma c_i(\zeta_{1,1} \ldots \zeta_{1,r-1}) p_{F_1}(x_r)^i$, where $p_{F_1}(x_i) = \zeta_{1,i} \in F_1$ has not changed if $i < r$. This polynomial is non-trivial, that is, not all $c_i(\zeta_{1,1} \ldots \zeta_{1,r-1})$ vanish, because, by definition, the coefficients of $f_{1,r}(x_r)$ are non-zero modulo $P_{i_1} \subset A$ and P_{i_1} is the center of p_{F_1} on A.

We see that when we made our finite extension from K to L, four paragraphs back, we should have added two more elements, $\sqrt{\delta - x_r}$ and $\sqrt{x_r - \gamma}$, where $\gamma, \delta \in F_1$, $\zeta_{1,r} - x_1 < \gamma < \zeta_{1,r} < \delta < \zeta_{1,r} + x_1$, and the interval (γ,δ) in F_1 contains only the one root $\zeta_{1,r}$ of $\Sigma c_i(\zeta_{1,1} \ldots \zeta_{1,r-1})T^i = 0$. (With the conditions indicated, the elements $\sqrt{\delta - x_r}$ and $\sqrt{x_r - \gamma}$ are infinitesimally small relative to R.) But there is no problem doing this retroactively, and the conclusion is, even with our new order or $E(f_{1,r},z) = E(x_r,y)$, we have $p_{F_1}(x_r) = \zeta_{1,r} \in F_1$. In particular, x_r is still infinitesimally small relative to R. Moreover, if one writes out the polynomials $f_{j,r} = f_{j,r}(x_r)$, $1 < j \leq s$, one sees that they are still infinitesimally small relative to F_j, since x_i, $i < r$, have not moved at all, and x_r has not moved relative to $F_1 \supset F_j$, by our reordering. This establishes our original hypotheses for our new order on $E(x_r,y)$, now with $D_E(x_r)$ algebraic, and hence completes the proof of 8.10.4. □

We can now prove the strong chain analogue of Proposition 8.10.3.

Proposition 8.10.5. Suppose A is a finite integral domain over R and $\mathfrak{P} = \mathfrak{P}_W[g_j]$ is a finite refinement of the weak order. Then any strong $\mathfrak{P}$-chain of prime ideals $P_{i_1} \subsetneqq \cdots \subsetneqq P_{i_s} \subset A$ can be refined to a strong $\mathfrak{P}$-chain $(0) \subsetneqq P_1 \subsetneqq \cdots \subsetneqq P_r \subset A$ of maximal length $r = \text{tr.deg}(A)$.

Proof: The proof consists of arguments used previously. First, by adjoining suitable $\sqrt{g_i}$ to A, we are reduced to the case where there are no g_j. Secondly, by extending the ground field, we are reduced to the case where $P_{i_s} \subset A$ is zero dimensional, that is, $A/P_{i_s} = R$. Both these reductions are just as in the proof of 8.6.10. We then apply Proposition 8.10.4 to get a maximal strong $\mathfrak{P}$-chain refinement of the P_{i_j} below P_{i_s}. This gives a discrete rank $r - r_s$ signed place p': $K \to K', \pm\infty$, where K is the fraction field of A, $r_s = \text{tr.deg}(A/P_{i_s})$, and K' is a finite algebraic extension of the fraction field of A/P_{i_s}. Finally, 8.4.3 gives a discrete rank r_s signed place p'': $K' \to R, \pm\infty$, which completes the strong $\mathfrak{P}$-chain above P_{i_s}. □

Remark: If A is a finite integral domain, $\mathfrak{P} = \mathfrak{P}_w[g_j] \subset A$, and $P \subset A$ a prime $\mathfrak{P}_d$-convex ideal, then the RHJ-property of $(A/P, (\mathfrak{P}_d/P)_d)$ follows rather easily from 8.10.5. In fact, in 8.10.5 one can add the conclusion that if $g \in A$, $g \notin P_{i_s}$, then the refinement chain can be constructed with $g \notin P_r$. Of course, P_r is zero dimensional, hence is a point $x \in X(A,\mathfrak{P})$, and moreover, x is highly non-degenerate. This argument would complete the proof of Proposition 8.6.9 which we omitted previously.

8.11. Orders on Function Fields

Let K be a real function field over the real closed ground field R. Write $K = R(x_1 \ldots x_n)$, $A = R[x_1 \ldots x_n]$. Thus A is an integral domain, $X = X(A,\mathfrak{P}_w) \subset R^n$ is an algebraic variety and $X_d = X(A,(\mathfrak{P}_w)_d)$ is the (semi)-algebraic variety of non-degenerate points of X. Note that $(\mathfrak{P}_w)_d = A \cap \mathfrak{P}_w(K)$, where $\mathfrak{P}_w(K)$ is the weak order on K. We know that if $g_i \neq 0$, then $\mathfrak{P}_w[g_i]$ is an order on A if and only if $U\{g_i\} \cap X_d \neq \emptyset$, where $U\{g_i\} = \{x \in X | g_i(x) > 0\}$. All non-empty open sets $U\{g_i\} \cap X_d$ are Zariski dense in X_d.

We will find it convenient to forget the degenerate points of X altogether, and work in X_d. Thus, the notation $U\{g_i\}$ now means $\{x \in X_d | g_i(x) > 0\}$. We will also find it convenient to work with the larger open sets $V\{g_i\} = \mathring{\overline{U}}\{g_i\} = \mathring{W}\{g_i\} \subset X_d$. (In X_d, $W\{g_i\} - U\{g_i\}$ has no interior.)

Let $\mathscr{V}$ be the collection of non-empty sets of the form $V\{g_i\}$. Thus $V\{g_i\} \in \mathscr{V}$ if and only if $\mathfrak{P}_w[g_i]$ is an order on A. The $V\{g_i\}$ are regular open sets, that is, $V\{g_i\} = \mathring{\overline{V}}\{g_i\}$, since, in fact, $\overline{V}\{g_i\} = \overline{U}\{g_i\}$. By Tarski-Seidenberg, the $V\{g_i\}$ are open, semi-algebraic sets. In affine space (hence in any semi-algebraic set) the $V\{g_i\}$ form a base for the same "topology" as the $U\{g_i\}$. That is, each V is a finite union of U's and each U is a finite union of V's. This is certainly not obvious. In fact, it is pretty much the same statement as Unproved Proposition 8.1.2. We will not use this fact, although there are strong indications that the $V\{g_i\}$ provide a more natural base than the $U\{g_i\}$.

By a *prefilter* $\mathscr{F} \subset \mathscr{V}$ we mean a non-empty subset of $\mathscr{V}$, closed under finite intersections. A prefilter is a *filter* if $V \in \mathscr{F}$, $V \subset V' \in \mathscr{V}$ implies $V' \in \mathscr{F}$. A filter is an *ultrafilter* if it is not properly contained in any other filter in $\mathscr{V}$. Note that $V\{g_i\} \subset V(h)$ if and only if $h(V\{g_i\}) \geq 0$. This is the property of the $V\{g_i\}$ which makes them more convenient than the $U\{g_i\}$ for the purposes of this section.

The set of filters in $\mathscr{V}$ is partially ordered by inclusion, an arbitrary intersection of filters is a filter (every filter contains $X_d = V(1)$), and a union of a chain of filters is a filter. Thus every filter is contained in an ultrafilter. Each prefilter $\mathscr{F} \subset \mathscr{V}$ is contained in a smallest filter $\mathscr{F}_d \subset \mathscr{V}$.

Let $\mathfrak{P} \subset A$ be any partial order. Define $\mathscr{F}(\mathfrak{P}) \subset \mathscr{V}$ by $V\{g_i\} \in \mathscr{F}(\mathfrak{P})$ if $g_i \in \mathfrak{P}$. $\mathscr{F}(\mathfrak{P})$ is a prefilter since $\mathfrak{P}_w[g_i]$ is an order on A if and only if $V\{g_i\} \in \mathscr{V}$, which implies the desired finite intersection property for $\mathscr{F}(\mathfrak{P})$. If $\mathfrak{P}_1 \subset \mathfrak{P}_2$, then $\mathscr{F}(\mathfrak{P}_1) \subset \mathscr{F}(\mathfrak{P}_2)$.

Let $\mathscr{F} \subset \mathscr{V}$ be an arbitrary prefilter in $\mathscr{V}$. Define $\mathfrak{P}(\mathscr{F}) \subset A$ by $g \in \mathfrak{P}(\mathscr{F})$ if there is $V \in \mathscr{F}$ with $g(V) \geq 0$. Equivalently, this says $V \subset V(g)$. Since the sets V are Zariski dense, it is clear that $\mathfrak{P}(\mathscr{F})$ is an order on A. In fact, $\mathfrak{P}(\mathscr{F})$ is a derived order, since if $f^2 g = h \in \mathfrak{P}(\mathscr{F})$,

$f \neq 0$, and $h(V) \geq 0$, then g must be non-negative on V. Otherwise, f would vanish on the Zariski dense set $U\{g_i\} \cap U(-g)$, where $V = V\{g_i\}$. If $\mathscr{F}_1 \subset \mathscr{F}_2$, then $\mathfrak{P}(\mathscr{F}_1) \subset \mathfrak{P}(\mathscr{F}_2)$. Also $\mathfrak{P}(\mathscr{F}) = \mathfrak{P}(\mathscr{F}_d)$, where $\mathscr{F}_d$ is the filter generated by $\mathscr{F}$.

We will study the compositions $\mathfrak{P}(\mathscr{F}(\mathfrak{P}))$ and $\mathscr{F}(\mathfrak{P}(\mathscr{F}))$. First, $\mathfrak{P} \subset \mathfrak{P}(\mathscr{F}(\mathfrak{P}))$ and $\mathscr{F} \subset \mathscr{F}(\mathfrak{P}(\mathscr{F}))$ are completely obvious from the definitions.

Proposition 8.11.1.

(a) If $\mathscr{F} \subset \mathscr{V}$ is any prefilter, then $\mathscr{F}(\mathfrak{P}(\mathscr{F})) = \mathscr{F}_d \subset \mathscr{V}$.

(b) If $\mathscr{F}$ is an ultrafilter, then $\mathfrak{P}(\mathscr{F}) \subset A$ is a total order. Also, if $\mathfrak{P} \subset A$ is a total order, then $\mathscr{F}(\mathfrak{P}) \subset \mathscr{V}$ is an ultrafilter.

(c) If $\mathfrak{P} \subset A$ is any order, then $\mathfrak{P}(\mathscr{F}(\mathfrak{P})) = \mathfrak{P}_d$.

Proof:

(a) Since $\mathfrak{P}(\mathscr{F}) = \mathfrak{P}(\mathscr{F}_d)$, we may as well assume $\mathscr{F}$ is a filter. We know $\mathscr{F} \subset \mathscr{F}(\mathfrak{P}(\mathscr{F}))$. Conversely, if $g_i \in \mathfrak{P}(\mathscr{F})$, so that $V\{g_i\} \in \mathscr{F}(\mathfrak{P}(\mathscr{F}))$, let $g_j(V_j) \geq 0$, $V_j \in \mathscr{F}$. Then $V_j \subset V(g_j)$ and since $\mathscr{F}$ is a filter, $V(g_j) \in \mathscr{F}$. Thus $V\{g_i\} \in \mathscr{F}$ and $\mathscr{F}(\mathfrak{P}(\mathscr{F})) \subset \mathscr{F}$.

(b) Suppose $\mathfrak{P} = \mathfrak{P}(\mathscr{F})$ admits a proper refinement, $\mathfrak{P} \subset \mathfrak{P}[g]$. Then $\mathscr{F}(\mathfrak{P}) = \mathscr{F} \subset \mathscr{F}(\mathfrak{P}[g])$ and since $\mathscr{F}$ is an ultrafilter, $\mathscr{F} = \mathscr{F}(\mathfrak{P}[g])$. But now $g \in \mathfrak{P}(\mathscr{F}(\mathfrak{P}[g])) = \mathfrak{P}(\mathscr{F})$, contradiction.

Secondly, assume $\mathfrak{P} \subset A$ is a total order and suppose $\mathscr{F} = \mathscr{F}(\mathfrak{P})$ is properly contained in a filter $\mathscr{G}$. Let $V = V\{g_i\} \in \mathscr{G} - \mathscr{F}$. Since $\mathfrak{P} \subset \mathfrak{P}(\mathscr{F}(\mathfrak{P})) \subset \mathfrak{P}(\mathscr{G})$, we have $\mathfrak{P} = \mathfrak{P}(\mathscr{G})$. But $g_i \in \mathfrak{P}(\mathscr{G})$, hence $V\{g_i\} \in \mathscr{F}(\mathfrak{P})$, contradiction.

(c) Finally, $\mathfrak{P} \subset \mathfrak{P}(\mathscr{F}(\mathfrak{P}))$ and $\mathfrak{P}(\mathscr{F}(\mathfrak{P}))$ is a derived order, so $\mathfrak{P}_d \subset \mathfrak{P}(\mathscr{F}(\mathfrak{P}))$. Also, $\mathfrak{P}_d = \cap \mathfrak{P}_\alpha$ where the intersection is taken over the total order refinements of $\mathfrak{P}$. Write $\mathfrak{P}_\alpha = \mathfrak{P}(\mathscr{F}_\alpha)$ where $\mathscr{F}_\alpha = \mathscr{F}(\mathfrak{P}_\alpha)$. Then $\mathscr{F}(\mathfrak{P}) \subset \mathscr{F}_\alpha$, $\mathfrak{P}(\mathscr{F}(\mathfrak{P})) \subset \mathfrak{P}(\cap \mathscr{F}_\alpha) \subset \cap \mathfrak{P}(\mathscr{F}_\alpha) = \mathfrak{P}_d$. □

We have thus established a natural, bijective, refinement preserving correspondence between orders on the function field K (which are the same as derived orders on A) and filters in the family $\mathscr{V}$ of subsets of X_d.

This result makes it clear how an arbitrary order on a function field K is related to the particular orders defined by evaluating polynomial functions on Zariski dense subsets of any affine model $X \subset R^n$ for K. Total orders correspond to evaluating functions on infinitesimally small open subsets, that is, germs of open sets.

Of course, the totality of all orders on K is a birational invariant of X. However, with respect to the particular model X, we can distinguish orders obtained by evaluating functions on bounded sets and orders obtained at infinity. With reference to the embedding $X \subset R^n$, which amounts to writing $A = R[x_1 \ldots x_n]$, we say a filter $\mathscr{F} \subset \mathscr{V}$ is *bounded* if it contains an element $V\{g_i\}$ which is a bounded subset of R^n. Otherwise, we say $\mathscr{F}$ is *unbounded*. If $\mathscr{F}$ has no bounded refinement, we say $\mathscr{F}$ is *at infinity*.

If $V\{g_i\} \subset \mathscr{V}$, we have the closure $\overline{V}\{g_i\} = \overline{U}\{g_i\}$ in X_d. If $\mathscr{F} \subset \mathscr{V}$ is a filter, by the *center* of $\mathscr{F}$ we mean

$$C(\mathscr{F}) = \bigcap_{V\{g_i\} \in \mathscr{F}} \overline{V}\{g_i\} \subset X_d.$$

If $\mathscr{F} = \mathscr{F}(\mathfrak{P}_w[g_i])$, then $C(\mathscr{F}) = \overline{V}\{g_i\}$. In particular, $C(\mathscr{F}(\mathfrak{P}_w)) = X_d$. The center of $\mathscr{F}$ could be empty, even if $\mathscr{F}$ is bounded. In the case of filters at infinity, homogeneous coordinates and hemispherical models of semi-algebraic sets makes these notions quite analogous to the centers of places in algebraic geometry over algebraically closed fields. This is especially true for total orders, because of the intimate relations between total orders and real places. In the real case, we have the added geometry provided by *partial* orders on function fields.

Proposition 8.11.2. If $\mathfrak{P} \subset K$ is an order with associated filter $\mathscr{F} \subset \mathscr{V}$ of subsets of X_d, then $x \in C(\mathscr{F})$ if and only if there is a total order refinement $\mathfrak{P}' \supset \mathfrak{P}$ with x a $\mathfrak{P}'$-convex maximal ideal of A. If $\mathscr{F}' = \mathscr{F}(\mathfrak{P}')$, then $C(\mathscr{F}') = x$. In general, if $\mathscr{F}$ is an ultrafilter, $C(\mathscr{F})$ consists of one point or is empty, and is always empty if $\mathscr{F}$ is at infinity.

Proof: The main point here is that if $g \in A$, $x \in V(g)$, then g is necessarily positive in any total order on K for which x is a convex ideal. This assertion can be deduced from either 8.6.5 or 8.11.1. For example, if $\mathfrak{P}_w[-g] \subset A$ is an order and $x \in V(g)$, then x is clearly not in $X \cap W(-g)$ if $g(x) > 0$ and even if $g(x) = 0$, x is a degenerate point of $X \cap W(-g)$.

The proposition is then proved as follows. If $x \in C(\mathcal{F})$, we choose a family of elements $g_\varepsilon \in A$, all $\varepsilon > 0 \in R$, with $g_\varepsilon(x) > 0$, $\bigcap_\varepsilon V(g_\varepsilon) = \{x\}$. This is easy using the technique of 8.1.1. The assumption $x \in C(\mathcal{F})$ implies $\mathfrak{P}[g_\varepsilon, \text{all } \varepsilon]$ is an order, clearly with center $\{x\}$. We can then take for $\mathfrak{P}'$ any total order refinement of $\mathfrak{P}[g_\varepsilon]$. Conversely, if a total refinement $\mathfrak{P}' \supset \mathfrak{P}$ exists with x $\mathfrak{P}'$-convex. Then $x \in \overline{V}\{g_i\} = \cap \overline{V}(g_i)$ all $g_1,\ldots,g_k \in \mathfrak{P}$. Otherwise, $x \in V(-g_j)$, some j, and g_j would be negative in $\mathfrak{P}'$.

The other statements of the proposition are also easily deduced from the remark in the first paragraph of the proof. □

Remark. If our ground field is $\mathbb{R}$, the real numbers, then a compactness argument implies that the center of a bounded filter is never empty.

Remark. If the order $\mathfrak{P} \subset A$ is finitely constructed using the operations of finite extension, finite intersection, and the operators $\mathfrak{P}_s, \mathfrak{P}_m, \mathfrak{P}_p, \mathfrak{P}_d$ of 3.12, then $C(\mathcal{F}(\mathfrak{P})) \subset X_d$ will be semi-algebraic. In general, however, this will not be expected.

Remark. If $\mathfrak{P} \subset K = R(x_1 \ldots x_n)$ is a total order and $p_R\colon K \to \Delta, \pm\infty$ the associated signed place, with Δ Archimedean over R, then $\mathcal{F}(\mathfrak{P})$ has center $\{(a_1 \ldots a_n)\}$ if and only if $p_R(x_i) = a_i \in R \subset \Delta$.

It is pretty easy to see intuitively (but not necessarily easy to prove) what all orders are like on function fields in one variable. An affine model,

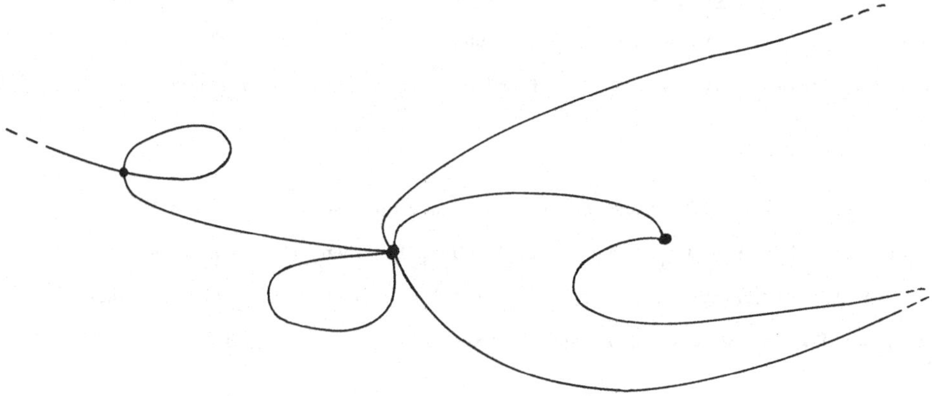

with degenerate points excised, will look like a smooth curve, with finitely many singular points. At each point some even number of branches comes in. A branch at x is a connected component (see 8.13) of $U - \{x\}$ where U is a very small open neighborhood of x. The total orders correspond precisely to the branches at all affine points, together with a finite number of branches at infinity. The individual sets $V\{g_i\}$ are finite unions of open intervals on the curve, with disjoint closures.

In 7.7 we studied places $p: K \to \Delta, \infty$ of real fields K. In particular, if Δ is real and Δ is given any total order, then by 7.7.2 p can be refined to a signed place, $p: K \to \Delta, \pm\infty$, thus inducing an order on K compatible with the order on Δ. To conclude this section, we study a finite variant of this problem of lifting orders in the case of function fields.

Suppose A is a real finite integral domain, $P \subset A$ a prime ideal, $\overline{A} = A/P$, and $\overline{\mathfrak{P}} \subset \overline{A}$ a total order. We ask when there exists a total order $\mathfrak{P} \subset A$ such that $\pi(\mathfrak{P}) = \overline{\mathfrak{P}}$, where $\pi: A \to \overline{A}$ is the projection. In particular, P will be $\mathfrak{P}$-convex, hence necessarily P is $(\mathfrak{P}_w)_d$-convex. This condition will be included in our proposition below.

Let $X = X(A,\mathfrak{P}_w)$, $X_d \subset X$ the non-degenerate points, and similarly $\overline{X} = X(\overline{A},\mathfrak{P}_w)$, $\overline{X}_d \subset \overline{X}$ the non-degenerate points. We first prove a lemma of independent interest.

Proposition 8.11.3. Suppose A is a real finite integral domain $X = X(A,\mathfrak{P}_w)$, $E \subset X$ a semi-algebraic subset. Then E is Zariski dense in X (that is, $I(E) = (0) \subset A$) if and only if $E \cap X_d$ has non-empty interior.

Proof: The "if" statement is clear since open sets in X_d are Zariski dense. Write $E = \cup E_i$, $E_i = Z\{f_{ij}\} \cap U\{g_{ik}\}$. Then if E is Zariski dense, some E_i is Zariski dense. Thus, $f_{ij} = 0 \in A$ and $E_i = U\{g_{ik}\} \cap X$. Since $I(X - X_d) \neq (0)$, we must have $U\{g_{ik}\} \cap X_d$ non-empty as claimed. □

Proposition 8.11.4. If $P \subset A$ is prime, $\bar{\mathfrak{P}} \subset \bar{A} = A/P$ a total order, then there exists a total order $\mathfrak{P} \subset A$ with $\pi(\mathfrak{P}) = \bar{\mathfrak{P}}$ if and only if for all finite sets $\{\bar{f}_i\} \subset \bar{\mathfrak{P}}$, $\bar{f}_i \neq 0$, the set $U\{\bar{f}_i\} \cap \bar{X}_d \cap X_d$ has interior in $\bar{X}_d$.

Proof: First a comment on notation. We have $\bar{X} \subset X$ from $\pi\colon A \to \bar{A}$. Also, we write $U\{f_i\} \subset X$, $U\{\bar{f}_i\} \subset \bar{X}$, if $f_i \in A$, $\pi(f_i) = \bar{f}_i$. Now, let $F = \{f \in A | \bar{f} \in \bar{\mathfrak{P}}, \bar{f} \neq 0\}$. Suppose $\mathfrak{P}_1 = \mathfrak{P}_w[F] \subset A$ is an order and $P \subset A$ is $(\mathfrak{P}_1)_d$-convex. Obviously, $\pi(\mathfrak{P}_1) = \bar{\mathfrak{P}} \subset \bar{A}$. From 7.7.3, we can choose a total order refinement $\mathfrak{P} \supset \mathfrak{P}_1$ such that P is $\mathfrak{P}$-convex. Then also $\pi(\mathfrak{P}) = \bar{\mathfrak{P}}$, since $\bar{\mathfrak{P}} \subset \bar{A}$ is already a total order. Conversely, if our desired total order $\mathfrak{P} \subset A$ exists, certainly $F \subset \mathfrak{P}$, hence $\mathfrak{P}_1 = \mathfrak{P}_w[F] \subset A$ is an order and P is $(\mathfrak{P}_1)_d$-convex.

We now prove the hypotheses of the proposition are sufficient. We must prove $\mathfrak{P}_1 \subset A$ is an order and P is $(\mathfrak{P}_1)_d$-convex. If $\mathfrak{P}_1$ is not an order, then some $\mathfrak{P}_w[f_i]$ is not an order, for finitely many $f_i \in F$. This is equivalent to $U\{f_i\} \cap X_d = \emptyset$, so $U\{\bar{f}_i\} \cap \bar{X}_d \cap X_d = \emptyset$. (Note $U\{\bar{f}_i\} \cap \bar{X}_d = U\{f_i\} \cap \bar{X}_d$.) If $\mathfrak{P}_1$ is an order, but P is not $(\mathfrak{P}_1)_d$-convex, then again there is a finite set of $f_i \in F$ with P *not* $(\mathfrak{P}_w[f_i])_d$-convex. By 8.6.5, $(A, (\mathfrak{P}_w[f_i])_d)$ is an RHJ-algebra, with $X(A, (\mathfrak{P}_w[f_i])_d) = \overline{U\{f_i\} \cap X_d}$. (Here, the bar denotes closure.) But if $U\{f_i\} \cap X_d \cap \bar{X}_d$ has interior in $\bar{X}_d$, then P is exactly the ideal of functions which vanishes on $U\{f_i\} \cap X_d \cap \bar{X}_d \subset \overline{U\{f_i\} \cap X_d} \cap \bar{X} = \overline{U\{f_i\} \cap X_d} \cap Z(P)$. But in this case, P would be $(\mathfrak{P}_w[f_i])_d$-convex.

Finally, we prove the hypotheses are necessary. Assuming P is $(\mathfrak{P}_w\{f_i\})_d$-convex, we know that $P = I(\overline{U\{f_i\} \cap X_d} \cap Z(P))$. That is, $\overline{U\{f_i\} \cap X_d} \cap \overline{X}$ is Zariski dense in $\overline{X}$. Since $f = \Pi f_i$ vanishes on $\overline{U\{f_i\} \cap X_d} - U\{f_i\} \cap X_d$ and $f \notin P$, we must have $U\{f_i\} \cap X_d \cap \overline{X}$ Zariski dense in $\overline{X}$. Applying 8.11.3 to $\overline{A}$, we conclude that $U\{f_i\} \cap X_d \cap \overline{X}_d$ has interior in $\overline{X}_d$, as desired. □

Corollary 8.11.5. In the situation above, if $X = X(A,\mathfrak{P}_w) = X_d$, that is, if X has no degenerate points, then any total order $\overline{\mathfrak{P}}$ on $\overline{A}$ lifts to a total order on A.

Proof: The point is if $\overline{f}_i \in \overline{\mathfrak{P}}$, $\overline{f}_i \neq 0$, then $U\{\overline{f}_i\} \cap \overline{X}_d \neq \emptyset$. If $X = X_d$, then $U\{\overline{f}_i\} \cap \overline{X}_d \cap X_d = U\{\overline{f}_i\} \cap \overline{X}_d$. □

For example, if $A = R[X_1 \ldots X_n]$, the polynomial ring, then 8.11.5 applies. However, the conclusion admits a trivial proof in this case. Let Δ denote the fraction field of $\overline{A} = A/P$. Then $\overline{\mathfrak{P}}$ induces a total order on Δ and we give the polynomial ring $\Delta[X_1 \ldots X_n]$ any total order with $(X_i - \zeta_i)$ infinitesimally small relative to Δ. We then restrict this order to the subring $R[X_1 \ldots X_n] \subset \Delta[X_1 \ldots X_n]$. We see this order lifts $\overline{\mathfrak{P}}$, by writing out the Taylor series of any $f \in R[X_1 \ldots X_n]$ in powers of the $(X_i - \zeta_i)$. Here $\zeta_i = X_i \pmod{P} \in \Delta$.

The geometry of 8.11.4 is roughly illustrated by the picture below, where

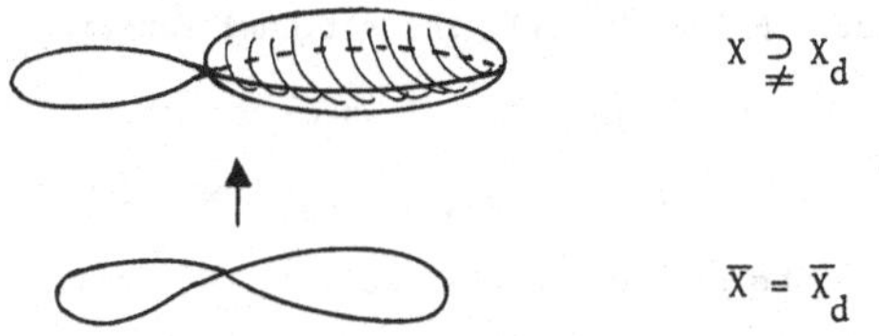

we have $\overline{X} = \overline{X}_d$ for simplicity. Using the ultrafilter interpretation of total orders $\overline{\mathfrak{P}} \subset \overline{A}$, we can think of $\overline{\mathfrak{P}}$ as picking out infinitesimally small open subsets of $\overline{X}_d$. If these small sets do not contain enough points in X_d, there can be no ultrafilter of sets in X_d giving an order on A lifting $\overline{\mathfrak{P}}$.

In the figure, orders $\bar{\mathfrak{P}}$ "centered" in the right half of $\bar{X}_d$ will lift, while orders centered in the left half of $\bar{X}_d$ will not.

We will state a slight generalization of 8.11.4.

Proposition 8.11.6. Suppose A is a real finite integral domain, $\mathfrak{P}_w[g_j] \subset A$ a finite refinement of the weak order and $P \subset A$ a prime $\mathfrak{P}_w[g_j]$-convex ideal. Suppose $\bar{\mathfrak{P}} \subset \bar{A} = A/P$ is a total order refining $\mathfrak{P}_w[\bar{g}_j] \subset \bar{A}$. Then there exists a total order $\mathfrak{P} \subset A$ with $g_i \in \mathfrak{P}$ and $\pi(\mathfrak{P}) = \bar{\mathfrak{P}}$, if and only if for all finite sets $\{\bar{f}_i\} \subset \bar{\mathfrak{P}}$, $\bar{f}_i \neq 0$, the set $\overline{U\{f_i,g_j\} \cap X(A,\mathfrak{P}_w[g_j])_d} \cap X(\bar{A},\mathfrak{P}_w[\bar{g}_j])_d$ has interior in $X(\bar{A},\mathfrak{P}_w[\bar{g}_j])_d$.

Proof: First, 8.11.3 generalizes routinely to sets $X = X(A,\mathfrak{P}_w[g_j])$. Then the proof of 8.11.6 is exactly like that of 8.11.4. The reason one must work with the *closure* of $U\{f_i,g_j\} \cap X(A,\mathfrak{P}_w[g_j])_d$ is that some g_j may belong to P, that is, $\bar{g}_j = 0$. If, in fact, all $g_j \notin P$, then one can just require that $U\{f_i,g_j\} \cap X(A,\mathfrak{P}_w[g_j])_d$ meets $X(\bar{A},\mathfrak{P}_w[\bar{g}_j])_d$ in a set with non-empty interior. For example, if $g_j \notin P$ and $X(A,\mathfrak{P}_w[g_j])_d = X(A,\mathfrak{P}_w[g_j])$, this always holds, giving a generalization of 8.11.5. □

8.12. Discussion of Total Orders on $R(x,y)$

Suppose given a total order $\mathfrak{P}$ on $R(x,y)$, the rational function field in two variables over the real closed field R. Let us assume that $\mathfrak{P}$ is centered at the origin in R^2, that is to say, $p_R(x) = p_R(y) = 0 \in R \subset \Delta$, where $p_R\colon R(x,y) \to \Delta, \pm\infty$ is the signed place associated to our order on $R(x,y)$ and the subfield $R \subset R(x,y)$.

The results of the preceding section show that the order $\mathfrak{P}$ is necessarily describable by infinitesimal behavior of functions $f \in R[x,y]$ near the origin. On the other hand, the order $\mathfrak{P}$ is also described by the signed place $p_R\colon R(x,y) \to \Delta, \pm\infty$. In this section we will reconcile these two descriptions by comparing invariants of p_R (the residue field Δ, value group Γ, rank of p_R, etc.) with the particular geometric behavior of functions f near the origin which determines whether f is positive or negative (rel $\mathfrak{P}$). We do not attempt to state a theorem, and our discussion is meant to be enlightening, not rigorous. Our discussion of signed places on $R(x,y)$ is a watered down version of the discussion of valuations on function fields of two variables given by Zariski in his work on resolution of singularities of surfaces.

There are two possibilities for the residue field Δ of the signed place p_R: $R(x,y) \to \Delta, \pm\infty$. Namely, Δ is either a function field in one variable, Archimedean ordered over R, or Δ is isomorphic to R. Assume first that $\Delta \neq R$. There is then $t = f(x,y)/g(x,y) \in R(x,y)$, with $R(t)$ Archimedean over R, and Δ is algebraic over $R(t)$. Let $\tau = D_R(t)$ be the transcendental cut of R defined by t and let (a_i', a_i'') be a nested family of intervals in R converging to τ. This means $a_i' < \tau < a_i''$ and every element of R is either smaller than some a_i' or larger than some a_i''. (The index set $\{i\}$ may be very large. Also, it does not follow that the differences $a_i'' - a_i'$ converge to 0.) In any event, with respect to the order $\mathfrak{P} \subset R(x,y)$, we have $a_i' < t < a_i''$.

Consider the neighborhood in R^2 defined by

$$U_{i,\varepsilon} = \{(x,y) \mid 0 < x^2 + y^2 < \varepsilon,\ a_i' g^2(x,y) < f(x,y)g(x,y) < a_i'' g^2(x,y)\}.$$

We decompose these $U_{i,\varepsilon}$ into their "connected components". These components are non-empty and have the origin in their closure. (In particular, $f(x,y)$ and $g(x,y)$ vanish at the origin.) Moreover, the behavior (number and approximate location) of these components stablizes for sufficiently small ε and a_i', a_i'' close enough to τ. The order $\mathfrak{P}$ is then determined by selecting one of the component families of the $U_{i,\varepsilon}$. (Connected components are studied in the next section, 8.13.) Specifically, $h(x,y) \in R[x,y]$ is positive in $\mathfrak{P}$ if $h(x,y)$ assumes only positive values on the selected component, infinitesimally close to the origin. We interpret this selection of a component of $U_{i,\varepsilon}$ as selecting a branch at the origin of the curve $f(x,y) = \tau g(x,y)$ over the larger field $R(\tau)$. If $\Delta = R(\tau,\mu)$, μ algebraic over $R(\tau)$, this choice is just the choice of one side of a real root of the minimal polynomial for μ.

As a concrete example, let R be the real algebraic numbers $t = x/y$, $\tau = \pi = 3.14159\ldots$.

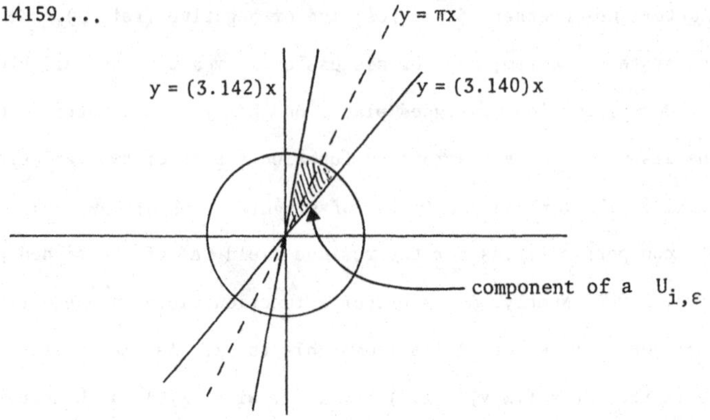

Given any polynomial $h(x,y)$, we can find numbers $r < \pi < s$, $\varepsilon > 0$ so that h has no zeros in the intersection of the cone $ry < x < sy$ and the puctured disc $0 < x^2 + y^2 < \varepsilon$. Then by requiring, say, $0 < y$, we single out a component of this region, hence $h(x,y)$ will have constant sign in this component.

In case $\Delta \neq R$, the value group Γ is always $\mathbb{Z}$. The value in $\mathbb{Z}$ of $h(x,y)$ is the *order* to which $h(x,y)$ vanishes at the origin along our chosen branch of the transcendental curve $f(x,y) = \tau g(x,y)$. That is, we find the first non-vanishing derivative of $h(x,y)$ on the branch at the origin.

Note that if $R = \mathbb{R}$, the real numbers, this example cannot exist since $\mathbb{R}$ admits no Archimedean extensions.

We now turn to the R-valued signed places $p_R: R(x,y) \to R, \pm\infty$. There are four possibilities for the value group Γ.

Case I. Γ non-Archimedean. Then $\Gamma = \mathbb{Z} \times \mathbb{Z}$, ordered lexicographically, $(m,n) < (m',n')$ if $m < m'$ or if $m = m'$, $n < n'$.

Case II. Γ Archimedean and discrete. That is, $\Gamma = \mathbb{Z}$.

Case III(a). Γ Archimedean, non-discrete, and containing two incommensurable elements. We can then assume $\Gamma = \{n + m\tau \mid n,m \in \mathbb{Z}\}$, where τ is an irrational real number. Γ is ordered as an additive subgroup of $\mathbb{R}$.

Case III(b). Γ Archimedean, non-discrete, but a subgroup of $\mathbb{Q}$, the rational numbers. The prime integers fall into two classes, p_i, q_j, where the p_i occur with arbitrarily high powers in denominators of elements of Γ and where the powers of q_j occuring in such denominators are bounded, say by b_j. Then

$$\Gamma = \{\tfrac{n}{m} | m = p_{i_1}^{\alpha_1} \cdots p_{i_r}^{\alpha_r} q_{j_1}^{\beta_1} \cdots q_{j_s}^{\beta_s}, \ 0 \leq \alpha_i < \infty, \ 0 \leq \beta_j \leq b_j\}.$$

We now analyze orders on $R(x,y)$ which yield signed places $p\colon R(x,y) \to R, \pm\infty$ with these value groups.

<u>Case I</u>. Here we have a rank 2 valuation. One of the variables, say y, defines an *algebraic* cut of the subfield $R(x) \subset R(x,y)$. In other words, there is an algebraic function $f(x,y) = f_0(x) + f_1(x)y + \cdots + f_m(x)y^m$, $f_i(x) \in R[x]$ with $f(x,y)$ infinitesimally small relative to $R(x)$, and irreducible in $R[x,y]$.

The curve $f(x,y) = 0$ thus has branches through the origin, as in our discussion in 8.11 or 8.3. The order $\mathfrak{P} \subset R(x,y)$ picks out not only a real

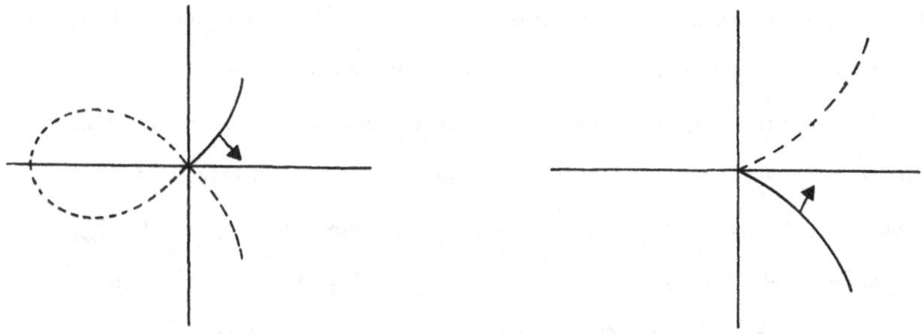

root of $f(x,y)$ over $R(x)$ (which is a branch of the curve $f(x,y) = 0$ in R^2 at the origin, as indicated by solid lines in the figures), but also a side of this real root in the real closure of $R(x)$ (as indicated by the normal arrows to our branches in the figures).

With these choices made, the $\mathfrak{P}$-sign of a function $g(x,y) \in R[x,y]$ is determined as follows. If $g(0,0) > 0$, then g is positive. If $g(0,0) = 0$, but $g(x,y)$ does not vanish on our branch C (that is, $f(x,y)$ does not divide $g(x,y)$), then g has the same sign as its values on C, infinitesimally close to the origin. Finally, if g vanishes on C, then g will not vanish on a small normal curve to C, near the origin, and the sign of g is the sign of the values of g on such a normal curve, on our preferred side of C. (In fact, this last "test" actually covers all cases.)

The value group Γ is $\mathbb{Z} \times \mathbb{Z}$. The value (m,n) assigned to g is determined by first writing $g = f^m h$, where h does not vanish on C (that is, h is not divisible by f). The invariant n is then the order to which h vanishes at the origin, along C.

<u>Case II</u>. $\Gamma = \mathbb{Z}$, $\Delta = R$. In this case, there is an "analytic curve" $0 = \sum_{i+j=1}^{\infty} a_{ij}x^i y^j$, $a_{ij} \in R$, with non-trivial real "branches" at the origin, and the $\mathfrak{P}$-sign of a function g is the sign of the values of g on a selected branch of the curve. Now, what does this mean in light of the fact that convergence may not even be *sensible*, and even if it is sensible, $\sum_{i+j=1}^{\infty} a_{ij}x^i y^j = h(x,y)$ may not converge? The answer is provided by looking at the honest curves $0 = h_r(x,y)$, where $h_r(x,y) = \sum_{i+j=1}^{r} a_{ij}x^i y^j$, as $r \to \infty$. The point is the infinitesimal behavior of branches of $h_r(x,y) = 0$ at the origin will *stabilize*. We can coherently select branches in the limit, as $r \to \infty$, and measure the sign of $g(x,y)$ on these branches, near the origin. The decision of whether g is positive or negative will be stable. It is precisely because $h(x,y)$ is *not* a polynomial, that our branch for $0 = h_{r+1}$ differs slightly from that for $0 = h_r$, and no $g(x,y) \neq 0$ can vanish on the branches chosen for all r. The value in $\Gamma = \mathbb{Z}$ assigned to g is the (stable) order to which g vanishes at the origin on the selected branches of the curves $0 = h_r$. (In other words, the degree of the first non-zero term in some power series.)

As a concrete example, consider $h(x,y) = y - \sum_{n=1}^{\infty} n!x^n$. Given $g(x,y)$, we look at the behavior of $g(x,y)$ infinitesimally close to the origin, along the honest curves $y = \sum_{n=1}^{r} n!x^n$. As r gets larger, we must look nearer and

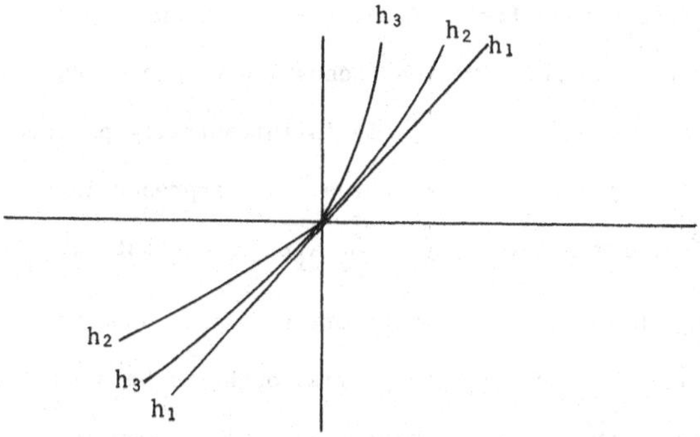

nearer the origin, but "convergence" of $h(x,y)$ is irrelevant.

Case III(a). $\Gamma = \{m+n\tau \mid m,n \in \mathbb{Z}, \tau \text{ irrational}\}$, $\Delta = R$. We give only one example here. Intuitively, the $\mathfrak{P}$-sign of $g(x,y) \in R[x,y]$ for an order yielding this value group is computed by restricting $g(x,y)$ to a "transcendental curve", for example, $y = x^{\tau}$. What one really does is find rationals r_i/s_i converging to τ and then restrict $g(x,y)$ to the curves $y^{s_i} = x^{r_i}$, or more precisely, to coherently chosen branches of this family of curves.

Case III(b). $\Gamma \subset \mathbb{Q}$, $\Delta = R$. Again, we give only one simple example. We might have a series representation $y = x^{m_1/n_1} + x^{m_2/n_2} + \ldots$, where $(m_1/n_1) < (m_2/n_2) < \cdots \in \Gamma$. This representation is purely formal. The finitely truncated formulas define honest algebraic curves with "stable" infinitesimal behavior near the origin. We test $g(x,y)$ by restricting to suitably selected branches.

It is clear that the "general" total order on $R(x,y)$ is complicated. This is consistent with our set theoretical characterization in the preceding section in terms of ultrafilters of certain open subsets. In general, the algebraist should only tolerate the orders of type I (discrete, rank r valuations on fields of functions in r variables). These are more in line with our philosophy of finite algebraic computability. The other types of orders are perhaps interesting to analysts.

For example, consider a first order differential equation $P(x,y)dx + Q(x,y)dy = 0$, $P,Q \in R[x,y]$, $Q(0,0) \neq 0$. We ask what interpretation can be made, for an *arbitrary* real closed field R, of the "solution curve", say through (0,0)? Although we do not expect an honest curve, we do have a procedure for deciding if a *polynomial* $f(x,y)$ is infinitesimally positive or negative at the origin, along our "phantom" curve. One approach is to just take the formal power series solution $y = \sum_{i=1}^{\infty} a_i x^i$ and construct an order, as in Case II above. However, it is much better to just use the differential equation itself to decide if a polynomial ought to be positive or negative along this curve, say in the positive x-direction, near the origin.

Specifically, first look at $f(0,0)$. If $f(0,0) = 0$, then we want to know $(df/dt)(0)$, where $y = y(t)$ is the "phantom" curve and $f = f(t,y(t))$ is the evaluation of f on the curve $t > 0$. This doesn't make sense, of course, but the *result* of applying the chain rule does make sense,

$$\frac{df}{dt}(0) = \frac{\partial f}{\partial x}(0,0) - \frac{\partial f}{\partial y}(0,0)\,\frac{P(0,0)}{Q(0,0)},$$

since the differential equation says $dy/dt = -P(t,y)/Q(t,y)$. If this computation of $(df/dt)(0)$ is positive, then f is positive in our total order. If $(df/dt)(0) = 0$, we compute $(d^2f/dt^2)(0)$, and so on until we finally reach a decision. Of course, if the differential equation has an algebraic solution $f(x,y) = 0$, we are in Case I above, rather than Case II.

This sort of interpretation of differential equations seems quite reasonable and worth further study. Equations of higher order and behavior near singular points are topics to be investigated. Of course, one wants to make sense not just of the germ of a solution curve at an initial point, but also the *continuation* of the curve. That is, given an initial point, one wants to associate an order $\mathfrak{D}_{x_0} \subset R[x,y]$, with $D_R(x) = x_0$, for all values of x_0 in some interval. This collection of orders will play the role of the solution curve of the differential equation through the initial point.

8.13. Brief Discussion of Structure Sheaves

The material in this section is partly in the form of an outline of a discussion to be worked out in detail elsewhere. On the other hand, the ideas are quite fundamental for our program of algebraizing topology.

Let S be a closed, semi-algebraic set, identified with the maximal convex ideal spectrum of an affine coordinate ring $(A,\mathfrak{P}) = (A(S),\mathfrak{P}(S))$. Recall from 8.9.3 that this situation is intrinsically characterized as follows. A is a reduced R-algebra of finite type. There are finitely many $\mathfrak{P}$-convex primes $P_i \subset A$ with $(0) = \cap P_i$, and finite refinements of the weak order $\mathfrak{P}_i = \mathfrak{P}_w[g_{ij}]$ on $A_i = A/P_i$ such that $\mathfrak{P} = A \cap \Pi(\mathfrak{P}_i)_d$, under the inclusion $A \to \Pi A_i$. The set S is the union of the sets $S_i = X(A_i,(\mathfrak{P}_i)_d)$, each of which is the closure of the set of algebraic simple points x of the real variety $X(A_i,\mathfrak{P}_w)$ with $g_{ij}(x) > 0$. In the semi-algebraic sets S_i, every non-empty strong open set $U\{h_j\} = \{y \in S_i | h_j(y) > 0\}$ is Zariski dense. We will refer to such $S_i = X(A_i(\mathfrak{P}_i)_d)$ as *irreducible components* of S.

In Chapter V, we constructed rather generally a structure sheaf relative to the *Zariski* topology on $\mathrm{Spec}(A,\mathfrak{P})$. The global sections turned out to be the ring $(A_{S(1)}, \mathfrak{P}_{S(1)})$ obtained from A by inverting all elements greater than 1. Because of the Nullstellensatz, this amounts to inverting all functions $f \in A$ with no zeros on S. Also because of the Nullstellensatz we can simplify the discussion by restricting this sheaf to the maximal convex ideals $X(A,\mathfrak{P}) = S$. If $f \in A$, the ring of sections over the basic Zariski open set $D(f) \subset S$ is $(A_{S(f)}, \mathfrak{P}_{S(f)})$, obtained by inverting all elements of A with no zeros in $D(f)$. Thus elements of $A_{S(f)}$ are functions on $D(f)$. The elements of $\mathfrak{P}_{S(f)}$ are those functions in $A_{S(f)}$ nowhere negative on $D(f)$. The stalks of this sheaf are the local rings $(A_x,\mathfrak{P}_x)$, $x \in S$. Elements of A_x may be regarded as germs of functions on Zariski open neighborhoods of x and such a germ belongs to $\mathfrak{P}_x$ if and only if it is nowhere negative on a Zariski open neighborhood of x.

On the semi-algebraic set S we also have the "strong topology", that is to say, the collection of open, semi-algebraic subsets $U \subset S$. We would like to study "sheaves" for this strong topology, but classical sheaf theory is intimately tied to the infinite procedures of point set topology, so we should proceed with some caution. Perhaps the most natural finiteness condition to impose is that we seek sheaves for the *Grothendieck* topology on the set of open subsets $U \subset S$ in which only *finite* covering families $\{U_i \to U\}$ are allowed. When discussing open, semi-algebraic sets and finite open coverings, it is obviously very convenient to assume Unproved Proposition 8.1.2. This assures us that, essentially, finite open covers of any $U \subset S$ just amount to writing U as a finite union of basic open sets $U\{f_i\} \subset S$. On the other hand, all the propositions proved in this section are independent of Unproved Proposition 8.1.2.

We have in mind three sheaves of rings, in fact, associated to any semi-algebraic set $E \subset R^n$. We call these sheaves the rational structure sheaf, the semi-algebraic structure sheaf, and the smooth structure sheaf. Each structure sheaf corresponds to a category of morphisms between semi-algebraic sets, although we do not study these morphisms here. These three categories can be interpreted, within real algebra, as delineating the three subjects, algebraic geometry, algebraic topology, and differential topology.

(I) Let $E \subset R^n$ be a semi-algebraic set, $A(E) = A(\bar{E}) = R[X_1 \ldots X_n]/I(\bar{E})$ the affine coordinate ring of E. Throughout this section the particular embedding of E in affine space can be suppressed. We can think of E invariantly as some dense subset $E = \cup E_i$, $E_i = Z\{f_{ij}\} \cap U\{g_{ik}\} \subset X(A(\bar{E}), \mathfrak{P}(\bar{E}))$, of the maximal ideal space of an RHJ-algebra $(A, \mathfrak{P})$ of a certain type. Nonetheless, it will sometimes be convenient to refer to the distance between points in R^n, $\|x - y\|$. Thus we do not go out of our way to avoid an affine embedding. The reader can reformulate for himself all statements in invariant form. For example, if we say a subset $S \subset E$ is closed and bounded in R^n, then this can be reformulated by saying that $S \subset X(A, \mathfrak{P})$ is closed and that all $h \in A$ are bounded as functions on S.

The rational structure sheaf is very similar to the structure sheaf for the Zariski topology. We first define it, then prove a few propositions which enable us to compute rings of sections, in some sense. If $U \subset E$ is (relatively) open, we define $I(U) = \{f \in A = A(E) | f(U) \equiv 0\}$. Then

$A/I(U)$ is a ring of functions on U and we define $A(U)$ to be the localization of this ring obtained by inverting all functions with no zeros on U. We define an order $\mathfrak{P}(U) \subset A(U)$, consisting of all functions nowhere negative on U. We obviously have a presheaf of partially ordered rings for our Grothendieck topology and the rational structure sheaf of E is the sheaf associated to this presheaf.

In order to make computations, let us first consider the case where $E = S = X(A,\mathfrak{P})$, where A is a finite integral domain over R, and $\mathfrak{P} = (\mathfrak{P}_w[g_i])_d$ is a derived order of a finite refinement of the weak order. Any non-empty basic open set $U = U\{h_j\} \subset S$ is Zariski dense, hence $A(U) = A$.

Proposition 8.13.1. Let $U = U\{h_j\} \subset S$, as above. Then, first, the order $\mathfrak{P}' = (\mathfrak{P}[h_j])_d \subset A$ coincides with $(\mathfrak{P}_w[g_i,h_j])_d$, hence $(A,\mathfrak{P}')$ is an RHJ-algebra. Secondly, $U \subset X(A,\mathfrak{P})$ coincides with the Zariski open set $D(h) \subset X(A,\mathfrak{P}')$, where $h = \Pi\, h_j$. Finally, $(A(U),\mathfrak{P}(U)) = (A_{S(h)}\ \mathfrak{P}'_{S(h)})$, which is the ring of sections in the Zariski structure sheaf over $D(h) \subset X(A,\mathfrak{P}')$, obtained by inverting $S(h) = \{f \in A | 0 \leq h^{2m} \leq f,\ \text{rel}\ \mathfrak{P}',\ \text{some}\ m \geq 0\}$.

Proof: Certainly $(\mathfrak{P}_w[g_i,h_j])_d \subset ((\mathfrak{P}_w[g_i])_d[h_j])_d = (\mathfrak{P}[h_j])_d$. But also, it is easy to argue that any $f \in (\mathfrak{P}[h_j])_d$ is nowhere negative on $U\{g_i,h_j\}$, hence belongs to $\mathfrak{P}_w[g_i,h_j]_d$ by Proposition 8.6.5.

It is completely obvious that $U\{h_j\} \subset X(A,\mathfrak{P})$ coincides with $D(h) \subset X(A,\mathfrak{P}')$, since $h_j \geq 0$, $\Pi\, h_j \neq 0$ means all $h_j > 0$. The last statement follows from the Nullstellensatz. □

In the case under consideration here, a function $f \in A(U)$ of the presheaf is completely determined by its restriction to any non-empty smaller open set $V \subset U$. In particular, this presheaf has no respect for "topological components". The effect of the sheafification is exactly to allow independent rational functions over the components of each U.

More precisely, let us say U is *connected* if U is not the disjoint union of two open semi-algebraic subsets. Equivalently, if $\{U_i \to U\}$ is a finite cover of U by open sets, then for any two indices i,j there is a chain $i = i_0, i_1, \ldots, i_n = j$ such that $U_{i_{k-1}} \cap U_{i_k} \neq \emptyset$, $k = 1,2,\ldots,n$. Suppose we have functions $f_i/g_i \in A(U_i)$. Since $f_i/g_i = f_ig_i/g_i^2$, we may assume all $g_i \geq 0$, and $g_i > 0$ on U_i. Then $f/g = \Sigma f_i/\Sigma g_i$ belongs to $A(U)$, since $\Sigma g_i > 0$ on $U = \cup U_i$. Suppose $(f_i/g_i)|_{U_i \cap U_j} = (f_j/g_j)|_{U_i \cap U_j}$, all i,j. Then if $U_i \cap U_j \neq \emptyset$, we have $f_ig_j = f_jg_i$ in the integral domain A, since $U_i \cap U_j$ is Zariski dense. It follows that if U is connected, then $(f/g)|_{U_i} = f_i/g_i$, all i, hence the section of the rational structure sheaf defined by the $f_i/g_i \in A(U_i)$ already comes from the presheaf element $f/g \in A(U)$.

On the other hand, we will prove below that any semi-algebraic set is a finite union of connected semi-algebraic subsets. Thus, if an open $U \subset E$ is not connected, a section of the rational structure sheaf over U consists of a section over each connected component of U. Thus we have proved the following.

Proposition 8.13.2. If $U \subset S$ is (relatively) open, $S = X(A,\mathfrak{P})$ irreducible and non-degenerate as above, and $U = \cup U_i$ is the decomposition of U into disjoint, (relatively) open connected components, then the ring of sections of the rational structure sheaf over U is $\Pi(A(U_i),\mathfrak{P}(U_i))$. □

If E is any semi-algebraic set, we can write $\overline{E} = \cup S_i$, where $S_i = X(A_i,\mathfrak{P}_i)$ and each A_i is a domain with $\mathfrak{P}_i$ the derived order of a finite refinement of the weak order. If $A = A(E)$, we have the inclusion $A \to \Pi A_i$ and $\mathfrak{P}(E) = A \cap \Pi \mathfrak{P}_i$. If $U \subset E$ is (relatively) open, then $I(U) = \bigcap_{U \cap S_j \neq \emptyset} P_j$ since if $U \cap S_j \neq \emptyset$, then $I(U \cap S_j) = P_j$. Let $A_0 = A/I(U)$ with inclusion $A_0 \to \Pi A/P_j$, the product taken over those j with $U \cap S_j \neq \emptyset$.

Proposition 8.13.3. If $U = U\{h_i\} \cap E$ is a basic open set in E, then the presheaf sections $(A(U),\mathfrak{P}(U))$ are obtained as follows. $(A(U),\mathfrak{P}(U))$ is the localization of $(A_o,\mathfrak{P}_o)$ obtained by inverting all functions $f \in A_0$ such that $Z\{f\} \cap U\{h_i\} \cap \overline{E} \subset \overline{E} - E$, where $\mathfrak{P}_o \subset A_0$ is the order $A_o \cap \Pi(\mathfrak{P}_j[h_i])_d$. (That is, $\mathfrak{P}_o$ consists of the functions nowhere negative on U.) In particular, if $E = \overline{E}$ is closed, then $(A(U), \mathfrak{P}(U))$ is the ring of sections of the Zariski structure sheaf associated to the RHJ-algebra $(A_0, \mathfrak{P}_0)$, over the Zariski basic open set $D(h) \subset X(A_0, \mathfrak{P}_0)$, where $h = \prod h_i$.

Proof: This is essentially the same as 8.13.1. Note that from 8.6, each $\mathfrak{P}_j[h_i]_d \subset A_j$ is an order giving an RHJ-algebra, precisely if $U\{h_i\} \cap S_j \neq \emptyset$. □

In general, in the reducible case, it is not true that if $U \subset E$ is connected, then every section of the rational structure sheaf over U already comes from the ring $A(U)$. In the figure below, the affine coordinate ring

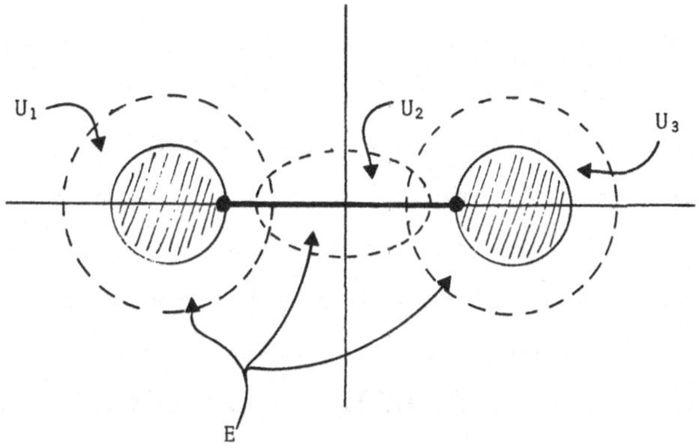

of the connected plane set E is $A = R[X,Y]$. But E is covered by U_1, U_2, U_3 and the rational functions $y \in A(U_1)$, $0 \in A(U_j)$, $j = 2,3$, fit together to give a globally defined rational function on E. A precise description of the global sections of the rational structure sheaf seems somewhat complicated.

Remark. As a general rule, when sheafifying a presheaf with respect to a non-topological sort of Grothendieck topology, the description of sections in terms of elements of stalks at each point which locally fit together is not appropriate. However, it turns out that the rational structure sheaf defined above actually coincides (on semi-algebraic open sets) with the classical *topological* sheafification of the presheaf defined by the $(A(U), \mathfrak{P}(U))$.

The stalks are clearly the rings $(A_x, \mathfrak{P}_x)$, where, if $x \in E$, A_x is the localization of A/I_x defined by inverting functions g with $g(x) \neq 0$. Here $I_x = \varinjlim_{x \in U} I(U)$ is the ideal of degeneracy at x (which coincides with $I(U)$ if U is sufficiently small). Thus an element of A_x is a germ of a function at $x \in E$. The positive elements $\mathfrak{P}_x \subset A_x$ are the germs which are locally non-negative near x.

Suppose given an element $s_x \in A_x$, for all $x \in U$, such that each x has a neighborhood U_x with elements $f_x/g_x \in A(U_x)$ which restrict to $s_y \in A_y$ for all $y \in U_x$. Such data exactly describes a section over U of the classical sheaf associated to the presheaf on question. Now, U is not compact, so we cannot assert that finitely many of the U_x cover U. However, let $V_x = D(g_x) \subset E$ denote the *Zariski* open set defined by $g_x \neq 0$. Thus f_x/g_x is defined on the much large set V_x. Moreover, if two functions f_1/g_1, f_2/g_2 agree on a semi-algebraic open set $U_1 \cap U_2$, then they will also agree on the much larger set $\cup(S_j - Z(g_1) - Z(g_2))$ where the S_j are those irreducible components of $\overline{E}$ with $S_j \cap U_1 \cap U_2 \neq \emptyset$. In this way, it is possible to replace the data defined by the $f_x/g_x \in U_x$ by *finitely many* functions f_{x_i}/g_{x_i} defined on larger neighborhoods of the form $U \cap V_{x_i} \cap S_i$. This says exactly that the sheafification of our presheaf $(A(U), \mathfrak{P}(U))$ with respect to the classical topology and the finite Grothendieck topology actually coincide.

II. We next define the sheaf of semi-algebraic functions on a semi-algebraic set $E \subset R^n$. Again, by "sheaf" we mean sheaf for the Noetherian-Grothendieck topology of finite, open, semi-algebraic covers of open subsets of E. We first define semi-algebraic functions and establish many basic properties.

Let $f\colon E \to R$ be a function. Then $E \times R$ is semi-algebraic, and we can consider the graph of f, $F \subset E \times R$.

Definition. The function $f\colon E \to R$ is a *continuous (semi-algebraic) function* if the graph $F \subset E \times R$ is a semi-algebraic set, and if for each semi-algebraic subset $S \subset E$, which is closed and bounded in R^n, the graph of f over S, $F \cap (S \times R)$, is a closed, bounded, semi-algebraic set in $R^n \times R = R^{n+1}$.

Note the definition involves both local and global properties of the graph of f. If we are given two such functions f_1: $E \to R$ and f_2: $E \to R$, with graphs $F_1 \subset E \times R$ and $F_2 \subset E \times R$, consider the graph $F_1 \underset{E}{\times} F_2$ of (f_1,f_2): $E \to R \times R$, that is, $\{(x,f_1(x),f_2(x)) | x \in E\} \subset E \times R \times R$. If $\pi_1(x,y_1,y_2) = (x,y_1)$ and $\pi_2(x,y_1,y_2) = (x,y_2)$, then obviously $F_1 \underset{E}{\times} F_2 = \pi_1^{-1}(F_1) \cap \pi_2^{-1}(F_2)$, which is semi-algebraic. If S is closed, bounded, and $S \subset E$, then $F_1 \underset{S}{\times} F_2$ is certainly bounded. Also, $F_1 \underset{S}{\times} F_2$ is closed in $R^n \times R \times R$, which is directly seen from the formula $F_1 \underset{S}{\times} F_2 = \pi_1^{-1}(F_1 \cap (S \times R)) \cap \pi_2^{-1}(F_2 \cap (S \times R))$. Since sum and product are polynomial maps $R \times R \to R$, we conclude from this paragraph and 8.9.4 the following result, by projecting $F_1 \underset{E}{\times} F_2 \subset E \times R \times R \to E \times R$, using sum or product $R \times R \to R$.

Proposition 8.13.4. Sums and products of continuous semi-algebraic functions are continuous semi-algebraic functions. □

As another immediate corollary of 8.9.4, we get a generalization of the remark following 8.9.4.

Proposition 8.13.5. A continuous semi-algebraic function assumes a maximum value on any closed, bounded, semi-algebraic set. □

We have chosen to define continuous functions without the usual ε's and δ's. The ε-δ definition turns out to be equivalent, although this is not completely trivial. Since the ε-δ definition is more convenient for certain arguments, we will now establish this equivalence. First, we need a lemma which guarantees that the local part of our definition of continuous function really carries some information. The following proposition is a sort of "curve selection lemma".

Proposition 8.13.6. Let E be a non-discrete semi-algebraic set, $x \in \overline{E}$ a point in the closure of E. Then there exist closed, bounded, one-dimensional, non-degenerate semi-algebraic sets $C = C_d$, with $x \in C \subset E \cup \{x\}$.

Proof: First, we may assume $E = Z\{f_i\} \cap U\{g_j\}$ since any E is a finite union of such sets. Then if $\overline{E} = \cup S_i$ is the decomposition of $\overline{E}$ into irreducible components, we have $E = \cup(S_i \cap E)$, and therefore we have, say, $x \in \overline{S_1 \cap E}$. This allows us to assume, in fact, that $E = U\{g_j\} \subset S = X(A,\mathfrak{P})$, where A is a domain, and $\mathfrak{P}$ the derived order of a finite refinement of the weak order. In fact, the g_j will actually be subsumed into the order $\mathfrak{P}$, that is, we may assume $\mathfrak{P} = (\mathfrak{P}_w[g_j])_d \subset A$. Also, we may as well assume $x = 0 \in R^n$.

Now we are in a position to apply the signed place perturbation theorem, 8.4.9 (or its extension in the Remark following the proof of 8.4.9) to the domain $A = R[x_1 \ldots x_n]$. We know that there is a total order $\mathfrak{P}'$ on A, refining $\mathfrak{P} = (\mathfrak{P}_w[g_j])_d$, such that the maximal ideal $(x_1 \ldots x_n) \subset A$, corresponding to the point $0 \in R^n$, is $\mathfrak{P}'$-convex. We may thus perturb this order and find a strong chain of $\mathfrak{P}$-convex ideals $0 \subsetneqq P_1 \subsetneqq \cdots \subsetneqq P_r = (x_1 \ldots x_n) \subset A$, where $\text{tr.deg}_R(A/P_i) = r - i$. Moreover, the Remark following 8.4.9 allows us to assume $g_j \notin P_{r-1}$.

The set of zeros of P_{r-1}, $x \in Z(P_{r-1}) \subset S$ is thus a one-dimensional semi-algebraic set, $x = 0$ is a non-degenerate point, and the g_j vanish on a discrete subset of $Z(P_{r-1})$. Also, $Z(P_{r-1})$ may contain some degenerate points. Nonetheless, we obtain our "curve" C of the proposition by intersecting $Z(P_{r-1}) \subset S$ with a closed ball sufficiently small to exclude all degenerate points of $Z(P_{r-1})$ and all zeros of the g_j, except $x = 0$. □

Next, we need to know that a semi-algebraic function satisfies a "polynomial equation" with the coordinate functions.

Proposition 8.13.7. If $E \subset R^n$ is a semi-algebraic set and $\varphi: E \to R$ a function such that the graph of φ, $F \subset E \times R$, is semi-algebraic, then there is a polynomial $P(x_1 \ldots x_n, y)$ such that $P(x_1 \ldots x_n, \varphi(x_1 \ldots x_n)) = 0$ for all $x = (x_1 \ldots x_n) \in E$.

Proof: Express $E = \bigcup_i Z\{f_{ij}\} \cap U\{g_{ik}\}$. If $x \in E$, $(x,\varphi(x)) \in Z\{f_{ij}\} \cap U\{g_{ik}\}$, then at least one f_{ij} occurs. Otherwise F would contain a whole interval of values (x,t), t near $\varphi(x)$, and thus F would not be the graph of a function. We can now set $P = \prod_i (\sum_j f_{ij}^2)$. □

Proposition 8.13.8. Let E be a semi-algebraic set, $f: E \to R$ a function. Then f is a continuous semi-algebraic function if and only if the graph of f, $F \subset E \times R$, is a semi-algebraic set and if for all $x \in E$, $0 < \varepsilon \in R$ there exists $0 < \delta = \delta(x,\varepsilon) \in R$, such that $\|x-y\| < \delta$, $y \in E$ implies $|f(x) - f(y)| < \varepsilon$.

Proof: First, we assume the ε-δ property. Suppose $S \subset E$, S a closed, bounded set. (We will frequently drop the adjective semi-algebraic, it being understood that all sets must be semi-algebraic.) We must show that $F \cap (S \times R)$ is closed, bounded. If $(x,t) \in \overline{F \cap (S \times R)} \subset R^n \times R$, then certainly $x \in S$, since S is closed. Now, if $t \neq f(x)$, the ε-δ property easily implies there are no points of F near (x,t) and this gives a contradiction. Thus $t = f(x)$, and $(x,t) \in F \cap (S \times R)$, as desired.

We will prove that $F \cap (S \times R)$ is bounded by induction on $\dim(S)$. The result is obvious if $\dim(S) = 0$, since then S is a finite set of points. Also, we may assume S is "irreducible", that is, $S = X(A,\mathfrak{B})_d$, where A is a domain $\mathfrak{B} \subset A$ a finite refinement of the weak order. This is justified because from 8.9 any closed S is a finite union of such irreducible pieces and if f is unbounded on S, then f would be unbounded on one of these pieces.

Suppose f satisfies the polynomial equation $0 = P(x_1 \ldots x_n, f(x_1 \ldots x_n))$, where $P(x,t) = p_0(x)t^m + \cdots + p_m(t)$. We may assume $p_0(x) \neq 0 \in A = A(S)$, since otherwise we could just drop this term from $P(x,t)$. Thus $Z(p_0(x)) = S' \subset S$ is a closed, bounded subset of lower dimension. We may thus assume $f(x') < b \in R$, for all $x' \in S'$. Consider $U = \{x \in S | f(x) < b\}$. Then $S - U$ is a closed, bounded, semi-algebraic set, disjoint from all zeros of $p_0(x)$. Thus $(p_0(x))^2$ assumes a minimal, *non-zero* value on $S - U$. This puts a bound on all zeros of $P(x,t)$, for $x \in S - U$, henc` a bound on $f(x)$, for $x \in S - U$. As f is also bounded by definition on U, we have f bounded on S.

Next, we go back and assume f continuous and prove the ε-δ property. Suppose for some $x \in E$, $0 < \varepsilon \in R$ and any $0 < \delta \in R$, there is a point $y = y(\delta) \in E$ with $\|x-y\| < \delta$ but $|f(x) - f(y)| \geq \varepsilon$. Then $x \in \overline{E}_0$, where $E_0 = \{y \in E \mid |f(x) - f(y)| \geq \varepsilon\}$. Since the graph F of f is semi-algebraic, E_0 is indeed semi-algebraic, as the image under projection $E \times R \to E$ of a semi-algebraic subset of F. By Proposition 8.13.6, choose a closed, bounded, one-dimensional, non-degenerate S with $x \in S \subset E_0 \cup \{x\} \subset E$. Since f is continuous, $F \cap (S \times R) = F_0$ is closed, bounded.

In general, if $F_0 \subset R^{n+1}$ is a closed, bounded set, and $B \subset R^{n+1}$ is closed, the squared distance function $d_B^2\colon F_0 \to R$, defined by $d_B^2(y) = \min_{b \in B} \|y-b\|^2$, obviously has the ε-δ property. (For fixed y, $\|y-b\|^2$ is large for $b \in B$ far from y and does assume a minimum on any closed, bounded subset of B, hence definitely assumes an absolute minimum on B.) We take

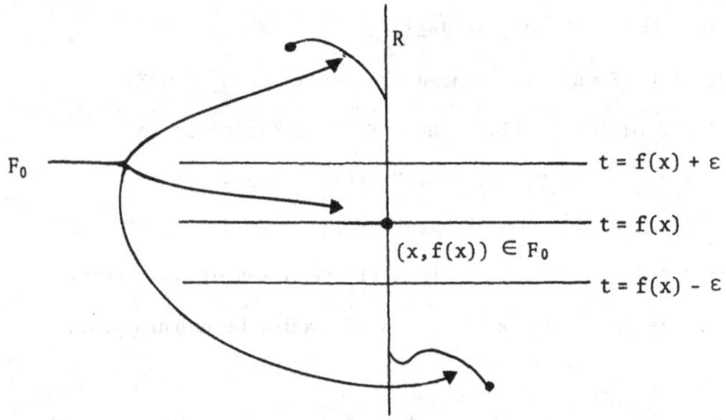

F_0 as above, $x \in E$ our fixed point, and $B = \{(x,t) \mid |f(x)-t| \geq \varepsilon\}$, part of the t-axis above $x = (x_1 \dots x_n)$. In this case, d_B^2 is easily computed explicitly:

$$d_B^2(x_1' \dots x_n', f(x_1' \dots x_n')) = \begin{cases} \Sigma(x_i' - x_i)^2 & \text{if } (x_1' \dots x_n') \neq (x_1 \dots x_n) \\ \varepsilon & \text{if } (x_1' \dots x_n') = (x_1 \dots x_n) . \end{cases}$$

In particular, it is clear that the graph of d_B^2, above F_o, is semi-algebraic, and, as already observed, has the ε-δ property. Thus d_B^2: $F_o \to R$ is *continuous* by the first part of the present proof. It must therefore assume a minimum value on F_o, which it obviously does not do. This contradiction establishes the ε-δ property for our original continuous function f. □

We have some immediate corollaries of 8.13.8.

Corollary 8.13.9. If $f: E \to R$ is a continuous semi-algebraic function and $U \subset R$ is an open, semi-algebraic set, then $f^{-1}(U) \subset E$ is (relatively) open, semi-algebraic in E. □

Corollary 8.13.10. If E is a *connected* semi-algebraic set, $f: E \to R$ a continuous semi-algebraic function, $x,y \in E$, and $f(x) < t < f(y) \in R$, then there exists $z \in E$ with $f(z) = t$. □

The following result will be useful in the last part of this section.

Proposition 8.13.11. Suppose E is a semi-algebraic set, $E_o \subset E$ a semi-algebraic subset such that $E \subset \overline{E}_o$. Suppose f_o: $E_o \to R$ is a continuous semi-algebraic function. Then there is at most one extension of f_o to a continuous, semi-algebraic function $f: E \to R$. Moreover, any function f which extends f_o and which has the ε-δ property on E is, in fact, a semi-algebraic function.

Proof: The uniqueness of f is obvious from the ε-δ property of continuous functions. Also, given an extension f of f_o with the ε-δ property, it is clear that the graph of f, $F \subset E \times R$, is just the closure in $E \times R$ of the graph of f_o, $F_o \subset E_o \times R \subset E \times R \subset \overline{E}_o \times R$. □

We made use above of the distance function to a closed set $B \subset R^n$, say d_B: $R^n \to R$, $d_B(y) = \min_{b \in B} |y-b|$. We now want to prove that d_B is always continuous, semi-algebraic.

Proposition 8.13.12. If $B \subset R^n$ is closed, semi-algebraic, then d_B: $R^n \to R$ is continuous, semi-algebraic.

Proof: The ε-δ property of d_B is immediate from the triangle inequality for distances. Thus we only need to prove that the graph of d_B in $R^n \times R$ is semi-algebraic. To see this, begin with the subset

$$D = \{(x,b, \|x-b\|) \mid x \in R^n,\ b \in B\} \subset R^n \times B \times R .$$

D is obviously semi-algebraic, hence by Tarski-Seidenberg, so is the image of D under projection π: $R^n \times B \times R \to R^n \times R$. Now, $\pi(D) \subset R^n \times [0,\infty)$, and the subset

$$D' = \{(x,t) \mid x \in R^n,\ 0 \leq t \in R,\ [0,t] \cap \pi(D) = \emptyset\}$$

is also semi-algebraic. This again follows from Tarski-Seidenberg since D' is defined by an elementary sentence. Note that the graph of d_B: $R^n \to R$ can now be described as $(\overline{D'} - D') \cup (B \times \{0\}) \subset R^n \times R$. Thus d_B is a semi-algebraic function. □

As a corollary of this argument and 8.13.8, we can prove that continuous semi-algebraic functions are uniformly continuous on closed, bounded sets.

Corollary 8.13.13. If $S \subset R^n$ is a closed, bounded semi-algebraic set, and f: $S \to R$ a continuous semi-algebraic function, then for all $0 < \varepsilon \in R$, there is a $0 < \delta \in R$ such that if $x,y \in S$ and $\|x-y\| < \delta$, then $|f(x)-f(y)| < \varepsilon$.

Proof: For each $x \in S$, the set $S_{x,\varepsilon} = S_x = \{y \in S \mid |f(x)-f(y)| \geq \varepsilon\}$ is a closed, bounded semi-algebraic set, by Corollary 8.13.9 (possibly $S_x = \emptyset$, but the set of such $x \in S$ with $S_x = \emptyset$ is *open* and can be discarded, since obviously *any* modulus of continuity δ will work for our ε-value at these points x. Anyway, the rest of this proof goes through even if some $S_x = \emptyset$.) Let $B_0 = \{(x,y) \mid x \in S,\ y \in S_x\} \subset S \times S$. Then B_0 is semi-algebraic, hence so is $B = \overline{B}_0 \subset S \times S$. It is not hard to prove that $S \cap B = \emptyset$, where $\Delta S = \{(x,x) \mid x \in S\} \subset S \times S$, say using the ε-δ property of f. Thus, the distance function d_B: $\Delta S \to R$ assumes a minimum, strictly positive value δ' on ΔS. We are here computing distances in $R^n \times R^n$. Obviously, δ' is no greater than the distance from x to S_x in R^n, for any $x \in S$. Thus, if $0 < \delta < \delta'$, then δ is a uniform modulus of continuity for f on S for our given ε. (If $B = \emptyset$, we define $d_B \equiv +\infty$.) □

We now define the sheaf of semi-algebraic functions on E. If $U \subset E$ is (relatively) open in E, we define $C^o(U)$ to be the ring (by 8.13.4) of continuous R-valued semi-algebraic functions on U, and $\mathfrak{P}^o(U) \subset C^o(U)$ the subset of nowhere negative functions. It is obvious that this presheaf is a sheaf, that is, if $U = \cup U_i$ is a *finite* open cover of U and $f_i \in C^o(U_i)$ with $f_i|_{U_i \cap U_j} = f_j|_{U_i \cap U_j}$, then there is a unique $f \in C^o(U)$ with $f|_{U_i} = f_i$. Moreover, $f \in \mathfrak{P}^o(U)$ if and only if all $f_i \in \mathfrak{P}^o(U_i)$.

Remark: Semi-algebraic functions are *not* infinitely collatable in general.

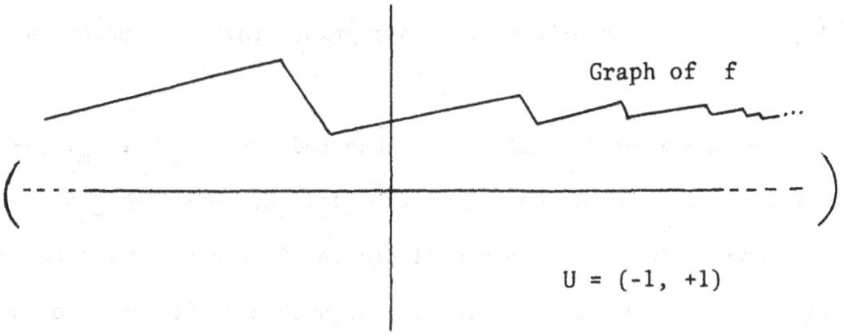

In the figure, f is a function on an open interval which is locally but not globally semi-algebraic.

Remark: Very little work is required to extend the notion of continuous R-valued semi-algebraic functions to vector valued functions f: $E \to R^m$, where $E \subset R^n$. First, the graph $F \subset E \times R^m$ must be semi-algebraic. Then one can give either the ε-δ definition or the closed, bounded property of the graph over any closed, bounded subset $S \subset E$. With either definition, one proves $f = (f_1 \dots f_m)$ continuous exactly if all the coordinate functions f_i: $E \to R$, $1 \leq i \leq m$, are continuous. Thus, the two definitions are again equivalent, by 8.13.8. It is also easy to prove that compositions $E \to E' \to E''$

of continuous, semi-algebraic functions are continuous, semi-algebraic, where $E \subset R^n$, $E' \subset R^{n'}$, $E'' \subset R^{n''}$, and that $f\colon E \to E'$ is semi-algebraic continuous, if and only if $g \circ f \in C^o(E)$, for all $g \in C^o(E')$.

We end this subsection with a proof of the fact that any semi-algebraic set E has only finitely many components. Recall that a semi-algebraic set E is connected if E has no proper semi-algebraic subsets which are both open and closed in E. We want to prove that any E is a finite union of connected semi-algebraic subsets which are both open and closed in E. These subsets are obviously maximal connected subsets, which can then be called the connected components of E.

Although we have notions of open sets, closed sets, connected sets, and so on, we must be careful not to assume facts from general topology which use infinite techniques inappropriate in our case. For example, we cannot routinely make use of the union of all connected semi-algebraic subsets containing a point $x \in E$ (which is how connected components are constructed "topologically") because there is no reason a priori that this union is semi-algebraic.

If E is any union of finitely many connected sets $E_1,\ldots,E_m$, then we can consider maximal connected subsets of the form $E_{i_1} \cup \cdots \cup E_{i_k}$. These are routinely shown to be both open and closed in E, hence are the connected components of E. If an E exists which is *not* such a finite union of connected sets, we can write $E = U_1 \cup U_1'$ where $U_1 \cap U_1' = \emptyset$ and both U_1 and U_1' are open and closed in E. Then, say, U_1 splits $U_1 = U_2 \cup U_2'$, and so on, producing an infinite strictly decreasing chain $U_1 \supset U_2 \subset U_3 \supset \cdots$ of open and closed subsets of E. The sets $V_i = U_i - U_{i+1}$ then give an infinite collection of pairwise disjoint, open and closed, semi-algebraic subsets of E. We refer to this process as a splitting process.

Clearly, we need only consider E of the form $Z\{f_i\} \cap U\{g_j\} \subset R^n$, where $1 \leq j \leq k$, say. Then E is the image of the real *variety* $V = Z\{f_i, y_j^2 g_j - 1\} \subset R^n \times R^k = R^{n+k}$, under projection $\pi\colon R^{n+k} \to R^n$, where $f_i, g_j \in R[x_1\ldots x_n]$, and we adjoin new variables $y_1\cdots y_k$. Thus it suffices to study varieties, since if $V_i \subset E$, $i \geq 1$, are disjoint, open and closed subsets, then so are

$\pi^{-1}(V_i) \subset V$. If some V is not a finite union of connected sets, we may assume V is irreducible and of least dimension with this property. Note that any 0-dimensional semi-algebraic set is finite, so $\dim(V) \geq 1$.

Let $V_0 \subset V$ be the algebraic simple points and $\Sigma V = V - V_0$ the singular set. Then $\dim(\Sigma V) < \dim(V)$, so we can write $\Sigma V = \Sigma_1 V \cup \cdots \cup \Sigma_k V$, where $\Sigma_i V$ is connected. We split $V = U_1 \cup U_1'$ into disjoint, open and closed subsets. Each $\Sigma_i V$ is either in U_1 or U_1'. This splitting process continues with, say, $U_1 = U_2 \cup U_2'$. Those $\Sigma_i V \subset U_1$ are either in U_2 or U_2'. Now, the conclusion we want is that V_0 actually contains infinitely many pairwise disjoint (relatively) open, semi-algebraic subsets V_i, $i \geq 1$, which are actually *closed in* R^n.

Note, if U_1' contains all $\Sigma_i V$, then $U_1 \subset V_0$ is one such set. If the splitting process continues infinitely, starting with $U_1 = U_2 \cup U_2'$, then we get all our V_i as above. If not, we must start over and go back and split U_1', but at least we have $V_1 = U_1$. Another case to consider is if some $\Sigma_i V$ are in U_1 and some in U_2. Then we haven't even found V_1, but there are less than k of the $\Sigma_i V$ in whichever of U_1, U_1' begins the infinite splitting process. It follows, then, that sooner or later the splitting process gives infinitely many $V_i \subset V_0$, which are semi-algebraic, pairwise disjoint, relatively open in V_0, and closed in R^n.

We are now in a position to use Whitney's proof [44] of the finiteness of the number of components. We know $\dim(V_i) = \dim(V_0) > 0$, since, in fact, V_0 is homogeneous, in the sense that it is a manifold. We choose a point $y \in R^n$ and a point $x_0 \in V_1$ so that the sphere of radius $\|y - x_0\|$ centered at y cuts the manifold V_1 transversally at x_0. In fact, we can assume that the normal vector to this sphere at x_0 lies in the tangent plane to V_1 at x_0. Let $g\colon\ V \to R$ be the function $g(x) = \|y - x\|^2$. If $I(V) = (f_i)$ and $\operatorname{codim}(V) = r$, we consider the subvariety $V' = \{x' \in V | \operatorname{rank}(df_i(x), dg(x)) = r\}$. Since each V_i is a closed, semi-algebraic set in R^n, $g|_{V_i}$ assumes a minimum value at, say, $x_i \in V_i$. Then $x_i \in V'$, hence V' is not a finite union of connected sets. On the other hand, by our choice of y and x_0, we have arranged that $x_0 \notin V'$, hence V' is a proper subvariety of V, necessarily of lower dimension. This contradiction proves our desired result.

Proposition 8.13.14. Every semi-algebraic set is a finite union of pairwise disjoint, connected, open and closed, semi-algebraic subsets. □

Remark: A further result is that if $x \in E$, then the connected component of E which contains x is the set of $y \in E$ such that there is a continuous, semi-algebraic "path" $p: [0,1] \to E$ with $p(0) = x$, $p(1) = y$, where $[0,1] \subset R$ is the unit interval. We will not prove this result here, although the results of this chapter are sufficient for constructing a proof. Essentially, one must look closely at global stratification and local geometry near a point.

Note that if our ground field is the field of real numbers, it is not obvious from the considerations leading to 8.13.14 that a *semi-algebraically* connected set is *topologically* connected. However, since $[0,1] \subset \mathbb{R}$ is both topologically and semi-algebraically connected, the fact that components are path components in general, does imply this fact about real numbers.

III. Finally we discuss briefly smooth semi-algebraic functions. For simplicity, we restrict our attention to open subsets $U \subset R^n$. We begin by giving a little more structure to the graph of a continuous semi-algebraic function $f: U \to R$.

Proposition 8.13.15. Suppose $f: U \to R$ is continuous, semi-algebraic and suppose $P(x_1 \ldots x_n, y)$ is a polynomial of least degree in y such that $P(x, f(x)) \equiv 0$, $x \in U$. Let $P = P_1 \cdot \ldots \cdot P_r$ be the factorization of P into irreducible factors in $R[X_1 \ldots X_n, Y]$, then:

(a) The P_i are distinct, that is, P is square free.

(b) If $\hat{P}_i = P/P_i$ and $U_i = \{x \in U \mid \hat{P}_i(x, f(x)) \neq 0\}$, then the U_i are non-empty and disjoint open subsets of U, and $P_i(x, f(x)) \equiv 0$ if $x \in U_i$.

(c) The principal ideals $(P_i) \subset R[X_1 \ldots X_n, Y]$ are convex prime ideals.

(d) $\dim(U - \cup U_i) < \dim(U) = n$. In particular, $U \subset \cup \overline{U}_i$.

Proof: If some P_i^2 divided P, we would find a polynomial of lower degree vanishing on the graph of f by dividing by P_i. Similarly, if some $U_i = \emptyset$, then $\hat{P}_i$ would vanish on the graph of f. The U_i are obviously disjoint, and P_i obviously must vanish on the graph of f over U_i.

The convexity of the ideals (P_i) follows now from 8.8.6, since the (P_i) have lots of zeros in R^{n+1}. Moreover, the other results of 8.8 imply that the algebraic simple zeros of P_i are dense on the graph of f over U_i. In particular, no polynomial can vanish on any open subset of the graph of f over U_i unless P_i divides it.

Suppose $U - \cup U_i$ contained an open set V. Let $F \subset R^{n+1}$ denote the graph of f over V. Then $I(F) \subset R[X_1 \dots X_n, Y]$ is convex and must have some associated prime of dimension n, which is therefore principal, say (Q). But all $\hat{P}_i$ vanish on F, and applying the arguments of the paragraph above to Q, we would deduce that all $\hat{P}_i \in (Q)$. As this is clearly impossible since the P_i are distinct irreducible polynomials, we conclude $\cup U_i$ is dense in U and (d) follows. □

We now want to define the subring $C^1(U) \subset C^0(U)$ of continuously differentiable semi-algebraic functions $f\colon U \to R$. We first assume $f \in C^0(U)$, then assume the limits

$$(\partial f/\partial x_i)(x) = \lim_{\varepsilon \to 0} \frac{f(x_1 \dots, x_i + \varepsilon, \dots x_n) - f(x_1 \dots x_n)}{\varepsilon}$$

exist in R for all $x \in U$ and define functions $(\partial f/\partial x_i)\colon U \to R$ with the ε-δ property. We can apply the chain rule to our relation $0 \equiv P(x,f(x))$, $x \in U$ of lowest degree and this gives

$$\begin{aligned} 0 &\equiv \partial P(x,f(x))/\partial x_i \\ &= (\partial P/\partial x_i)(x,f(x)) + ((\partial P/\partial y)(x,f(x))((\partial f/\partial x_i)(x)) \end{aligned}$$

Since $(\partial P/\partial y)(x,f(x) \not\equiv 0$ on U, and because no P_i divides $\partial P/\partial y$ where $P = P_1 \dots P_r$ as in 8.13.14, we deduce that $V = \{x \in U | (\partial P/\partial y)(x,f(x)) \neq 0\}$ is dense in U. The function $\partial f/\partial x_i$ is obviously semi-algebraic over V, since it can be written $(-(\partial P/\partial x_i)/(\partial P/\partial y))(x,f(x))$ if $x \in V$. Thus by 8.13.11, $\partial f/\partial x_i\colon U \to R$ also belongs to $C^0(U)$.

We refer to such functions $f\colon U \to R$ as C^1-semi-algebraic functions. The set of such $C^1(U) \subset C^0(U)$ is a subring. By iterating the procedure, we can define C^r-semi-algebraic functions $f \in C^r(U) \subset C^{r-1}(U) \subset \cdots \subset C^0(U)$, by requiring that f have continuous partial derivatives of order up to and including r, $1 \leq r \leq \infty$. Clearly, C^r-functions are finitely collatable, hence we have a sheaf of C^r-functions, defined on (variable) open subsets $U \subset R^n$.

There is no difficulty whatever now in extending the inverse function theorem, Proposition 8.7.1, to the case of a C^1-map $Y = (Y_1 \ldots Y_n)$: $(R^n,0) \to (R^n,0)$ with non-singular derivative $((\partial Y_i/\partial X_j)(0))$. The local injectivity of Y is proved by standard estimate arguments using the hypothesis of differentiability. The local surjectivity of Y is proved just as in the earlier proof of 8.7.1, using the minimum value property of continuous semi-algebraic functions on closed, bounded sets.

In fact, we will sketch another proof, similar to a standard proof in the classical case of real numbers. Beginning with any C^1-coordinate system $x_1 \ldots x_n$ near 0, and C^1-functions $y_1 \ldots y_n$ of the x_j, with $y(0) = 0$ and $(\partial y_i/\partial x_j)(0) = \mathrm{Id}$, one shows that $y_1, x_2, \ldots, x_n$ is a C^1 coordinate system. This argument uses "completeness" in the form of the intermediate value property on intervals for continuous functions. Since an interval is connected (in our sense), we have the intermediate value property from 8.13.10 at our disposal. The inverse function theorem is then proved by iterating this substitution procedure.

Note that even if originally the x_i are the standard coordinate functions and the y_i are polynomials, at the *second step* of this proof, the $y_2, \ldots, y_n$ will not generally be polynomials in $y_1, x_2, \ldots, x_n$, but will be C^1-semi-algebraic functions of $y_1, x_2, \ldots, x_n$. Thus, this proof would not have been feasible in the special case dealt with in 8.7.

The C^∞ theory is quite different from the C^r theory, $1 \leq r < \infty$. For example, if $U \subset R^n$ is open, $r < \infty$, then $C^r(U)$ is not an integral domain. Specific examples are easy. Let $f, g \in C^3((-1,+1))$ be defined by $f(x) = x^4$ if $x \geq 0$ and $f(x) = 0$ if $x \leq 0$, $g(x) = 0$ if $x \geq 0$ and $g(x) = x^4$ if $x \leq 0$. Then $f \cdot g = 0$.

If U is connected, then $C^\infty(U)$ is an integral domain. (In general, if $U = \cup U_i$ is the decomposition of U into disjoint open connected components, then obviously $C^r(U) = \Pi\, C^r(U_i)$, $0 \leq r \leq \infty$.) The C^∞-functions on U are known as Nash functions and have been widely studied in the case of ordinary real numbers, [30] through [42]. The basic result here is the following. We use the notation C_x^∞ for the ring of germs of C^∞ semi-algebraic functions at $x \in R^n$, $C_x^\infty = \varinjlim_{x \in U} C^\infty(U)$.

Proposition 8.13.16. Suppose $0 \in U \subset R^n$, U a connected, open, semi-algebraic set, and suppose $f \in C^\infty(U)$. Then:

(a) There is an *irreducible* polynomial $P(X_1 \ldots X_n, Y) \in R[X_1 \ldots X_n, Y]$ such that $P(x, f(x)) \equiv 0$, $x \in U$.

(b) The function f is *determined* by its germ $[f] \in C_0^\infty$, that is, $C^\infty(U) \to C_0^\infty$ is injective.

(c) The germ $[f] \in C_0^\infty$ is *determined* by its formal power series

$$\{f\} = \sum_{k=0}^{\infty} \sum_{\Sigma i_j = k} \frac{1}{(i_1)!\ldots(i_n)!} \frac{\partial^k f}{\partial X_1^{i_1}\ldots\partial X_n^{i_n}}(0) X_1^{i_1} \ldots X_n^{i_n} \in R[[X_1 \ldots X_n]],$$

that is, $C_0^\infty \to R[[X_1 \ldots X_n]]$ is injective.

(d) The power series $\{f\}$ is a formal solution of the equation $P(x_1 \ldots x_n, \{f\}) = 0$.

(e) Given an irreducible polynomial $P(X_1 \ldots X_n, Y)$ and a formal power series $\{f\}$ such that $P(X_1 \ldots X_n, \{f\}) = 0$, then there is a germ $[f] \in C_0^\infty$ with underlying power series $\{f\}$.

Sketch of Proof: Let $P(x, f(x)) \equiv 0$, $P = P_1 \ldots P_r$ factored as in 8.13.15 into distinct irreducible factors, $U_i \subset U$ the open subset where P/P_i is non-zero. To prove (a), we must prove that no $x \in U$ belongs to $\overline{U}_i \cap \overline{U}_j$, $i \neq j$, because it then follows from connectedness that there is only one U_i. We may as well prove $0 \notin \overline{U}_i \cap \overline{U}_j$, $i \neq j$.

If $0 \in \overline{U}_1$, then we can find $x_0 \in U_1$ arbitrarily near 0 such that $(\partial P_1/\partial y)(x_0, f(x_0)) \neq 0$. Near $(x_0, f(x_0))$, the graph of f coincides with the zeros of P_1. The non-vanishing of $(\partial P_1/\partial y)(x_o,f(x_o))$ allows us to compute all partial derivatives $(\partial^I f/\partial x^I)(x_o)$, $I = (i_1 \ldots i_k)$, in terms of the coefficients of P_1, by simply iterating the chain rule and using the identity $0 \equiv P_1(x,f(x))$, near x_o. For example, if there is only one variable x, then

$$0 = \frac{\partial P_1}{\partial x}(x_o,f(x_o)) + \frac{\partial P_1}{\partial y}(x_o,f(x_o))\,\frac{\partial f}{\partial x}(x_o),$$

$$0 = \frac{\partial^2 P_1}{\partial x^2}(x_o,f(x_o)) + 2\,\frac{\partial^2 P_1}{\partial y\partial x}(x_o,f(x_o))\,\frac{\partial f}{\partial x}(x_o) + \frac{\partial^2 P_1}{\partial y^2}(x_o,f(x_o))\left(\frac{\partial f}{\partial x}(x_o)\right)^2$$

$$+ \frac{\partial P_1}{\partial y}(x_o,f(x_o))\,\frac{\partial^2 f}{\partial x^2}(x_o),$$

and so on. In fact, these expressions are just the coefficients (up to constant factor) of powers of $(x - x_o)$ obtained by formally computing $P_1(x,f(x))$, where $f(x)$ is replaced by the power series $\sum_{k=0}^{\infty} (1/k!)f^{(k)}(x_o)(x-x_o)^k$. The multivariable analog is also true, of course.

Now, we want to assert that the formal power series of f *at the origin* $\{f\}$, as in parts (c), (d) of the proposition, is a formal solution of $P_1(X_1 \ldots X_n, \{f\}) = 0$. But each coefficient in this formal expansion is a finite expression which, by continuity of all $\partial^I f/\partial x^I$, and also of course by continuity of P_1 and its derivatives, can be evaluated as $\lim_{x_o \to 0}$ of the corresponding coefficients of the formal computation about $x_o \in U_1$. But we *know* these coefficients vanish, hence so do the coefficients at the origin. This proves (d).

The polynomial ring and power series ring are integral domains $R[X_1 \ldots X_n] \subset R[[X_1 \ldots X_n]]$. Thus, if some power series f is algebraic over $R[X_1 \ldots X_n]$, there is a *unique* irreducible polynomial (up to factor in $R[X_1 \ldots X_n]$), $P \in R[X_1 \ldots X_n][Y]$, with $P(X_1 \ldots X_n, f) \equiv 0$. We therefore conclude from the paragraphs above that we cannot have $0 \in \overline{U}_i \cap \overline{U}_j$ if $i \neq j$. This proves (a). But also, we conclude that if the power series $\{f\} \equiv 0$, then the irreducible polynomial $P(X_1 \ldots X_n, Y)$ is just the polynomial Y. This means that the function $f \equiv 0$, and (b) and (c) are proved.

Part (e) is quite interesting. However, we prefer to defer the proof until we make a more detailed study of power series rings in a subsequent chapter. □

Remark: The C^∞ structure sheaf is analogous to the rational structure sheaf, in the sense that a function $f\colon U \to R$ which is a C^∞-semi-algebraic function in each germ, that is, $[f] \in C^\infty_x$, all $x \in U$, is in fact already in $C^\infty(U)$. In other words, C^∞ functions are infinitely collatable.

Appendix

THE TARSKI-SEIDENBERG THEOREM

In this appendix we give Paul Cohen's proof [62] of the Tarski-Seidenberg theorem [56], [57], which we state in the following geometric form.

Proposition A.1. Let $E \subset R^n$ be a semi-algebraic set, $\pi: R^n \to R^{n-1}$ the projection onto the last $(n-1)$ coordinates. Then $\pi E \subset R^{n-1}$ is semi-algebraic.

Actually, the proof provides an algorithm for writing down πE as a semi-algebraic set, in terms of the polynomials used to define E. However, we will leave to the reader the details of keeping track of this algorithm through the various steps of the proof.

Proposition A.1 can be reformulated in logical language as "elimination of quantifiers". Specifically, by a polynomial relation $A(x_n \dots x_n, t_1 \dots t_m)$ we mean a finite sentence built up from basic relations $p(x_1 \dots x_n, t_1 \dots t_m) > 0$, where p is a polynomial with coefficients in R, using the logical connectives "and", "or", "not". Thus, the set of points $(x_1 \dots x_n, t_1 \dots t_m) \in R^{n+m}$ such that the sentence $A(x_1 \dots x_n, t_1 \dots t_m)$ is true, is just a general semi-algebraic set in R^{n+m}. By an elementary sentence, we mean a sentence of the type $(Q_1 x_1) \cdots (Q_n x_n) A(x_1 \dots x_n, t_1 \dots t_m)$, where the Q_i are quantifiers, $\exists$ or $\forall$, and A is a polynomial relation. Consider the set $\{(t_1 \dots t_m) \in R^m | (Q_1 x_1) \cdots (Q_n x_n) A(x_1 \dots x_n, t_1 \dots t_m)\}$. Another form of the Tarski-Seidenberg theorem is that this set is semi-algebraic.

Proposition A.2. There is a polynomial relation $B(t_1 \dots t_m)$ (in fact, an algorithm for producing B), such that $B(t_1 \ldots t_m) \Leftrightarrow (Q_1 x_1) \cdots (Q_n x_n) \cdot A(x_1 \ldots x_n, t_1 \ldots t_m)$.

Since $(\forall x_n)A$ is equivalent to $\sim(\exists x_n)(\sim A)$, it is clear by induction that Proposition A.2 needs to be proved only for $n = 1$ and $Q_1 = \exists$. This is what we actually prove below. In this case, the set $\{(t_1 \ldots t_m) \in R^m | (\exists x_1)A(x_1, t_1 \ldots t_m)\}$ is just the projection to R^m of the semi-algebraic set $\{(x_1, t_1 \ldots t_m) \in R^{m+1} | A(x_1, t_1 \ldots t_m)\}$. We see therefore that Propositions A.1 and A.2 are equivalent.

As an example of the sort of sets described by elementary sentences, we mention the topological closure of a semi-algebraic set $E \subset R^m$, $\overline{E} = \{(t_1 \ldots t_m) \in R^m | (\forall x_0)(\exists x_1) \cdots (\exists x_m)(x_0 = 0 \text{ or } (x_1 \ldots x_m) \in E \text{ and } \sum_{i=1}^{m} (x_i - t_i)^2 < x_0^2\}$.

In order to prove Proposition A.2, we need to introduce temporarily a new notion. Consider functions f with domain a subset $X \subset R^m$ and values in R. Note that the set of polynomials in one variable of degree $\leq d$ is an affine space over R of dimension d+1, with coordinates $(a_0 \ldots a_d)$, the coefficients of a polynomial $q = a_0 x^d + \cdots + a_d$. Thus, it makes sense to talk about semi-algebraic subsets of the space of polynomials of degree $\leq d$.

Definition A.3. The function $f: X \to R$ is *psemi-algebraic* if for all D and all semi-algebraic subsets $E \subset R$, the set $\{(q,x) \in R^{d+1} \times R^m | q(f(x)) \in E\}$ is semi-algebraic.

Note that the definition guarantees that the domain of a psemi-algebraic function is a semi-algebraic set. If $t \in R$, let $\text{sign}(t) = +1, 0, -1$ according to whether $t > 0$, $t = 0$, or $t < 0$. Since semi-algebraic subsets of R are rather simple, it is easy to see that f is psemi-algebraic if for all d and $\lambda = +1, 0, -1$ the sets $\{(q,x) \in R^{d+1} \times R^m | \text{sign } q(f(x)) = \lambda\}$ are semi-algebraic. It is also clear that polynomials are psemi-algebraic. We will need to know that differences of psemi-algebraic functions are psemi-algebraic. However, it is not much harder to prove the following.

Lemma A.4. If $g_1, \ldots, g_k$ are psemi-algebraic functions of m variables and f is a polynomial in k variables, then $f(g_1(x),\ldots,g_k(x))$ is a psemi-algebraic function of $x = (x_1\ldots x_m)$.

Proof: We must show that the sets $\{(q,x) \in R^{d+1} \times R^m | \text{sign } q(f(g_1(x),\ldots$ $\ldots,g_k(x))) = \lambda\}$ are semi-algebraic. Now $q(f(y_1\ldots y_k)) = Q(y_1\ldots y_k) =$ $Q_0 y_k^e + \cdots + Q_e$, where the $Q_i = Q_i(y_1\ldots y_{k-1})$ are polynomials whose coefficients are polynomials in the coefficients $(a_0\ldots a_d)$ of q, depending only on the coefficients of f. Therefore, since $g_k(x)$ is psemi-algebraic, it follows that the set $\{(q,y_1\ldots y_{k-1},x) | \text{sign } Q(y_1\ldots y_{k-1},g_k(x)) = \lambda\}$ is semi-algebraic, hence can be expressed as a Boolean combination of sets of the form $\text{sign } F(a_0\ldots a_d,y_1\ldots y_{k-1},x) = \mu$, where F is a polynomial. By similar reasoning, the set $\{(q,y_1\ldots y_{k-2},x) | \text{sign } F(a_0\ldots a_d,y_1\ldots y_{k-2},g_{k-1}(x),x) = \mu\}$ is semi-algebraic, and it follows that $\{(q,y_1\ldots y_{k-2},x) | \text{sign } Q(y_1\ldots y_{k-2},$ $g_{k-1}(x),g_k(x)) = \lambda\}$ is semialgebraic. Continuing this type of argument, we eventually conclude that $\{(q,x) | \text{sign } Q(g_1(x),\ldots,g_k(x)) = \lambda\}$ is semi-algebraic, as desired. □

Let $p(x) = b_0 x^n + b_1 x^{n-1} + \cdots + b_n$ be a polynomial with real roots $\xi_1(b) < \xi_2(b) < \cdots$, $b = (b_0\ldots b_n)$. Of course, the ξ_i are not everywhere defined. The main result needed to prove Proposition A.2 is the following.

Proposition A.5. The $\xi_i(b)$ are psemi-algebraic functions of b.

Proof: We argue by induction on n, the Proposition being trivial if $n = 1$. Since the derivative $p'(x)$ has degree $\leq n-1$, with coefficients $(nb_0,(n-1)b_1,\ldots,b_{n-1})$, we have by induction that the roots of $p'(x)$, say $\eta_1(b) < \eta_2(b) < \cdots < \eta_{k(b)}(b)$, are psemi-algebraic functions of $b = (b_0\ldots b_n)$. The data $\text{sign}(b_0),\ldots,\text{sign}(b_n),\text{sign } p(\eta_1(b)),\ldots,\text{sign } p(\eta_k(b))$ determines the number of roots $\xi_i(b)$ of $p(x)$ as well as locating each $\xi_i(b)$ in one of the intervals $(-\infty,\eta_1(b))$, $(\eta_j(b),\eta_{j+1}(b))$, $(\eta_k(b),\infty)$, or among the points $\eta_1(b),\ldots,\eta_k(b)$. This is so because $p(x)$ is monotonic between any two roots of $p'(x)$ and on $(-\infty,\eta_1)$, (η_k,∞), and because the $\text{sign}(b_j)$ determine the behavior of $p(x)$ at $\pm\infty$.

We must show that the sets $\{(q,b) \in R^{d+1} \times R^{n+1} | \text{sign } q(\xi_i(b)) = \lambda\}$ are semi-algebraic. We can divide q by p and write $q(x) = s(x)p(x) + r(x)$, where degree(r) < degree(p). Then $q(\xi_i(b)) = r(\xi_i(b))$, hence $\{(q,b) | \text{sign } q(\xi_i(b)) = \lambda\} = \{(q,b) | \text{sign } r(\xi_i(b)) = \lambda\}$. Moreover, the coefficients of r are polynomials in the coefficients $a = (a_0 \ldots a_d)$ and $b = (b_0 \ldots b_n)$ of q and p. Thus, by induction, the roots $\gamma_1 < \cdots < \gamma_m$ of $r(x)$ psemi-algebraic functions of (a,b). (Strictly speaking, there are cases to be distinguished in the formula for $r(x)$, namely if $b_0 \neq 0$ or if $b_0 = 0$, $b_1 \neq 0$, and so on. This causes no problem.)

Consider the sign data consisting of $\text{sign}(b_i)$, $\text{sign } p(\eta_i(b))$, $\text{sign}(\eta_i(b) - \gamma_j(a,b))$, $\text{sign } p(\gamma_j(a,b))$. By Lemma A.4, $\eta_i(b) - \gamma_j(a,b)$ is a psemi-algebraic function of (a,b). Therefore, any one of the finitely many possibilities, Φ, for this sign data determines a semi-algebraic set $E_\Phi = \{(a,b) \in R^{d+1} \times R^{n+1} | (a,b) \text{ realizes sign data } \Phi\}$. We claim that $\text{sign } r(\xi_i(b))$ is constant on E_Φ. Therefore, the set $\{(q,b) | \text{sign } r(\xi_i(b)) = \lambda\}$ is a finite union of some of the E_Φ, hence is semi-algebraic.

To see that $\text{sign } r(\xi_i(b))$ is determined by Φ, we observe first that we know from Φ which interval $(-\infty, \eta_1(b))$, $(\eta_j(b), \eta_{j+1}(b))$, or $(\eta_k(b), \infty)$ contains $\xi_i(b)$, or whether $\xi_i(b) = \eta_j(b)$, some j. Secondly, we know from Φ which of the roots $\gamma_j(a,b)$ of $r(x)$ lie in this interval. Thirdly, $p(x)$ is monotonic on this interval, hence the values of $\text{sign } p(\gamma_j(a,b))$ determine the interval $(-\infty, \gamma_1(a,b))$, $(\gamma_j(a,b), \gamma_{j+1}(a,b))$, or $(\gamma_m(a,b), \infty)$ which contains $\xi_i(b)$, or whether $\xi_i(b) = \gamma_j(a,b)$, some j. But $\text{sign } r(x)$ is constant on the intervals $(-\infty, \gamma_1)$, (γ_j, γ_{j+1}), (γ_m, ∞), and our claim is established. □

Proof of Proposition A.2: We can now complete the proof of the Tarski-Seidenberg theorem. It clearly suffices to show that given finitely many polynomials $f_i(x_1 \ldots x_n)$ and signs λ_i, the set $\{(x_2 \ldots x_n) \in R^{n-1} | (\exists x_1) \text{ sign } f_i(x_1 \ldots x_n) = \lambda_i\}$ is semi-algebraic. Regard the f_i as polynomials in x_1 with coefficients which are polynomials in $(x_2 \ldots x_n)$. The real roots of the f_i, say $\xi_{ij}(x_2 \ldots x_n)$, are psemi-algebraic functions of $(x_2 \ldots x_n)$. Consider the sign data $\text{sign}(\xi_{ij} - \xi_{i'j'})$, which orders the roots ξ_{ij}, say $\xi_1 \leq \xi_2 \leq \cdots \leq \xi_t$. There are finitely many possible orderings, which partitions R^{n-1} into disjoint, semi-algebraic subsets.

If $(x_2 \ldots x_n)$ is fixed, the f_i have constant sign on the intervals $(-\infty, \xi_1)$, (ξ_j, ξ_{j+1}), (ξ_t, ∞). (In fact, these signs are computed by the sign data sign $f_i(\xi_1 - 1)$, sign $f_i(\frac{\xi_j + \xi_{j+1}}{2})$, sign $f_i(\xi_t + 1)$. Alternatively, one could use the sign data sign $(\partial^k f_i / \partial x_1^k)(\xi_j)$, $k > 0$. See 7.2.5.) It is obvious that the truth of $(\exists x_1)$ sign $f_i(x_1 \ldots x_n) = \lambda_i$ depends only on the data sign $f_i(-\infty, \xi_1)$, sign $f_i(\xi_j)$, sign $f_i(\xi_j, \xi_{j+1})$, sign $f_i(\xi_t, \infty)$. Hence $\{(x_2 \ldots x_n) | (\exists x_1) \text{sign } f_i(x_1 \ldots x_n) = \lambda_i\}$ is semi-algebraic. □

We can now use elimination of quantifiers to eliminate the notion of psemi-algebraic functions. This observation is due to Efroymson [36].

Proposition A.6. A function $f: X \to R$, $X \subset R^m$, is psemi-algebraic if and only if the graph of f, $F \subset R^m \times R$, is a semi-algebraic set.

Proof: A very special case of the definition of a psemi-algebraic function f shows that the set $\{(t,x) | f(x) - t = 0\}$ is a semi-algebraic set. But this is F.

Conversely, if F is semi-algebraic, let $F = \{(x,t) | A(x,t)\}$ for some polynomial relation $A(x,t)$. Then $\{(q,x) | \text{sign } q(f(x)) = \lambda\} = \{(q,x) | (\exists t)(A(x,t)$ and $\text{sign } q(t) = \lambda\}$ which is semi-algebraic by Proposition A.2. □

Bibliography

I. Hilbert's 17th Problem

1. D. Hilbert, Über die Darstellung definiter Formen als Summe von Formenquadraten, Math. Ann. 32 (1888), 342-350.
2. E. Landau, Über die Darstellungen definiter binärer Formen durch Quadrate, Math. Ann. 57 (1903), 53-64.
3. E. Landau, Über die Darstellung definiter Funktionen durch Quadrate, Math. Ann. 62 (1906), 272-285.
4. E. Artin, Über die Zerlegung definiter Funktionen in Quadrate, Abh. Math. Sem. Univ. Hamburg 5 (1927), 100-115.
5. J.W.S. Cassels, On the representation of rational functions as sums of squares, Acta Arith. IX (1964), 79-82.
6. A. Pfister, Zur Darstellung definiter Funktionen als Summe von Quadraten, Invent. Math. 4 (1967), 229-237.
7. D.W. Dubois, Note on Artin's solution of Hilbert's 17th problem, Bull. Amer. Math. Soc. No. 4, 73 (1967), 540-541.
8. T.S. Motzkin, The Arithmetic-Geometric Inequality, and, Algebraic Inequalities, Inequalities, Vol. 1 (O.Shisha, Ed.) Academic Press, New York, 1967, 205-254 and 199-203.

II. Ordered Fields and Real Varieties

9. E. Artin and O. Schreier, Algebraische Konstruktion reeller Körper, Abh. Math. Sem. Univ. Hamburg 5 (1926), 85-99.
10. E. Artin and O. Schreier, Eine Kennzeichnung der reell abgeschlossenen Körper, Abh. Math. Sem. Univ. Hamburg 5 (1927), 225-231.
11. R. Baer, Über nicht-Archimedisch geordnete Körper, S.-B.

Heidelberger Akad. Wiss. Math.-Natur. Kl. 8. Abh. 1927 (Beiträge zur Algebra. L.).

12. W. Krull, Allgemeine Bewertungstheorie, J. Reine Angew. Math. 167 (1931), 160-196.

13. S. Lang, The theory of real places, Ann. of Matn. 57 (1953), 378-391.

14. D. W. Dubois, A Nullstellensatz for ordered fields, Ark. Mat. 8 (1969), 111-114.

15. H. Gross and P. Hafner, Über die Eindeutigkeit des reellen Abschlusses eines angeordneten Körpers, Comment. Math. Helv. 44 (1969), 491-494.

16. J.J. Risler, Une caractérisation des idéaux des variétés algébriques réelles, C.R. Acad. Sci. Paris 271 (1970), 1171-1173.

17. D.W. Dubois and G. Efroymson, Algebraic theory of real varieties, I., Studies and Essays presented ot Y.W. Chen on his sixtieth birthday, Taiwan, 1970), 107-135.

18. P. Ribenboim, Le théorème des zéros pour les corps ordonnés, Séminaire d'Algebre et Théorie des Nombres, Dubreil-Pisot, 24e année, 1970-71, exp. 17.

19. M. Knebusch, On the uniqueness of real closures and the existence of real places, Comment. Math. Helv. 47 (1972), 260-269.

20. G. Efroymson, Local reality on algebraic varieties, J. Algebra 29 (1974), 133-142.

21. D.W. Dubois and G. Efroymson, A dimension theorem for real primes, Canad. J. Math. 26 (1974), 108-114.

22. G. Stengle, A Nullstellensatz and a Positivstellensatz in semialgebraic geometry, Math. Ann. 207 (1974), 87-97.

23. M. Knebusch, On algebraic curves over real closed fields. I., Math. Z. 150 (1976), 49-70.

24. M. Knebusch, On algebraic curves over real closed fields. II., Math. Z. 189 (1976), 189-205.

III. Rings of Functions and Algebraic Topology

25. R. Swan, Vector bundles and projective modules, Trans. Amer. Math. Soc. 105 (1962), 264-277.

26. C. E. Watts, Alexander-Spanier cohomology and rings of continuous functions, Proc. Nat Acad. Sci. 54 (1965), 1027-1028.

27. E.G. Evans, Jr., Projective modules as fiber bundles, Proc. Amer. Math. Soc. 27 (1971), 623-626.

28. K. Lønsted, Vector bundles over finite CW complexes are algebraic, Proc. Amer. Math. Soc. 38 (1973), 27-31.

29. R. Swan, Topological examples of projective modules, Trans. Amer. Math. Soc. 230 (1977), 201-234.

IV. Real Analytic-Algebraic Functions

30. J. Nash, Real algebraic manifolds, Ann. of Math. 56 (1952), 405-421.

31. M. Artin and B. Mazur, On periodic points, Ann. of Math. 81 (1965), 82-99.

32. K. Lønsted, An algebraization of vector bundles on compact manifolds, J. Pure Appl. Algebra 2 (1972), 193-207.

33. J.J. Risler, Un théorème des zéros en géométrie analytique réelle, C.R. Acad. Sci. Paris 274 (1972), 1488-1490.

34. R. Palais, Equivariant and real algebraic differential topology, Part I. Smoothness categories and Nash manifolds. Notes, Brandeis University 1972.

35. J.J. Risler, Sur l'anneau des fonctions de Nash globales, C.R. Acad. Sci. Paris 276 (1973), 1513-1516.

36. G. Efroymson, A Nullstellensatz for Nash rings, Pacific J. Math. 54 (1974), 101-112.

37. J.J. Risler, Résultats récents sur les fonctions de Nash, Séminaire Pierre Lelong (Analyse) Année 1974/75, Lecture Notes in Mathematics No. 524, Springer-Verlag, 1976.

38. J.J. Risler, Sur l'anneau des fonctions de Nash globales, Ann. Sci. École Norm. Sup. 8 (1975), 365-378.

39. J. Bochnak and J.J. Risler, Le thérème des zéros pour les variétés analytique réeles de dimension 2, Ann Sci. École Norm. Sup. 8 (1975), 353-364.

40. T. Mostowski, Some properties of the ring of Nash functions, Ann. Scuola Norm. Sup. Pisa III (1976), 245-266.

41. G. Efroymson, Substitution in Nash functions, Pacific J. Math. 63 (1976), 137-145.

42. J. Bochnak, Sur la factorialité des anneaux des fonctions de Nash, Comment. Math. Helv. 52 (1977), 211-218.

V. Topology of Semi-Algebraic Sets

43. O.A. Oleǐnik, Estimates of the Betti numbers of real algebraic hypersurfaces, Rec. Math. (Mat. Sb.) N.S. 28(70), (1951), 635-640.

44. H. Whitney, Elementary structure of real algebraic varieties, Ann. of Math. 66 (1957), 545-556.

45. J. Milnor, On the Betti numbers of real varieties, Proc. Amer. Math. Soc. 15 (1964), 275-280.

46. S. Lojasiewicz, Triangulations of semi-analytic sets, Ann. Scuola Norm. Sup. Pisa 18 (1964), 449-474.

47. R. Thom, Sur l'homologie des variétés algébrique réeles, Differential and Combinatorial Topology (Morse Symposium), Princeton University Press, 1965.

48. H. Hironaka, Triangulations of algebraic sets, Proc. Amer. Math. Soc., Symp. in Pure Math., 29 (1975), 165-185.

49. J. Bochnak, Quelques propriêtes quantitatives des ensembles semi-algêbrique, Ann. Scuola Norm. Sup. Pisa 2 (1975). 483-495

VI. Witt Rings and Reality

50. D.K. Harrison, Witt Rings, Lecture Notes, Dept. Math., Univ. of Kentucky, Lexington, KY, 1970.

51. J. Leicht and F. Lorenz, Die Primideale des Wittschen Ringes, Invent. Math. 10 (1970), 82-88.

52. M. Knebusch, Real closures of semi-local rings and extensions of real places, Bull. Amer. Math. Soc. 79 (1973), 78-81.

53. M. Knebusch, Real closures of commutative rings I, J. Reine Angew. Math. 274/275 (1975), 61-89.

54. M. Knebusch, Real closures of commutative rings II, J. Reine Angew. Math. 286/287 (1976), 278-213.

55. T.Y. Lam, Ten lectures on quadratic forms over fields, Conference on Quadratic Forms, 1976, Queen's Papers in Pure and Appl. Math. No. 46, Queen's University, Kingston, Ont., Canada, 1977.

VII. Real Algebra and Logic

56. A. Tarski, A decision method for elementary algebra and geometry, 2nd ed., revised, Berkeley and Los Angeles, 1951.

57. A. Seidenberg, A new decision method for elementary algebra, Ann. of Math. 60 (1954), 365-374.

58. A. Robinson, On ordered fields and definite functions, Math. Ann. 130 (1955), 257-271.

59. A. Robinson, Further remarks on ordered fields and definite functions, Math. Ann. 130 (1956), 405-409.

60. A. Robinson, Introduction to Model Theory and the Meta Mathematics of Algebra, North-Holland Publishing Co., Amsterdam, 1965.

61. G. Kreisel and J.L. Krivine, Elêments de logique mathêmatique, thêorie des modèles, Dunod., Paris, 1967.

62. P.J. Cohen, Decision procedures for real and p-adic fields, Comment. Pure Appl. Math. 22 (1969), 131-151.

VIII. Basic Algebra Texts

63. M.F. Atiyah and I.G. Macdonald, Introduction to Commutative Algebra, Addison-Wesley, 1969.

64. N. Jacobson, Lectures in Abstract Algebra, Vols. 1-3, Van Nostrand, 1951, 1953, 1964.

65. N. Jacobson, Basic Algebra I, W.H. Freeman, 1974.

66. S. Lang, Algebra, Addison-Wesley, 1965.

67. B.L. van der Waerden, Modern Algebra, Vols. 1,2, Frederick Ungar Publishing Co., 1953.

68. O. Zariski and P. Samuel, Commutative Algebra, Vols. 1,2, Van Nostrand, 1958, 1960.

Notation

(Introduction not included)

Index

(Introduction not included)